The purpose of this comprehensive text is to increase awareness of human reproduction and its consequences. The central theme links reproductive capacity, the social consequences of the multiple stresses this places on the environment and the ways this relates back to the reproductive health of humans and other animals. In the first section, the biology of human reproduction is discussed, including such topics as the treatment and causes of infertility, growth and maturation, parental behaviour and neonate biology. The effects of procreational biology on the foundation of human social structure are also examined. The second part deals with reproduction as it relates to health and social issues such as stress, fertility control, AIDS, teratogens and errors of sexual differentiation. It is an invaluable resource for all those wishing to update their knowledge of human reproductive biology.

A Guide to Reproduction

Social Issues and Human Concerns

A Guide to Reproduction

Social Issues and Human Concerns

IRINA POLLARD
Senior Lecturer, Biological Sciences, Macquarie University, Sydney, Australia

CAMBRIDGE UNIVERSITY PRESS

Published by the Press Syndicate of the University of Cambridge
The Pitt Building, Trumpington Street, Cambridge CB2 1RP
40 West 20th Street, New York, NY 10011-4211, USA
10 Stamford Road, Oakleigh, Melbourne 3166, Australia

First published 1994

Printed in Great Britain at the University Press, Cambridge

A catalogue record for this book is available from the British Library

Library of Congress cataloguing in publication data

Pollard, Irina.

A guide to reproduction : social issues and human concerns / Irina Pollard
p. cm.
ISBN 0-521-41862-3 (hardback) ISBN 0-521-42925-0 (pbk.)

1. Human reproduction. 2. Human reproduction – Social aspects.
3. Infertility. I. Title.
QP251.P65 1994
612.6 – dc20 93-38304 CIP

ISBN 0 521 41862 3 hardback
ISBN 0 521 42925 0 paperback

KW

To Morgan

Contents

Preface and acknowledgements

The purpose of this book is to provide a comprehensive, inter-disciplinary text for science and medical students, and others wishing to up-date their knowledge of reproduction and related social issues. Because the book follows the content of the third year university biology course I teach, some background knowledge in animal physiology and biochemistry is assumed. The text is also intended to be thought provoking for those with little biological background who may be challenged to pursue the subject further. At the end of each chapter the General references section has sufficient documentation for the reader to find the source of most of what is discussed.

The 1980s have been an exciting decade in the reproductive sciences because of the substantial contributions which have come from all its subdisciplines. This is reflected in the overall structure of the book, which is aimed at an increased awareness of human reproductive issues and their consequences across disciplines. Theories and principles are integrated to create a holistic overview.

I especially owe a dept to Roger Hiller for critically reading all my drafts, for generously giving expert advice and for engaging in probing discussions throughout the gestation of this book. Roger's well-judged criticism resulted in important modifications and improvements, and his insistence on clarity of expression saved me from many blunders. I also owe a debt to David Pollard who read the entire manuscipt and gave generously of his talents and time. Also, I extend my thanks to those colleagues who read and improved various chapters: Brian Atwell, Frank Burrows, Walter Ivantsoff, David Laing, Ian Pike and Michael Sinosich. I am indebted to Ron Oldfield and Jenny Norman for expertize in photography; special acknowledgements for individual micrographs appear with the relevant figures in the text.

The illustrations are an integral part of this book and all the drawings are originals. I am very fortunate to be associated with two exceptional artists whose complementary creativity resulted in clear, meaningful illustrations from my roughest sketches and ideas. Special thanks are due to both of them. Barbara Duckworth (from School of Biological Sciences, Macquarie University) did the labelling throughout and her special expertise in generating computer graphics, as well as in free-hand drawing, is reflected in her final products. Barbara is responsible for Figs. 2.1, 2.2, 3.2, 3.3, 4.3–4.5, 4.7, 4.8, 5.4, 6.1, 9.2, 10.1, 10.2, 13.1, 13.2, 16.1–16.3, 17.1, 19.1 and 19.2. Betty Thorn (formally from School of Biological Sciences, Macquarie University) has a special talent in bringing biological drawings to life. Betty is responsible for the cover illustration and Figs. 3.1, 4.1, 5.1, 5.5, 8.1, 8.2, 9.1, 11.1, 13.3 and 14.1. The drawings are the work of Barbara Duckworth and Betty Thorn.

Finally, the writing of this book was suggested by my editor Alan Crowden and I thank him and Cambridge University Press especially Tracey Sanderson and Jane Ward for giving me this opportunity and assisting its delivery: once I began to research and write, I found the project challenging and pleasurable.

Irina Pollard

Part one
Reproductive biology

1 Fertility and infertility

Since the 1950s, biologists have contributed greatly to advances in the reproductive sciences and medicine. The essence of this reproductive revolution has been its unifying character. Contributions from the disciplines of genetics, biochemistry, prenatal medicine, immunology, epidemiology, socioanthropology and ecological ethics have provided choices, accelerated social change and rewarded procreation through improved health. Despite these new opportunities, medical technology is not the only essential for solving many of our problems because health determinants, particularly reproductive health, are usually more dependent on increased political awareness and a rising socioeconomic level. The unhealthy aspects of reproduction are intimately connected through the stresses of severe crowding, poverty, inequality of resource distribution and political instability. Issues of human concern, including the consequences of unregulated fertility, infertility, infant/maternal mortality, intrauterine growth retardation, congenital abnormality, sexually-transmitted disease and environmental deterioration, are the central themes of this book and are integrated into the various chapters dealing with theories and principles of human reproductive function.

HEALTH DETERMINANTS

Reproductive health is predetermined by genetic endowment and by environment but it can be improved or undermined by individual behaviour, advanced by better socioeconomic conditions and changed by medical knowledge and services. Improved medical social services are essential to the raising of reproductive health but can only be effective if they facilitate the necessary changes in human behaviour. Health, as defined in the Constitution of the World Health Organization (WHO), is a state of 'complete physical, mental and social well-being not merely the absence of disease or infirmity', succinctly emphasizing the positive aspects of a realised potential. For reproductive health, three basic interconnected elements have to be satisfied. These are control of fertility, success in procreation and safety in their pursuit.

The regulation of reproductive potential requires the conscious separation of fertility from sexual pleasure and enjoyment of relationships. Men and women, since prehistoric times, have agreed that it is desirable that fertility be controlled. Not until the second half of the 20th century, however, has knowledge and technological development given humankind real control over fertility. Reproductive behaviour is an important health determinant which is essentially in individual hands. Planning of pregnancies to take place at optimal times and under optimal conditions makes a major contribution to better health for parents, children and society. Pregnancies too close, too early, too late and too many are high-risk behaviours that deny access to their potential for millions of people, who spend their lives uneducated, unhealthy and ill-housed. This loss of potential has serious physical, psychological, social and economic consequences. Numerous surveys show that many women, especially in developing countries, still have no control over their fertility. Aside from economic costs, unwanted fecundity transgresses on ecological issues. Species besides *Homo sapiens* have a right to exist, and *Homo sapiens* is a knowledgeable enough species to know that further environmental degradation is incompatible with survival.

A second aspect of reproductive health is the outcome. The birth of a healthy infant with good postnatal prospects, unencumbered by the effects of intrauterine deprivation, should be the norm. Socioeconomic improvement is a significant factor in alleviating high-risk pregnancies as it is the women and children under poverty who are the most vulnerable. In depressed societies, the status of women is also depressed and status raising factors, such as education, access to contraceptives, reasonable employment opportunities and equality under the law, are not generally available. For the millions of unplanned children, life is also predetermined by the poverty cycle which limits access to education and health services and denies them independence to withstand harmful traditional pressures. Traditionally, improved socioeconomic conditions have been linked to absolute economic growth. Significant improvements in reproductive health can, however, be achieved without corresponding economic growth because the absolute level of growth is less important than the manner in which the economic rewards are shared by members of the community and whether the existing economic resources are managed sustainably. For instance, improving the status of women in a society will make a significant contribution to reproductive health within the same level of economic development and a healthier human resource will increase future economic productivity.

The last aspect of reproductive health relates to the reasonable expectation that engaging in sexual intercourse, fertility regulation, pregnancy and childbirth is not participating in a high-risk activity with adverse implications for partners or offspring. Safe reproduction is also a matter of basic human rights, especially those concerned with the health of the next generation. Too little is known about the etiology of congenital anomalies, but the majority of malformations at birth have a dual cause, comprising a genetic predisposition triggered by environmental teratogens. These include infectious diseases, therapeutic drugs, recreational drugs, psychological stress, radiation and environmental pollutants, most of which are human in origin. Humane communities generally judge that failure to provide life-preserving treatment to a newborn constitutes child neglect. Surely the failure to make adequate envionmental provisions during reproduction that would minimize the need for intervention at or after birth is equally culpable.

Reproduction is a very sensitive indicator of health status and can be impaired by a modest failure in physiological homeostasis. The United Nations' medium projection predicts that from our present 5 billion, the population will stabilize at between 8 and 14 billion some time in the next century. It is also predicted that more than 90% of the increase will occur in the poorest countries whose populations already overextend their available environmental resources. These projections are based on assumptions of fecundity and life expectancy which are largely guesswork and necessarily ignore the consequences of this enormous population growth on the human reproductive capacity. Paradoxically, one of the biological consequences of unregulated fertility is infertility (Chapter 13). Humans may not be exempt, as a survey by the US Office of Technology and Assessment found that the rate of infertility among women aged 20–24 years rose significantly from 3.6% in 1965 to 10.6% in 1982. In addition, the number of couples with primary infertility doubled, increasing from 500 000 in 1965 to 1 million in 1988. A British study suggested that subfertility was another problem that many couples confront, since 20–35% of pregnancies take more than 1 year to conceive. The legacy of unregulated population growth coupled with appalling environmental deterioration may be human infertility on a mass scale.

INFERTILITY: THE PRICE OF EXCESS FECUNDITY?

Pregnancy provides the best basis for assessing a couple's fecundity and a couple is considered infertile if no conception has been achieved after 12 months or more of unprotected coitus of average frequency. Infertility is hard to assess accurately, but several studies have reported that up to 15% of couples in developed countries are involuntarily infertile, with a much higher percentage in developing countries (30–40% of women in parts of tropical Africa). The main causes of infertility in developing countries are sexually-transmitted diseases (the main culprits are gonorrhoea and chlamydia) and repeated pregnancies causing secondary infertility due to poor hygiene at the time of childbirth, abortion or miscarriage. Infertility due to primary gonadal failure, such as permanent azoospermia or congenital absence of ovaries, can be described as sterility.

According to WHO figures, of the infertile couples (up to 80 million people worldwide) 30–40% have an exclusively male factor, 25% have factors in both male and female, 40% have a predominant female factor and in 2–15% no diagnosis can be made after a complete investigation. It has also been reported that in as many as 35% of couples the infertility may have multiple origins. Female infertility is age dependent and its rate increases from 30 years of age to the menopause. Since the 1970s, an increasing number of couples, mainly from western societies, have been seeking treatment for infertility. This may, of course, be due to an increased candour that the evolving reproductive technologies has encouraged, or it may reflect increased infertility resulting from, among other things, exposure to low concentrations of a range of toxins. Thousands of potential toxins are released continually into the environment, the great majority of which have never been studied for their effects on mammalian reproduction. The increasing number of women in the workforce has prompted many studies on the potential effects of occupational exposure to toxins on female reproduction. These studies have also been the impetus for similar investigations in men. Male occupational exposure to heat, noise, vibrations, radiation, microwaves, toxic chemicals such as lead, anaesthetics, pesticides and synthetic oestrogens may affect semen quality, fertility and libido. According to a US Office of Technology Assessment study, people working in agriculture, laboratories, oil, chemical and atomic industries, pulp and paper manufacturing, and textiles are most likely to be exposed to substances that adversely affect reproduction. Some scientists claim that an increase in several testicular abnormalities including cancer, undescended testicles and low sperm count could be linked to rising levels of oestrogens in the environment. Environmental pollutants such as polychlorinated biphenyls (PCPs), some organochloride insecticides and detergents exhibit oestrogenic activity and may affect normal differentiation of male fetuses (Chapter 16). Lifestyle also modulates fertility; for example, an association between cigarette smoking and teratozoospermia (sperm morphology less than 50% normal) has been documented. Environmental factors may also increase impotency, which can be psychogenic, neurological or vascular in origin.

Male infertility

Men have almost always been assumed to be fertile so recognition of male infertility has been slow. Until the second half of the 20th century the blame for a couple's infertility was laid on the female partner. Now, scores of male infertility clinics exist around the world and the male reproductive science, andrology, has emerged and gained widespread acceptance. In fertile men, the sperm concentration varies between 60–80 million active sperm per ml of semen in an ejaculate volume of approximately 2.5–3.5 ml. The turning point between fertile and infertile sperm numbers is taken as 20 million active sperm per ml, but sperm density alone is insufficient to assess the quality of a semen sample. Other parameters such as sperm motility, viability and morphology are especially important. The least adequately assessed parameter of semen analysis is sperm morphology, which is, typically, examined on wet preparations or on smears of stained seminal fluid. Only morphological parameters, such as head size and shape, length and appearance of flagellum, presence or absence of multiple head or flagellae, are assessed because functional faults are beyond the limits of resolution of the light microscope. The transmission and scanning electron microscopes are useful tools in evaluating the ultramorphology of sperm as they permit a detailed assessment of structural integrity of subcellular components (Fig. 1.1). A more precise assessment of sperm motility can now be obtained with video micrographic, computer-assisted semen analysis systems that measure sperm velocity, progression, amplitude and frequency, and head movement. These measures permit a deeper evaluation of sperm function but such systems are expensive and are only available in sophisticated research centres.

Once sperm have arrived on site, other qualities needed for fertilization of oocytes, involving recognition, binding and penetration through the zona pellucida, and fusion with the vitelline membrane of the oocyte (Chapter 8), must be included. Diagnostic techniques and methods for

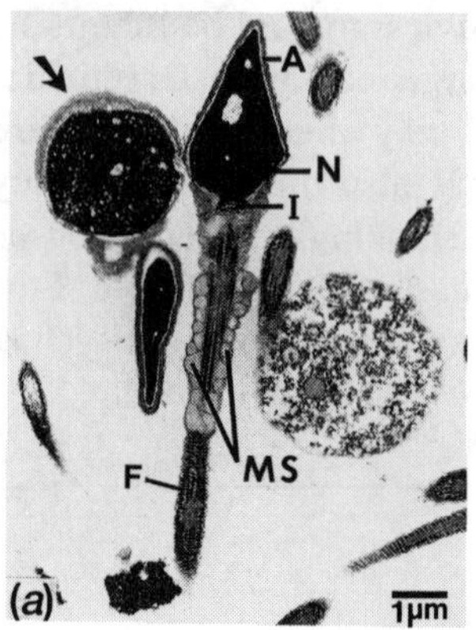

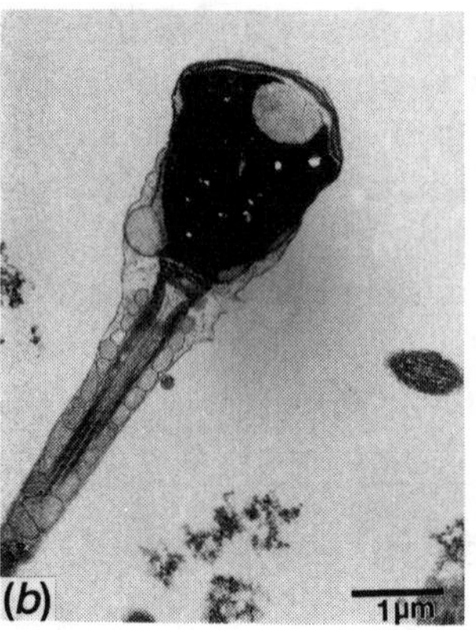

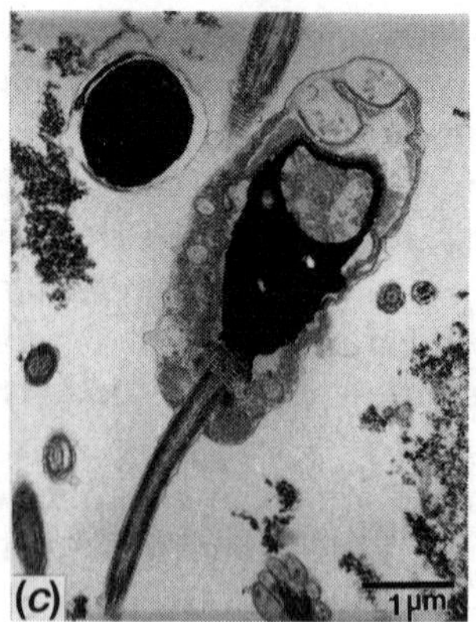

Fig. 1.1 Transmission electron micrograph showing normal and abnormal forms of human sperm. (*a*) A normal sperm showing the acrosome (A) covering the anterior part of the head or nucleus (N), implantation site (I), midpiece surrounded by the mitochrondrial sheath (MS) and the principal piece of flagellum (F). The round-headed abnormality (arrow) illustrated has a circular nucleus lacking an acrosome and has a detached flagellum. The round-headed syndrome is frequently described but represents less than 1% of sampled abnormal cells and occasionally can be found in 'normal' populations. (*b*) An abnormal sperm characterized by vacuolations in the nucleus and postacrosomal region with separation of the acrosomal membranes. (*c*) A grossly abnormal sperm characterized by extensive vacuolation of the nucleus, alteration of shape, incomplete formation of the acrosome and copious excess cytoplasm around the nucleus. I am indebted to R. A. Boadle, electron microscopist, Dept. of Anatomical Pathology, ICPMR Westmead Hospital, Sydney, for supplying this interesting photograph.

the *in vitro* assessment of the functional competence of human sperm include the sperm penetration assay (or the zona-free hamster oocyte test) and the hemizona assay. The sperm penetration assay employs hamster ova which have been separated from their surrounding zona pellucida before their incubation with sperm. Sperm capable of penetrating the vitelline membrane are considered capable of fertilization (the resulting 'humster' does not develop beyond the two-cell stage). The hemizona assay has been developed as a predictor of fertilization potential and uses matching halves of a human zona pellucida from a non-living oocyte to evaluate tight binding (as opposed to initial attachment). One half provides an internal control while the other half is observed for test binding. The assay index is derived from the percentage of bound sperm from a subfertile man over the percentage of bound sperm from a fertile control. Oocytes, fresh or frozen, come from surgically removed ovaries, post-mortem ovarian tissue and failures from IVF programmes. Sperm preparation protocols and reagents for the artificial enhancement of sperm function have also been developed. Sperm numbers and quality may be boosted in some men by large doses of vitamins B and/or C and frequent ejaculation.

Between 50 and 75% of all male infertility is asymptomatic and without demonstrable cause because, as already mentioned, functional defects place assessment of the sperm's integrity outside conventional semen analysis. Identified causes of infertility range from congenital defects to an inappropriate lifestyle (common modifiers of male fertility are further described in Chapters 5 and 8). The etiology of infertility can be classified under the following headings: testicular causes affecting spermatogenesis, post-testicular (ductal factors and sexual dysfunction), pretesticular (including endocrine disorders), immunological causes and bacterial, viral or parasitic urogenital infections. The frequency of known causes of infertility varies from country to country but common causes, in descending order of prevalence, include varicocele (dilation of the, commonly left, spermatic vein), testicular failure (idiopathic[1] oligospermia, generalized immotility or ineffective motility classed as asthenozoospermia and teratozoospermia), accessory gland infection, sperm autoimmunity, congenital abnormalities, systemic factors, sexual dysfunction and obstructive azoospermia. For example, a history of postpubertal mumps or sexually-transmitted disease, especially if not promptly treated, can result in seminiferous epithelial damage and ductal obstruction. Varicoceles occur in 10–20% of the general population but are present in approximately 40% of infertile men. Varicoceles result from inadequate venous drainage from the internal spermatic

[1] Of unknown cause.

vein. This results in varicosed areas and scrotal heating due to excessive venous pressure build-up. A real concern in this field is not so much infertility as subfertility and compromised reproductive capacity as this may be a source of congenital abnormalities of which 60% have no known cause and are thought to result from the interaction of genetic and environmental factors.

Female infertility

Fertility, often in the form of a goddess or phallic symbol, has been worshipped since ancient times because the 'barren woman' caused much concern. The etiologies of female infertility fall into similar categories to those listed for the male, although some kind of treatment for infertile women has always been available. Since the early 1960s, however, important breakthroughs, such as ovulation-inducing drugs and gynaecological microsurgery, have resulted in more women being successfully treated than their partners. Female infertility can be classified under the following broad categories: endocrine causes including anovulation and luteal phase defect, tubal occlusion, disorders of the cervix and uterus including malformation, abnormal growths (fibroids), endometriosis and hostile cervical mucus, immunological causes and infections. Both men and women can develop antisperm antibodies with some evidence that women infected with sexually-transmitted disease are more likely to develop antibodies against sperm because sperm come in contact with the immune system through genital lesions. Miscarriage is quite common in all pregnancies (15–20% of all human embryos die in the early stages of development, which serves as the mechanism for the selection of normal and near-normal embryos) but is significantly higher in communities with substandard hygiene, poor nutrition and a high rate of disease.

A major cause of infection in women is pelvic inflammatory disease (PID). It has been estimated that between 30–50% of all cases of female infertility and ectopic pregnancy are caused by PID-related tubal adhesions. Sexually-transmitted diseases underlie an estimated 75% of all cases of PID but infection after backstreet abortion and genital mutilation in African teenage girls at circumcision and/or infibulation also accounts for significant incidences of PID. The term PID was introduced because the condition is seldom confined to a single pelvic organ and is caused by an inflammatory infection in the upper genital tract. From the clinical point of view, the majority of women suffering from pelvic inflammatory infections may not even be aware of the original infection or recover completely, but persistent morphological damage to various parts of the genital tract may result. In many cases the decreased or abolished reproductive capacity may not become apparent until years later. PID can involve one or both fallopian tubes (salpingitis), one or both ovaries (oophoritis), the uterus, and the peritoneum in the pelvis and abdominal wall. Many opportunistic infections may be introduced during coitus, especially if the immune system and the cervix (the natural barrier against the invasion of microorganisms) is compromised after miscarriage, abortion, cervical surgery, the presence of certain types of intrauterine contraceptive devices and primary infections, such as, *Neisseria gonorrhoea* and *Chlamydia trachomatis*. Since the 1980s chlamydial infection and gonorrhoea have increased in epidemic proportions with, globally, 200 million new cases of gonorrhoea reported to the WHO each year. Up to 20% of women thus infected will develop PID.

Complete or partial tubal occlusion may be diagnosed by hysterosalpingography which involves introducing radio-opaque dye through the uterus into the fallopian tubes under X-ray fluoroscopy visualization. When the dye moves into the fallopian tubes, occlusions or other abnormalities can be seen: if the fallopian tube is open, the dye fills the tube and spills into the abdominal cavity. If the hysterosalpingogram is abnormal, laparoscopy allows direct visualization of the pelvic organs through an endoscope. Complete or partial tubal blockage and adhesions can sometimes be corrected by microsurgery.

MEDICAL TECHNIQUES FOR ASSISTED REPRODUCTION

Efforts to overcome infertility by engaging in all manner of sexual practices and manipulations date back at least to Biblical times. Even to-day not all infertile couples seek infertility treatment with, in industrialized communities, an estimated 51% of couples with primary infertility and 22% with secondary infertility seeking treatment. The demand is, however, rising, probably as a result of a more conducive social milieu and familiarity with the available reproductive technologies. It may also be rising as a result of increased primary infertility (a failure to reach fertility), greater availability of services for secondary infertility (malfunction of the mature reproductive system) and a decreased number of infants available for adoption. Among infertile couples seeking medical treatment, 50% will achieve a viable pregnancy and 85–90% of these will be the result of conventional therapy. Ovulation induction, surgery and artificial insemination are the most widely used and successful approaches to overcoming infertility.

Children conceived by artificial insemination

Artificial insemination (AI) as a technique has been known for centuries; it is said that in the 14th century Arab tribesmen impregnated enemy mares with semen of inferior stallions! In the early 1890s, Russian scientists established AI as a tool in livestock production and its application is now broadening to include conservation programmes for endangered species. The first recorded human birth after artificial insemination with husband's sperm (HI) was in 1790 when the brilliant Scottish physician John Hunter inseminated a woman with epididymal sperm from her husband who had an urethral defect. In recent times, HI has been useful in cases of paraplegia (sperm is collected by electroejaculation), obstructed vas deferens or epididymis (sperm is aspirated from the epididymis) and forced separation of couples (prisoners on long-term sentences). It is also widely used for idiopathic infertility. Between 15–30% of women become pregnant during six insemination (menstrual) treatment cycles. The pregnancy success rate is increased if the sperm are first subjected to a separation procedure such as 'swim-up'. This involves layering media directly upon a semen sample. After incubation, the top fraction containing the most motile (and presumably the healthiest) sperm is removed for insemination. A modification of the 'swim-up' technique involves the use of albumin columns which also selects sperm by their inherent motility. In this procedure, a sample of semen is placed on top of the column, incubated and then the bottom fraction, containing the most active sperm, is removed.

Donor insemination (DI) is the insemination of a woman with sperm from a donor other than her husband. DI is much more successful than HI, with a reported 60% birth rate after six insemination cycles and is, therefore, one of the major treatments for male infertility. The technique is routinely used in Australia, Europe and the USA; over 30 000 DI births were registered in the USA in 1987. The procedure does not carry an increased risk of spontaneous abortion or congenital anomalies and has advantages over adoption in that the child is genetically related to the mother and the couple can experience conception, pregnancy and delivery. The disadvantage of half-siblings, related through the father, unknowingly marrying and bearing children is minimized as most AI clinics limit the number of pregnancies from each donor. It has been estimated that the chance of genetic incest occurring is one in 40 000 if sperm from one donor is used to produce six offspring in a city with a population of 3 million. Insemination centres restrict recruitment of donors to men who are healthy, free from transmissible genetic disorders and sexually-transmitted diseases and have, as far as can be determined by laboratory analysis, semen with a high fertilization potential. Access to personal details that may affect sperm quality, including *in utero* exposure to DES (diethylstilbestrol, see Chapter 16), the use of prescription and recreational drugs, occupational exposure to excessive heat, radiation and toxic chemicals, may also be required. In addition, all semen is frozen and quarantined in sperm banks for six months after which the donor is retested for antibodies against HIV, the human immunodeficiency virus, because antibodies may not show up for 3–6 months after infection.

Two other techniques of HI related to *in vitro* fertilization and gamete intrafallopian transfer (see next section) but far simpler are also popular: intrauterine insemination (IUI) and direct intraperitoneal insemination (DIPI). Patients are sometimes superovulated because the rationale behind both techniques is to increase the number of available oocytes and thus maximize the chances of conception. Intrauterine insemination works because placing the prepared semen high up in the uterus, allows sperm to bypass the cervix and its mucus which might possibly contain antisperm antibodies. For DIPI, the semen sample is placed into the pouch of Douglas (a peritoneal space behind the vagina) by means of a needle-guided catheter after puncturing the wall of the posterior fornix of the vagina. Sperm are picked up from the pouch of Douglas by the fimbriae and enter the ampullae of the oviducts in similar manner to oocyte pick-up.

Children conceived by *in vitro* fertilization and related technologies

The basic and, to the public, most familiar reproductive technologies are *in vitro* fertilization (IVF) and gamete intrafallopian transfer (GIFT). These techniques either alone or in conjunction with the other technologies offer additional hope to those infertile couples for whom conventional treatment has been unsuccessful. Although most appropriate as a method of achieving conception in women with absent or occluded oviducts (fallopian tubes), IVF has also proved a useful form of treatment in many other causes of female and male infertility. In oligospermia, for instance, the small number of viable sperm recovered are spared negotiation of the female reproductive tract and can concentrate around the oocyte. Other conditions for which IVF is useful are ovaries enclosed in adhesions, defective oocyte pick-up, male or female immunity to sperm and idiopathic infertility. The world's first IVF baby, Louise Brown, conceived under the control of gynaecologist Patrick Steptoe and embryologist Robert Edwards, was born in England in 1978. This achievement revolutionized the treatment of human infertility and opened up prospects for the plight of endangered species. There has been much progress since then. Australia's test-tube pioneers have earned several medical triumphs including breakthroughs in quality control for culture, administration of fertility drugs, successful embryo freezing and assisted fertilization. These contributions originated mostly from the Monash University–Queen Victoria Medical Centre–Epworth Hospital IVF Program (Melbourne) research teams. Under the direction of Wood and Trounson, these research successes have helped the centre to become one of the world's most recognised IVF clinics.

Because of the speed of new scientific inventions and technical improvements, the field of assisted reproduction has earned the trade a high degree of trust and popularity. In the 1990s the techniques are a major growth industry, providing employment opportunities for reproductive biologists, phsicians and technicians. IVF does not, however, supply all the answers; the overall live delivery rate is low (approximately 5–15%), there are inherent risks of complications and the programme is relatively costly. The extent to which women are emotionally stressed by treatment schedules and pregnancy anticipation may influence the outcome of IVF, particularly if excessive elevation of prolactin and cortisol is involved, and in some men the stress of undergoing an IVF treatment cycle may also impair the sperm quality. Counselling against heightened emotional distress has become an essential component of the IVF procedure.

In general terms the IVF programme involves six consecutive manipulations: ovarian stimulation, oocyte or egg retrieval, sperm collection, *in vitro* fertilization, embryo culture and an embryo transfer (ET) into the uterus. The aim of ovarian stimulation is to increase the number of mature oocytes available for retrieval (the pregnancy success rate is improved by the simultaneous transfer of two to three embryos; freezing of extra embryos for future use further augments the pregnancy rate per stimulated cycle). The technique of superovulation involves treatment with gonadotrophic hormones to increase the number of oocytes matured in a given menstrual cycle. To achieve ovarian stimulation, specific combinations of the various 'fertility drugs', such as the weak synthetic oestrogen clomiphene citrate (CC), human menopausal gonadotrophin, human chorionic gonadotrophin, follicle stimulating hormone and gonadotrophin releasing hormone analogues, are routinely used. Circulating ovarian steroid levels and ultrasound images of the ovaries are used to monitor the development and maturation of preovulatory follicles and the time when oocytes are ready to be collected by aspiration (oocyte pick-up). Various laparoscopic or ultrasound-guided techniques can be used for oocyte collections but the optimal technique is the transvaginal ultrasound-guided approach. This entails passing a collection needle into the ovary along a defined path visualized on a television monitor and transmitted from an ultrasonic transducer which is designed for introduction into the vagina. The method minimizes risk of morbidity because, unlike the older methods, it is relatively non-invasive and can be done under local anaesthesia or light intravenous sedation.

The addition of washed sperm to cultures containing the retrieved mature oocytes results, within a few hours, in fertilization. Although the technique is simple, this part of the IVF procedure is the most delicate, requiring careful control of laboratory conditions if embryo viability is not to be compromised. Embryos subjectively scored as having good viability predictors, such as the ability to divide evenly, regularly and rapidly, are then replaced *in utero*. Transfer is usually carried out after 48 hours in culture when the embryo is at the 4-cell stage, although 2-cell and 8-cell embryos can be transferred equally successfully. IVF also makes it possible to transfer the embryo derived from the gametes of genetic parents to a separate gestational mother. Recently IVF technology has been extended to include menopausal and perimenopausal women who have been implanted with embryos from oocytes obtained from younger, fertile donors. Embryo transfer technology will eventually empower the veterinary and medical professions with the ability to select for sex, desirable physical characteristics and lack of deleterious genes; it also will permit genetic manipulation of embryos. The manipulation of heredity in medicine and agriculture (gene therapy and genetic engineering) for carefully defined purposes and under appropriate supervision can both be ethically acceptable and socially desirable. In addition, the application of embryo technology for the preservation of endangered species holds broad appeal (Chapter 14).

GIFT became a significant treatment alternative in 1986 for patients with idiopathic infertility and differs from IVF in that fertilization occurs in the gestational mother and the gamete recipient must have at least one normal oviduct. Like IVF, GIFT requires ovarian hyperstimulation and is followed by a two-step procedure: oocyte collection and subsequent transfer of oocytes and sperm into the ampulla of the oviduct. The technique has been used in cases of obstructive azoospermia when sperm aspirated from the caput region of the epididymis are used. Live delivery rates following GIFT are higher than those for IVF, ranging from 18–26%. ZIFT or zygote intrafallopian transfer involves the transfer of the fertilized conceptus before cleavage has commenced. It has been introduced as an alternative to GIFT for patients with semen abnormalities where the screening for successful fertilization is advantageous. The procedure has an approximate success rate of 17%.

Superovulation by exogenous gonadotrophins is not risk free and can impair fertility by several mechanisms. The multifactorial etiology may include genetic deficiencies in the superovulated oocyte or embryo and/or hostile factors in the maternal environment. For example, insufficient progesterone or excessive oestrogenic stimulation of the reproductive tract may lead to an asynchrony between embryonic and uterine development. Birth registers reveal that there is a significant association between superovulation and increased neural tube and other defects in infants. The interpretation of such findings is controversial because they may be the result of either the drugs used in the treatment of infertility or an inherent defect in the subfertile patient or both. Only large prospective studies, including the offspring from fertile donors, can elucidate the roles of subfertility and its treatment in the etiology of malformations. Ovarian hyperstimulation may also result in iatrogenic complications with serious health consequences for the woman. The ovarian hyperstimulation syndrome refers to an enlargement of the ovaries with the development of multiple ovarian cysts and, in severe cases, an acute, potentially fatal, fluid shift out of the intravascular space resulting in generalized oedema.

Recently, deaths from iatrogenically acquired Creutzfeldt–Jakob disease (CJD) have received much publicity. CJD is a degenerative, fatal viral disease of the brain that was inadvertently transmitted to recipients treated with extracts from human cadaveric pituitary glands. The use of human pituitary-derived hormones for the treatment of specific pituitary hormone deficiencies was established in the early 1960s and discontinued in 1985 when an association of CJD with previous treatment using human-derived growth hormone was established. Pituitary-derived therapeutic hormones were then replaced by genetically engineered ones, but not before an estimated 30 000 individuals had received human pituitary-derived hormones. These were given mostly to children of short stature (growth hormone) or infertile women (gonadotrophins). CJD normally occurs sporadically or in familial form and may have an incubation period of as long as 35 years.

Technical advances in *in vitro* fertilization technology

Research into male infertility treatment is proceeding at an accelerating rate. Patients with sperm abnormalities, which normally prevent oocyte fertilization, can now have children as a consequence of sperm micromanipulaton or assisted fertilization. Biologically selected barriers which prevent abnormal gametes from penetrating the zona pellucida are low density, poor structure, abnormal or weak motility, immotility and dysfunctions at the level of binding with the zona; some of these may now be circumvented by micromanipulation in conjunction with IVF technology. Micromanipulation allows manipulation of the zona (zona drilling or partial zona dissection) so as to facilitate sperm passage, insertion of sperm under the zona (subzonal sperm insertion) or injection of sperm into the ooplasm (sperm microinjection). Pregnancies have been reported utilizing these techniques and have been hailed as a possible breakthrough in treating male infertility.

The safety of micromanipulation and assisted fertilization technologies is of concern, especially in the case of the most invasive method of direct ooplasm injection where penetration of the plasma membrane is more likely to result in damage to the oocyte. So far babies conceived as a result of micromanipulation seem healthy, but it is too early to determine the long-term consequences.

The chance that a spontaneous pregnancy will occur in an infertile couple is relatively high but depends on the cause of the infertility. In one survey, more than 50% of the pregnancies in couples with diagnosed cervical factors, endometriosis (in which endometrial tissue grows, misplaced in the oviduct, ovary or peritoneal cavity), partial tubal disease and moderate sperm defects occurred without treatment; whereas couples with anovulation and total tubal occlusion had higher conception rates following treatment. In addition to learning about the available treatments for infertility and their efficacy, couples may wish to explore other alternatives, including adoption and surrogate parenthood, or they may choose not to become parents at all.

SURROGATE PARENTHOOD

Whatever its antiquity and cultural pedigree, surrogacy, when used in conjunction with the reproductive technologies, is a relatively recent phenomenon. In human reproduction a surrogate is defined as a woman who agrees to bear a child on behalf of an infertile couple. Unlike adoption, an agreement is made before conception for the surrogate to become pregnant and bear a child for a specific couple and, in commercial cases, to be paid for her services. In a general sense, using surrogate mothers to overcome childlessness is an old concept: Genesis 16 versus 1–6 relates how Hagar, the servant girl, bore a son for Abraham because his wife Sarah was infertile. The most common form of technologically based surrogacy, known as genetic or partial surrogacy, occurs when the surrogate is inseminated using sperm from the husband of the infertile woman or, in situations where the husband is also infertile, by donor sperm. In either case, since she provided the oocyte to be fertilized, the surrogate is the biological mother. A couple would be likely to contract for partial surrogacy where the woman is unable or unwilling (in cases of inheritable genetic disease) to provide either the genetic (ova) or the gestational (uterus) components for childbearing. Gestatory, gestational or total surrogacy involves the surrogate being implanted with an embryo produced *in vitro* from donor gametes and the surrogate provides only the gestational component of reproduction. This situation is less common than genetic surrogacy because it is technically more complicated and requires extra resources such as IVF availability. An interesting early case was that of the South African 'granny' surrogate, 48-year-old Pat Anthony, who gave birth to triplets in 1987 from embryos genetically belonging to her own daughter and son-in-law. A couple would be likely to contract for gestatory surrogacy where the woman has a missing or malformed uterus or has a severe medical condition (hypertension, diabetes) contraindicating pregnancy. Modern surrogacy challenges traditional assumptions about parenthood because artificial reproduction procedures make it possible to separate out the various phases of the reproductive process. It is now possible for a child to be subject to multiple parenting with two different men (genetic and adoptee) fulfilling the differing functions of 'father' and up to three different women (genetic, gestational and adoptee) fulfilling the differing functions of 'mother'. The birth of such children also holds wider kin implications for grandparents or other relatives.

A surrogate arrangement can take one of two forms. The most common form is commercial surrogacy and usually involves drawing up a contract with specific items to be agreed to and fulfilled by both parties. The distinguishing feature, however, is payment of a fee agreed to prior to conception and, usually, paid after the child has been safely surrendered to the commissioning couple. The second form of surrogate arrangement is altruistic surrogacy where a woman agrees to bear a child on behalf of a couple but requires no payment or reward in return. Most reported cases of altruistic surrogacy have involved private agreements between family members and/or close friends. Altruistic surrogacy is not an unusual concept as in all societies there are instances where a woman gives her baby to another individual soon after it is born; for example, children can be brought up by aunts, uncles or grandparents. As a commercial contract, surrogacy has developed into big business since the late 1970s when the practice quickly spread from the USA to Canada, western Europe, Australia and thence globally. Along the way all forms of surrogacy have attracted controversy, praise, condemnation and government scrutiny. However, compared to the social issues alluded to at the beginning of this chapter, surrogacy has attracted a disproportionate amount of publicity and concern.

General references

Barrett, C. L., Chauhan, M. M. & Cooke, I. D. (1990). Donor insemination—a look to the future. *Fertility & Sterility*, **54**, 375–387.

Carp, H. J., Toder, V., Mashiach, S., Nebel, L. & Serr, D. M. (1990). Recurrent miscarriage: a review of current concepts, immune mechanisms, and results of treatment. *Obstetrical & Gynecological Survey*, **45**, 657–669.

Cohen, J. (1991). Assisted hatching of human embryos: review. *Journal of in vitro Fertilization and Embryo Transfer*, **8**, 179–190.

Collins, A., Freeman, E. W., Boxer, A.S. & Tureck, R. (1992). Perception of infertility and treatment stress in females as compared with males entering *in vitro* fertilization treatment. *Fertility & Sterility*, **57**, 350–356.

Davies, M. C. (1991). Pregnancy without ovaries. *Contemporary Reviews in Obstetrics & Gynaecology*, **3**, 119–125.

Gagnon, C. (1988). The role of environmental toxins in unexplained male infertility. *Seminars in Reproductive Endocrinology*, **6**, 369–376.

Keel, B. A. & Webster, B. W. (eds.) (1990). *CRC Handbook of the Laboratory Diagnosis and Treatment of Infertility*. CRC Press, Boca Raton, FL.

Lazarus, L. (1991). Creutzfeldt–Jakob disease and human pituitary-derived hormone therapy. *Fellowship Affairs, Royal Australasian College of Physicians*, **10**, 21–28.

Macnab, A. J. & Zouves, C. (1991). Hypospadias after assisted reproduction incorporating *in vitro* fertilization and gamete intrafallopian transfer. *Fertility & Sterility*, **56**, 918–922.

Oehninger, S. & Alexander, N. (1991). Male infertility: the focus shifts to sperm manipulation. *Current Opinion in Obstetrics & Gynecology*, **3**, 182–190.

Pride, S. M., James, C. J. & Yuen, B. H. (1990). The ovarian hyperstimulation syndrome. *Seminars in Reproductive Endocrinology*, **8**, 247–260.

Robert, E., Pradat, E. & Laumon, B. (1991). Ovulation induction and neural tube defects: a registry study. *Reproductive Toxicology*, **5**, 83–84.

Sathananthan, A. H. (1992). Ultrastructural pathology of *in vitro* fertilization. In: *Diagnostic Ultrastructure of Non-neoplastic Diseases*, eds. D. W. Papadimitriou, D. W. Henderson & D. V. Spagnolo, pp. 435–450. Churchill Livingstone, New York.

Sauer, M. V., Paulson, R. J. & Lobo, R. A. (1990). A preliminary report on oocyte donation extending reproductive potential to women over 40. *New England Journal of Medicine*, **323**, 1157–1160.

Seibel, M. (ed.) (1990). *Infertility: A Comprehensive Text*. Apple Lange, Norwalk, CT.

Seidel, G. E. (1991). Embryo transfer: the next 100 years. *Theriogenology*, **35**, 171–180.

Tadir, Y., Wright, W. H., Vafa, O., Liaw, L. H., Asch, R. & Berns, M. W. (1991). Micromanipulation of gametes using laser microbeams: review. *Human Reproduction*, **6**, 1011–1016.

2 Sex determination and gamete maturation

OUR ORIGINS

The foundation for normal adult reproduction is set during fetal life. Sex, as are all other functions, is the product of genetics and the environment working as one informational unit which is so complex that, under normal conditions, their separate contributions cannot be isolated. Prenatal and postnatal development takes place in an environmental context, that is, the combination of the individual's changing milieu and life experience. Epigenetic influences, that are all the external variables which modulate the activity of genes, become more evident the older an individual becomes. Age, education, experience, conditioning and disease all shape appearance, thought, emotion and behaviour. For example, babies look more like each other than do fully formed adults, and genetically distinct adults may have more in common with their peers than with themselves at an earlier age. The process of development starts in the ovaries and testes of the parents where the cells destined to become the future gametes are first differentiated as a distinct germ line and, on fertilization, forge the generational link between parent and offspring.

Since epigenetic effects can be operative at any time during the differentiation of the gametes, the health and living conditions of both parents from the time of gamete formation to the conception of the offspring are crucial, as is the mother's situation during pregnancy (Chapters 8 and 9). There are aspects of early development which, if uncorrected perinatally, may have long-term effects upon the individual. Fetuses have mechanisms by which they adapt to deteriorating environmental conditions brought about by drug abuse, disease, nutritional deprivation and non-adaptive lifestyles. Adaptation is subject to genetic variability and, if the limits of genetic adaptability are exceeded, fetal growth retardation and/or organ damage may result. Should epigenetic influences adversely affect the sexual differentiation of the fetal germ cells, then a changed genetic programme may be perpetuated in the offspring of a subsequent generation. Conversely, reproduction under good conditions creates a positive force in shaping human identity.

SEXUAL DIFFERENTIATION

The development of an individual follows an orderly process consisting of:

(a) Chromosomal sex at the time of fertilization
(b) Gonadal sex
(c) Formation of the sexual phenotype

The above sequence of events is sometimes referred to as the Jost paradigm, named after Alfred Jost's classical experiments carried out in the 1940s. Jost demonstrated that in mammals it is the presence, or absence, of fetal testes which determines the sex of the developing fetus. The subsequent differentiation of the accessory sex structures follows the primary differentiation of the gonad, whose endocrine secretions in turn influence their development. Thus, each step in the process of sexual differentiation is dependent on the preceding one, and, under normal circumstances, chromosomal sex agrees with the phenotypic sex. Occasionally, however, chromosomal sex and phenotypic sex do not agree, or the sexual phenotype may be ambiguous.

Chromosomal sex

Chromosomal sex is determined at fertilization when the sperm provides either an X chromosome resulting in a 46, XX zygote or a Y chromosome resulting in a 46, XY zygote. Scientists have known the chromosomal basis of sex since the early 1900s, but not until 1959 was it shown that the presence of the Y chromosome is essential for the development of males. This followed observations on the karyotype of a male with Klinefelter syndrome (XXY) and a female with Turner syndrome (XO). For the past 30 years, ever-smaller regions of the Y chromosome have been the focus in the search for the genetic source of masculinity. Finally, early in 1991, British researchers announced that a gene capable of switching mammalian development from its default female pattern to a male one had been identified. In humans (as in all mammals) no matter how many X chromosomes are present, a testis will develop as long as a Y chromosome is also present. However, in both men and women, infertility can arise from an incorrect number of X chromosomes. For example, Klinefelter's syndrome individuals (XXY, affecting about one boy in 1000) are male in general appearance with underdeveloped testes and enlarged breasts due to impairment of normal development by products of the extra X chromosome. Turner's syndrome individuals (the most frequent human chromosomal disorder of recognised pregnancies with most affected XO conceptuses aborting spontaneously) are infertile, sexually undeveloped women. It is of interest to note that during early fetal life of the XO individual seemingly normal ovaries develop, which then progressively lose their oocytes so that, at birth, the infant's ovaries are reduced to connective tissue remnants resulting in permanent sexual immaturity. During the evolution of the XX/XY sex-determining system, the development of X-dosage compensation, which works to equalize the expression of X chromosomes in normal males and females, was necessary.

The human Y chromosome is divided into two portions: the euchromatic portion comprises the entire short arm and one third of the proximal long arm. The remaining portion is heterochromatic (condensed, inactive in DNA transcription). The euchromatic segment on the short arm contains the testis-determining gene (termed *TDF* in the human and *Tdy* in the mouse) and other genes required for normal spermatogenesis. The British team, led jointly by Lovell-Badge from the MRC National Institute for Medical Research and Goodfellow from the Imperial Cancer Research Fund, London, isolated the single gene that diverts the bipotential embryonic gonad from the default ovarian pathway to that of testis differentiation and masculinization. This testis-determining gene they refer to as *SRY* (sex-determining region of the Y chromosome) in humans and *Sry* in mice. Final proof of the role of *Sry* came from transgenic mice possessing the normal female XX chromosomes plus *Sry*. These developed into males with normal testes and subsequent male behaviour. Identification of *Sry* does not totally elucidate the process of differentiation, but it does mark the end of an exciting chapter in the search for maleness. The complete process of sexual differentiation is a complicated one as many genes are implicated in human sex determination, some of which are located on the sex chromosomes and some on the autosomes. It seems that normal male phenotypic differentiation depends on the presence of *TDF* together with the participation of at least 19 X-linked and autosomal genes common to both male and female embryos.

Genes controlling ovarian differentiation are located on both X chromosomes because normal differentiation of the primitive gonad into an ovary occurs only in the presence of two intact X chromosomes; the XO woman (Turner's syndrome) is an undeveloped female possessing infantile sex organs and is frequently sterile. Additionally, as the embryo differentiates, functional differences may result from alternative female-specific splicing of RNA transcribed from common X-linked and autosomal genes.

Gonadal sex

Until a human embryo is about 5 weeks old it lacks any kind of sex organs, and 12 weeks later male sexual development is completed. At approximately 5 weeks of gestation when the embryo is about 8 mm long, a ridge of cells, the gonadal ridge or genital crest, destined to develop into the future testis or ovary grows out of each mesonephros, the temporary excretory organ of the embryo. The gonadal ridge cells will differentiate into the somatic cells of the future gonad: in the male, the Sertoli and interstitial Leydig cells and in the female the granulosa and theca cells. The germ cells themselves, destined to differentiate into oocytes and spermatozoa, arise from an entirely different location outside the embryo proper, the epiblast or dorsal endoderm of the yolk sac. They actively migrate, by means of amoeboid movement guided by chemotactic attractive forces, timing their arrival to coincide with the formation of the gonadal ridges which they invade. Once inside the gonadal ridges, the germ cells multiply by mitotic divisions. Now the undifferentiated gonads have all the cellular components necessary to develop into either testes or ovaries.

Around 6–7 weeks of gestation, human fetal gonads show the first microscopic signs of tissue differentiation, but only in gonads destined to develop into testes. Ovarian differentiation is retarded by several weeks compared to that of the testes. Cortical and medullary regions of the primitive gonad are discernible at this time and if a testis is to develop it will arise from the medulla while the cortex regresses. If an ovary is to develop, the cortical elements are the ones to differentiate while the medullary portion undergoes regression. The sudden appearance of a large number of Sertoli cells and the subsequent organization of seminiferous cords (later to become the seminiferous tubules of the adult testis) are the first microscopic signs of testicular differentiation. Starting from the central part of the testis and spreading towards the periphery, the seminiferous cords are formed by the alignment of connected Sertoli cells enclosing the germ cells. The proliferating sex cords establish contact with the ingrowing mesonephric connections, the rete testis cords, later to constitute the rete testis leading to the efferent ducts and to the epididymis (Chapter 5).

It can be assumed that the *TDF* gene is expressing at, or prior to, the differentiation of the male-supportive Sertoli cell lineage, but from the time of conception there are clear differences between potential ovaries and potential testes. The XX and XY gonads differ in their rate of development even before the Sertoli cells commence to differentiate. For example, in the rat at 13.5 days of gestation, the gonads of XY fetuses are on average 40% larger than those of their XX litter mates, and male newborns are normally heavier and bigger than their sisters. This growth discrepancy commences soon after fertilization as early cleavage is faster in male conceptuses. In humans, ultrasound scans reveal that male fetuses are already ahead of their female counterparts by 6 weeks of age when visible signs of sexual differentiation have only just begun. It has been suggested that the crucial function of the male-determining gene on the Y chromosome is to accelerate the development of the male fetus in order to set the conditions in which the Sertoli cells can develop. It is essential for the testes to differentiate early in development because unless the germ cells are enclosed inside the seminiferous cords they either degenerate or enter the first stages of meiosis. If this earliest stage in the differentiation of egg cells occurs, an ovary will develop in a genetically male fetus. Factors retarding fetal growth incorporate the inherent risk of a desynchronization of sequences critical to male sexual differentiation. While a female fetus can afford to develop at leisure, becoming a male is a race against time, highlighting perhaps the first competitive event in a male's life.

Phenotypic sex

The somatic cells of the fetal testes produce two classes of hormones. The first is Müllerian inhibiting hormone (MIH), a glycoprotein of Sertoli cell origin which demolishes the Müllerian duct or potential female reproductive tract and conditions the seminiferous epithelium. The second is testosterone, a steroid of Leydig cell origin which stabilizes the Wolffian or potential male reproductive tract and directs the development of the external genitalia.

The reproductive tract

Phenotypic sex denotes the development of the reproductive tract through which, in the adult, the germ cells will eventually travel. Each embryo originally develops two sets of potential reproductive tracts, the mesonephric or Wolffian duct, and the paramesonephric or Müllerian duct. The Wolffian duct has the potential to develop into the male reproductive tract, including vas deferens, epididymis, seminal vesicle and ejaculatory ducts, while the Müllerian duct can give rise to the female reproductive structures, including the oviducts (fallopian tubes), uterus and upper part of the vagina. In any embryo, depending on the genetic sex, one set of ducts develops while the other regresses.

Dissolution of the basement membranes and condensation of the mesenchymal cells around the ducts is an early event in Müllerian regression. Not much is understood about the mechanism of action of MIH although the gene expressing MIH has been identified. MIH is not solely responsible for Müllerian duct regression as steroid hormones also influence the process. When remnants of Müllerian duct persist it is usually accompanied by cryptorchidism, a failure of the testes to descend. Another physiologically important function of MIH is connected with the further morphological differentiation of Sertoli cells and the formation of junctional complexes that form a crucial part of the Sertoli cell–testis barrier. The strategic location of adherent junctions between adjacent Sertoli cells allows the development of anatomical and functional subdivisions of the seminiferous epithelium. This barrier is absent at birth, but it becomes effective at puberty, concomitant with spermatogenesis. All aspects of sexual differentiation depend on joint interactions between Sertoli and germ cells which regulate a set of autocrine (within the same cell) and paracrine (on adjacent cells) developmental substances. It is not so much the circulating hormones, but rather the endocrine environment on site effected by cell to cell communication which is crucial in early development.

With completion of the basement membrane along the base of the seminiferous cords, a second stage of development in the fetal tests commences. Starting at approximately 8 weeks of gestation, the steroid-secreting interstitial or Leydig cells differentiate between the sex cords. Maximal cell numbers are reached at 14 weeks when Leydig cells occupy 50% of the relative area of the human fetal testes. Subsequently the numbers decrease with only a few present at term. Corresponding fluctuation in testosterone secretion causes the virilization of the Wolffian duct, urogenital sinus and genital tubercle (see next section).

The external genitalia

Until the 8th week, the external genitalia are identical in both sexes and have the capacity to develop, from rudiments common to male and female embryos, in either direction. Under the influence of testosterone these rudiments are converted into male genitalia; in the absence of testosterone the female genitalia are formed. Figure 2.1 depicts successive stages in the development of undifferentiated human external genitalia through the crucial 12–14 week point at which male and female fetuses can be distinguished by inspection of the external genitalia. In the male, testosterone is responsible for the development of the genital ducts, that is, Wolffian duct virilization, while the development of the external genitalia depends on a metabolite of testosterone, 5α-dihydrotestosterone (DHT). In the absence of testosterone and DHT the Müllerian duct develops, apparently autonomously, and the endocrine differentiation of the fetal ovary, as evidenced by the beginning of oestradiol synthesis, also occurs. Oestradiol is not essential for female phenotypic development but is involved in the differentiation of the primordial follicles. The granulosa cell of ovaries from 8–20-week-old fetuses convert testosterone to oestradiol which promotes a further cascade of ovarian development (see next section).

Prior to the synthesis of any hormone by a tissue, enzyme differentiation is necessary. For instance, as illustrated in Fig. 2.1, the urogenital sinus and genital tubercle can, on acquiring 5α-reductase, convert circulating testosterone to DHT. Similarly, in the fetal testis the progressive increase in 3β-hydroxysteroid dehydrogenase activity is reflected in increased testosterone synthesis from its precursors pregnenolone and androstenedione. In the fetal ovary oestradiol production reflects increased aromatase activity. It follows, then, that changes in the rates of only a few enzymatic reactions at a critical time in development may have profound consequences for sexual differentiation. Initially the Leydig cells synthesize testosterone as an autonomous function of the steroidogenic cells, but later in differentiation, when functional receptors for luteinizing hormone (LH) and follicle stimulating hormone (FSH) are present, testosterone and oestrogen synthesis is regulated quantitatively by the gonadotrophins, either gonadotrophin releasing hormone (GnRH) from the fetal pituitary and/or chorionic gonadotrophin (hCG) from the placenta (Chapter 9). It is during the development of the fetal hypothalamic–pituitary–gonadal axis that male and female fetuses are vulnerable to epigenetic influences. These influences have the power to modify fetal GnRH secretion and permanently modify the endocrine feedback axes in the adult (Chapters 13 and 16). Unlike the fetal ovary, where granulosa aromatase expression is quantitatively important, synthesis of aromatase in Sertoli cells of the testis is inhibited, probably by MIH also of Sertoli cell origin.

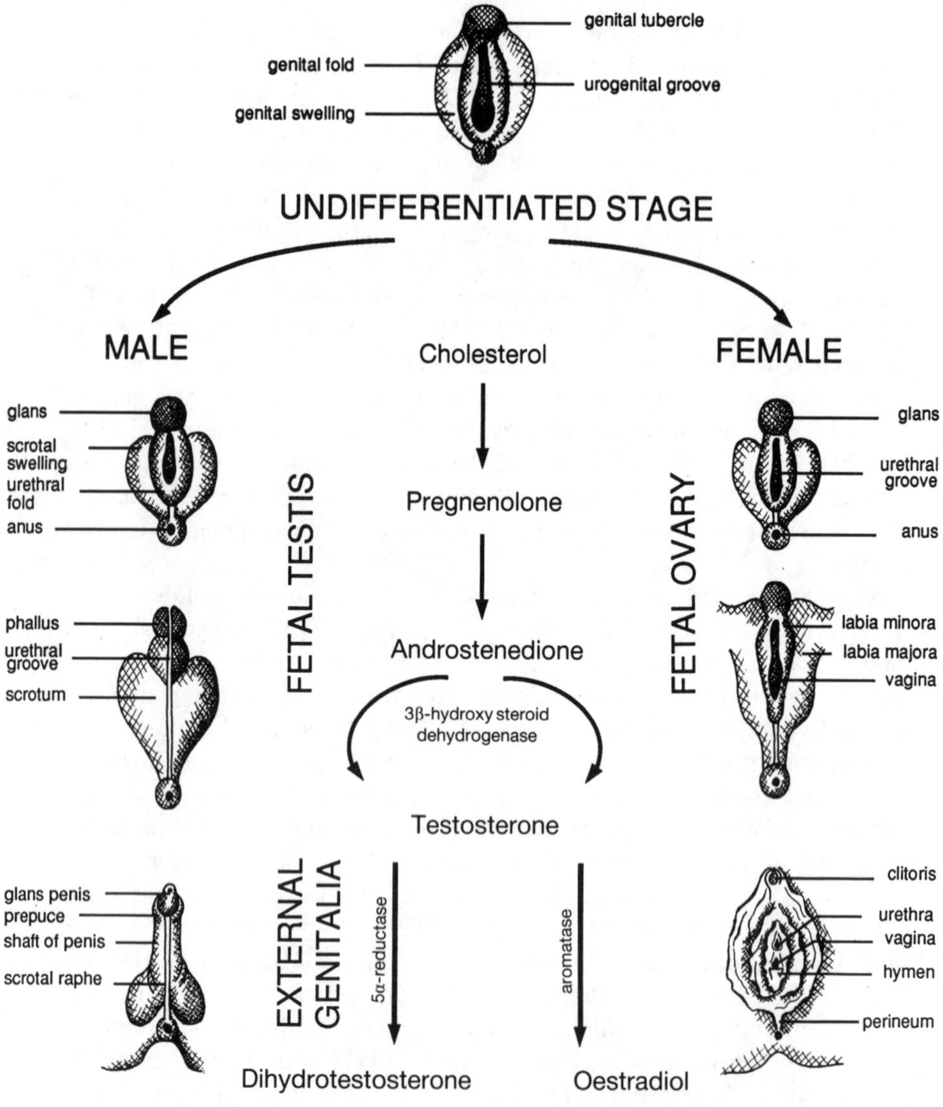

Fig. 2.1 The differentiation of the external genitalia from common primordia consisting of undifferentiated labio-scrotal folds that are located on either side of the urogenital groove. In the female these folds remain separate and become the labia minora and majora. In the male they fuse to form the corpus spongiosium and the scrotum. Typically by 12–14 weeks male and female fetuses can be distinguished by inspection of the external genitalia.

PSEUDOHERMAPHRODITISM

Gonadotrophin-dependent testosterone synthesis further regulates the growth and differentiation of the external genitalia and descent of the testes. DHT is an androgen that is 10 times more potent than testosterone and consequently local 5α-reduction in target tissues possessing 5α-reductase can greatly amplify the androgenic signal. Ambiguous genitalia in a newborn pose immediate problems of gender assignment, rearing and prolonged health concerns since many forms of hermaphroditism are accompanied by a high incidence of gonadal malignancy. Congenital anomalies which result in the failure to achieve normal sexual dimorphism may be due to intrinsic defects in gene expression and/or epigenetic modification of normal gene expression. Cases of ambiguous genitalia exist in which testosterone synthesis is normal but 5α-reductase is deficient due to a genetic error. The documentation of such an enzyme deficiency in the human population living in the Dominican Republic has attracted much publicity because, probably due to inbreeding, the incidence of male pseudohermaphroditism (normal karyotype 46,XY) is unusually high. Although pseudohermaphroditism, also called by the clumsy name pseudo-vaginal perineoscrotal hypospadia, is not a life-threatening malformation it is socially and clinically significant when moderate to severe. Hypospadia represents a continuum from small anomalies in penis morphology of no functional concern to severe cases of ambiguous genitalia. The severe cases are always raised as girls because at birth each child has a labial-like scrotum, a blind vaginal pouch and a clitoris-like phallus. The testes and derivatives of the Wolffian duct system, such as the epididymis and the vas deferens, are normally differentiated. At puberty, however, there is an absence of breast development and in a proportion of the boys the voice breaks, muscle mass increases, the phallus enlarges to become a penis and spermatogenesis is initiated. The postpubertal psychosexual orientation is often masculine although early female childhood conditioning can cause some conflicts. Late virilization in these boys may be caused by the combination of higher levels of plasma testosterone at puberty than during embryogenesis and by the presence of some residual 5α-reductase activity.

GAMETOGENESIS AND MATURATION

There are chronological and morphological differences between the processes of spermatogenesis and oogenesis, although both share the common goal of forming a haploid genome and have many aspects of maturation common to both the male and female germ cells. Male germ cell maturation involves the maintenance of a mitotic stem cell population in the adult, whereas in the female, germ cells enter into meiotic prophase during fetal life. Male and female germ cells also differ with respect to the stages of greatest morphogenesis: oocytes undergo their period of growth while arrested in meiosis, sperm are formed after meiosis when they are genetically haploid. The mature sperm is a highly polarized cell having a morphology that is distinctive in each mammalian species and has a singular function, the transmission of the haploid genome to the egg at fertilization. This mission has influenced selection for the sperm cell's architectural organization and metabolic processes. Maturation in the oocyte, on the other hand, is a process endowing it with the ability to complete meiosis, undergo fertilization and prepare male and female haploid chromatin for syngamy and embryo development. The structural and regulatory proteins needed to complete these processes result largely from transcription at an earlier growth stage of the oocyte with translation or post-translational modification of proteins occurring shortly before fertilization. Essential proteins, particularly those with structural or housekeeping functions, may be stored in the oocyte for a long period (Chapter 8).

The ovum: chronology of oogenesis

Although the ovary was recognized as an anatomical entity by Herophilus of Alexandria in about 300 BC, the mammalian egg or ovum was not identified until early in the 19th century. De Graaf, in the 1670s, had recognized that eggs came from the ovary but thought that the entire Graafian follicle was an egg. It was not until microscopes became more powerful and generally available that good morphological descriptions enabled further physiological understanding. In 1825 von Baer, an Estonian biologist, correctly described the exact anatomical relationship between the egg and the follicle in mammals.

Figure 2.2 summarizes the main events of ovum maturation or oogenesis.

Fig. 2.2 The various stages of oogenesis. Oogonia develop from primordial germ cells which, after a number of mitotic divisions, enter meiotic division I and are then called primary oocytes. Shortly before ovulation the primary oocyte completes meiosis I and becomes a secondary oocyte, or unfertilized egg,

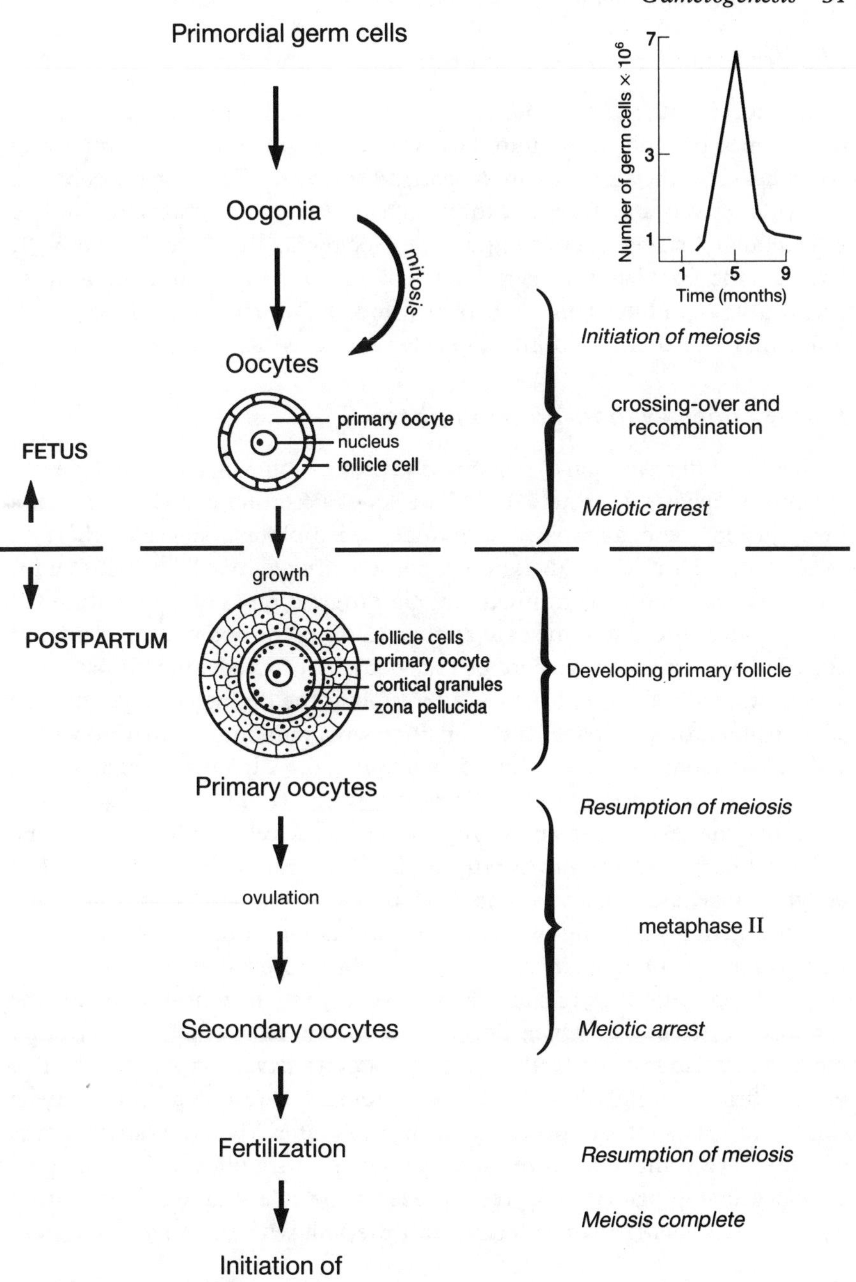

arrested in metaphase of meiosis II. The second meiotic division is completed only if the secondary oocyte is fertilized. *Inset* depicts the numbers of germ cells in the ovaries of the fetus.

The primordial germ cells or oogonia

At about 3 months of gestation, the large amoeboid primordial germ cells, now termed oogonia, move into the cortex where, together with supporting epithelial cells, they give rise to the cortical sex cords. The oogonia continue to divide actively and begin to exhibit a characteristic morphology including intercellular bridges connecting adjacent oogonia. By about 5 months of fetal life the population of germ cells has increased by mitotic division to about 6 or 7 million (Fig. 2.2, *inset*), and it is a subset of these which differentiate further into primary oocytes and, eventually, primary follicles.

Primary oocytes: non-growing phase

Following differentiation of oogonia, the proportion which do not degenerate enters meiosis at around 11–12 weeks of gestation and then proceeds through leptotene, zygotene, pachytene and diplotene stages of the first meiotic prophase. Once meiotic prophase commences, there is no endocrine requirement for continued meiotic progression as oogonia located at ectopic sites also enter meiosis at that time, as do male germ cells not enclosed in seminiferous cords. The oocytes proceed approximately in synchrony with meiosis. Meiosis is arrested at the diplotene stage, the last stage of meiotic prophase, and transforms oogonia into primary oocytes. Granulosa (somatic) cells begin to surround the diplotene stage oocytes forming primary (primordial) follicles at between 16–21 weeks of gestation. Towards the last trimester of pregnancy, thecal cells with steroidogenic potential surround the developing follicles. During differentiation many oocytes degenerate because the first meiotic prophase is a stage that involves genetic crossing-over and recombination, a high-risk procedure which may result in pairing anomalies, chromosomal breaks and other errors. Epigenetic factors such as cytotoxic agents may also influence the proportion of oocytes which degenerate. Germ cells which do not enter meiosis are also lost. At birth, nearly all oocytes have completed prophase of the first meiotic division and will remain arrested at the dictyate (diplotene) stage of late prophase until years later when ovulation occurs and they are stimulated to resume meiosis. It is significant that this pool of non-growing primary oocytes represents the sole source of unfertilized eggs in the sexually mature female and the link with the next generation.

Primary oocytes: growing phase

Each oocyte is contained within a cellular follicle that grows concomitantly with the oocyte and depends for its nutritional and functional control on the follicle cells. The granulosa and theca, and their paracrine/autocrine communication network generated by gap junctions, is crucial to the regulation of oocyte maturation. Atresia continues with approximately 50% of the follicles degenerating following the first 2 weeks after birth. However, while some follicles disappear, others are recruited into a phase of growth. Recruitment of oocytes into the growing pool is under gonadotrophin control and during the prepubertal period certain of the developing follicles may even form an antrum. However, since adequate GnRH stimulation is necessary for complete follicular development, these follicles also degenerate.

Recruited primary follicles progress through a series of co-ordinated and definable morphological stages while remaining arrested in the first meiotic prophase. Oocyte growth is concerned with the accumulation and storage of substances necessary to initiate early embryogenesis. The unfertilized egg contains an extensive store of material ready for mobilization following successful fertilization. For example, a fully grown mouse oocyte contains approximately 200 times the amount of RNA, 1000 times more ribosomes and 50–60 times more protein, excluding the zona pellucida, than is found in a typical somatic cell. The human oocyte likewise undergoes a 500-fold increase in volume and becomes one of the largest, if not the largest, cell of the body. The increase of the nucleus or germinal vesicle is due in particular to accelerated ribosomal-RNA synthesis. As ribosomal-RNA accumulates it can be used, on fertilization, for the immediate synthesis of ribosomes thus allowing increased protein synthesis. Concomitantly, the Golgi apparatus expands its activity of processing and concentrating many necessary secretory products such as the zona pellucida (a thick extracellular coat surrounding all mammalian eggs), glycoproteins and cortical granules (Chapter 8). Contact between the oocyte and the innermost layer of follicle cells is maintained via junctional complexes formed between the oocyte microvilli and the follicle cell extensions which penetrate the zona pellucida. The zona pellucida contains sperm receptors which, at fertilization, mediate sperm–egg interaction and act as a secondary block to polyspermy. A cytoplasmic lattice, or fibrillar arrays, in the cytoplasm is also layed down and probably serves to increase storage space for ribosomes and yolk.

In summary, the dictyate pool of primordial follicles consists of somatic and germinal constituents that grow co-ordinately through a series of well-characterized phases. At the completion of growth, the follicle contains an enlarged oocyte with prominent germinal vesicle, or nucleus and nucleolus, enclosed within an extracellular coat, the zona pellucida. The somatic component includes inner layers of columnar cells, the corona radiata or cumulus oophorus, a number of layers of granulosa cells with a fluid-filled cavity or antrum, and an outer layer of thecal cells in communication with blood vessels.

Oocyte maturation: the aquisition of meiotic competence

Meiotic maturation refers to the conversion of fully grown oocytes, present in antral follicles, into unfertilized eggs just prior to ovulation and following the preovulatory LH surge (Chapter 4). Meiotic maturation involves nuclear progression from dictyate of the first meiotic division to metaphase of the second meiotic division. Other metabolic changes necessary for egg activation at fertilization also take place. Maintenance of meiotic arrest is dependent on the integrity of the cumulus–oocyte complex and uncoupling of the oocyte from the cumulus provides the appropriate stimulus for resumption of meiosis. Although the details of the mechanism by which meiotic arrest is mediated are not clear, it has been postulated that inhibitory levels of cyclic AMP, generated by the oocyte or more likely transferred to it from the cumulus, are involved. An alternative theory postulates that a maturation-promoting factor (MPF) mediated by hypophysial and/or follicular hormones is responsible for meiotic maturation.

In culture, fully grown oocytes undergo meiotic maturation spontaneously in the absence of added hormones or other special factors. For instance, mammalian oocytes matured and fertilized *in vitro* exhibit normal preimplantation development and develop into viable fetuses on transplantation into receptive uteri. If this were not so, the whole technology surrounding IVF could not have developed. Normally there is, under increasing LH stimulation, a significant decrease in the number of gap junctions coupling the cumulus cells and the oocyte, and this may be the critical factor necessary to induce meiotic maturation *in vivo*. Spontaneous meiotic maturation *in vitro* may also take place as a result of expansion and sloughing off of the cumulus cells which become detached from the oocyte, thus removing inhibitory influences.

The mature sperm

In the normal male, sperm production, which is initiated at puberty, is a continuing process lasting into senescence. As in the female, there is a succession of critical steps in the development of the sperm until it finally becomes capable of fertilization. Spermatogonia remain quiescent throughout childhood but during early adolescence begin to proliferate into primary spermatocytes with meiosis occurring at puberty. The control of spermatogenesis is fully described in Chapter 5.

Surprisingly, it was only in the 19th century that the role of sperm in fertilization was discovered, a result of the combination of better microscopic lenses with the development of the cell theory. Sperm were first described in 1678 by two independent investigators, the famous Dutch microscopist Anton von Leeuwenhoek and Nicolas Hartsoeker. Leeuwenhoek believed sperm to be parasitic animals living in semen, hence their name spermato–zoa, animacules or sperm animals. Hartsoeker, on the other hand, believed that the entire embryonic individual, the homunculus, lay preformed within the head of the sperm. Despite Hartsoeker's more penetrating insight, the prevailing dogma for the next 200 years was that sperm had nothing to do with reproduction, because to accept the alternative hypothesis was to also accept the waste of a vast potential of male homunculae. Later in the 18th century, Spallanzani observed that filtered toad semen, devoid of sperm, could not fertilize eggs. However, the existing dogma was so entrenched that, despite this ingenious experiment, even Spallanzani himself thought that the sperm were parasites and that the viscous fluid retained by the filter paper was the agent of fertility. Eventually, in 1821, Dumas and Prevost claimed that sperm were the active agents of fertilization and not parasites and, further, that the sperm actively penetrated the egg, thus contributing materially to the next generation. To strengthen their argument they pointed out the absence of sperm in the sterile mule. This claim was disregarded – an unfortunately recurring theme in the history of science. Finally, in 1876 Hertwig demonstrated the penetration of the egg by the sperm and the union of their nuclei in the Mediterranean sea urchin.

The specialized structural features of the mature spermatozoon reflect its unique function and comprise two main components, a head and a tail. The head consists of the acrosome, the highly condensed nucleus and lesser amounts of cytoskeletal structures. The acrosome, which contains enzymes essential for fertilization, is a large secretory granule closely surrounding and overlying the anterior end of the nucleus. The flagellum or tail contains the energy sources necessary to produce motility and consists of a centrally placed axoneme, a highly ordered complex of microtubules surrounded by dense fibres extending from the head to near the posterior end of the axoneme, and mitochondria wrapped in a tight helix around the dense fibres. The tail, like the head, is closely wrapped by the plasma membrane and contains little cytoplasm. It is the plasma membrane which is unique in the sperm as it is regionally delineated into functional surface domains. This regional differentiation of the sperm surface results in components associated with specialized functions being located in specific domains. Substances involved in the acrosomal reaction are associated with the plasma membrane of the acrosome region and substances involved in flagellar activity are associated with the flagellum.

DIFFERENTIATION OF SEXUAL BEHAVIOUR

The brain itself is also a sexually differentiated organ. In the female, for example, the pattern of pituitary gonadotrophin secretion controls the normal ovulatory cycles. For the establishment and maintenance of male sexual function both morphologically and behaviourally co-ordinated hormonal interactions are necessary. The hypothalamus of both sexes evolves toward the female cyclic secretion of gonadotrophins unless the endocrine hypothalamus is exposed to androgens during prenatal and early postnatal development which results, after maturation, in an irreversible differentiation of the nervous system with a tonic, constant secretion of LH and FSH. In this, the hypothalamus mirrors the differentiation of the gonads. In the newborn female rat, for example, a single injection of testosterone results in the absence of cyclic gonadotrophin release and anovulation in adult life due to the establishment of a tonic (male) GnRH secretion pattern.

Sexual behaviour, likewise, is dependent for its expression on genetic coding but is vulnerable to modifying epigenetic influences during critical periods in differentiation and to conditioning. The sexual behaviour of rats can be altered by manipulating the testosterone level during sexual differentiation of the brain; for example, exposure to androgen is necessary for the development of male copulatory behaviour and prenatally androgenized female monkeys play in a manner more typical of males. Information concerning the neuronal basis of human sexuality is difficult to obtain as adequate animal models of complex human behaviours cannot be developed, consequently masculinization of the neurogenic mechanisms in humans is less well understood. While social influences and sex steroids play a major role in adult gender identity, there is also evidence for a role of early hormonal influences on sexual behaviour. However, in the area of sexual behaviour more than elsewhere, the individual must be studied in the context of the product of evolutionary history, individual genotype and environmental development. In the evolution of human sexual behaviour in particular, natural selection seems to have incorporated a flexible strategy relying for its expression not so much on hormonal but on cerebral control. This inherent flexibility may explain why, despite consistent effort, scientific attempts to explain diverse sexual behaviours biologically, have met with little success. For instance, no consistent hormonal or genetic differences have been found between heterosexual, bisexual and homosexual preferences, although a neurohormonal theory of sexual orientation may be supported by findings that prenatally androgenized women may have a higher incidence of homosexual feelings. These reports have generated a revived interest in evidence that homosexuality is, at least in part, a biological phenomenon.

Inconclusiveness is even more poignant in the less well understood and accepted experience of transsexuals. It is estimated that, one case of transsexualism is found in every 100 000 of the male population and every 400 000 of the female poulation in the USA (Johns Hopkins University). In Sweden and England, the incidence is significantly higher where 1 in 40 000 males and 1 in 100 000 females are reported to be transsexual. The majority (approximately 80%) who request and successfully go through gender reassignment are men. Similarly, little is understood of other sex-related human behaviours existing since antiquity as part of the sexual repertoire in all human cultures. These behaviours can be categorized as destructive (rape, incest and paedophilia) or wayward (voyeurism, exhibitionism, fetishism and prostitution). In western society there is a preoccupation with the need to categorize 'natural masculinity' and then apply this norm to all men. In the 1940s, soon after the sex steroids had first been synthesized, there were many attempts to prove that homosexual males were different from heterosexual males. Old and new findings have been equally inconclusive as no statistically valid study exists that is based on random sampling from the entire male population. Writings over the last few hundred years, however, indicate evolving socially accepted views as to what it is to be masculine and these represent a continuum from homosexuality to heterosexuality, or in biological language, a normal continuum of alternative phenotypes.

The Grafenberg spot

For women, as for men, there has been constant social pressure for an all-encompassing definition of female sexuality. The G or Grafenberg spot debate is a prime example. Wolffian and Müllerian ducts do persist, in remnant form, in the opposite sex. In the 1970s, these vestigial remains aroused interest in an acrimonious debate regarding the now celebrated G spot. The G spot in women, a vestigial homologue of the prostate, is located inside the anterior wall of the vagina. The debate revolved around how many genital foci of erotic arousal are possessed by women: the clitoris alone, or the clitoris together with the G spot. After Kinsey called attention to the clitoris as the main focus of sexual sensitivity, medical and laboratory studies concentrated on discovering the one universal mechanism of sexual response in women. When Masters and Johnson announced in 1966 that all orgasms were essentially the same, many sexologists felt relieved and adopted an even more dogmatic view than that of Masters and Johnson. This view was finally challenged in 1980 by sexologists Ladas, Whipple and Perry who, in their book *The G Spot*, demolished the prevailing either/or view and put forward the concept of orgasm as a continuum of sexual responses. Female orgasm, vaginally or clitorally induced, involves the whole body and may result in one or multiple climaxes. This heralded a new, less dogmatic era, acknowledging the orgasm to be only one part of female sexuality inseparable from other individual feelings that accompany the experience.

HORMONAL CONTROL OF BRAIN DIFFERENTIATION

The sexual differentiation of the mammalian brain comprises two stages:

(a) An early organizational stage integral to the development of the phenotype, that is, the development of the nervous system coinciding with sex determination
(b) A later activational stage which is activated at puberty.

As in the primary differentiation of the gonad, development of the central nervous system (CNS) is dependent upon an orderly sequence of hormonal influences. When phenotypic differentiation takes place, the brain also undergoes differentiation, coinciding with a critical period of maximal sensitivity to the effects of steroids. In mammals, timing of the sensitive period for the potentiation of male sexual behaviour by gonadal steroids varies from species to species. It occurs entirely prepartum in rhesus monkeys, humans and other mammals with long gestation periods, spans the perinatal period in rats and mice, and is postnatal in short-gestation species such as Syrian hamsters and marsupials like the possum. This variation is related to and depends on the timing of gonadal differentiation.

Brain sex in the male is not, however, solely determined by prenatal exposure to testosterone, as oestradiol-17β is also critical for differentiation of the CNS. This holds true for all mammals, including humans; however, dihydrotestosterone (a non-aromatizable androgen) may also have a role in the organization of masculine behaviour in primates. Thus, aromatization of testosterone is, at least in part, an important prerequisite for androgen action, but whether this applies to all effects of androgen in the CNS is not clear. Normally, early Leydig cell function is co-ordinated with the development of the male CNS and oestrogen is formed from testosterone by aromatization. Synchronized regulation of testosterone levels, the synthesis of steroid receptors and aromatase activity in the CNS, determines the development of neuronal mechanisms destined to regulate adult masculine behaviour. Female fetuses make no masculinizing quantities of testosterone so other pathways of biochemical action develop within the brain's organization.

Viviparity in mammals carries with it the potential danger that androgens and oestrogens produced by the pregnant female might cross the placenta and reach brain receptors of female fetuses and so induce a degree of masculinization. For centuries it has been known that when cattle carry twins of differing sex, the female is often sterile and shows various degrees of morphological and behavioural masculinization (or defeminization). These are the 'freemartin' females. However, an error in sexual differentiation occurs only if the female's placenta has been in anastomosis with that of her twin brother's. Laboratory experimentation has demonstrated that such developmental modification is caused by testosterone, MIH and, possibly, whole somatic cells passing from the male twin to the female twin via the anastomosis. After testosterone treatment in adulthood, female rats which have developed *in utero* between two males display more mounts than do females that have developed in the same uterine horn without males. These observations demonstrate that the early organizational phase of female brain differentiation is influenced by the environment and can, as a result, be primed with hormones to exhibit strong male behaviours. Normally a variety of protective mechanisms exist to diminish this danger. Steroids are transported primarily in the plasma and are, for the most part, bound to proteins. The free steroids, in constant equilibrium with the bound fraction, are the only active component. Corticosteroid-binding globulin (CBG) and sex hormone-binding globulin (SHBG) increase during pregnancy or following administration of oestrogens and thyroid hormones. Transport proteins vary from species to species: in rodents, unlike in humans, SHBG is absent and oestrogens are bound by α-fetoprotein. Male fetuses exposed prenatally to the synthetic oestrogen diethylstilboestrol (DES) appear to have been feminized/demasculinized (Chapter 16). DES was used in the 1960s for the treatment of at-risk human pregnancies and its potent hormonal activity may be due to its lack of binding to plasma protein.

General references

Bailey, M. J. & Pillard, R. C. (1991). A genetic study of male sexual orientation. *Archives of General Psychiatry*, **48**, 1089–1096.

Bianchi, N. O. (1991). Sex determination in mammals. How many genes are involved? A review. *Biology of Reproduction*, **44**, 393–397.

Blyth, B. & Duckett, J. W. (1991). Gonadal differentiation: a review of the physiological process and influencing factors based on recent experimental evidence. *Journal of Urology*, **145**, 689–694.

Döhler, K. D. (1991). The pre- and postnatal influence of hormones and neurotransmitters on sexual differentiation of the mammalian hypothalamus. *International Review of Cytology*, **131**, 1–57.

Gondos, B. (1991). Gonadal disorders in infancy and early childhood. *Annals of Clinical & Laboratory Science*, **21**, 62–69.

Haug, M., Brain, P. F. & Aaron, C. (eds.) (1991). *Heterotypical Behaviour in Man and Animals*. Chapman and Hall, London.

Koopman, P., Gubbay, J., Vivian, N., Goodfellow, P. & Lovell-Badge, R. (1991). Male development of chromosomally female mice transgenic for Sry. *Nature*, **351**, 117–121.

McLaren, A. (1991). Development of the mammalian gonad: the fate of the supporting cell lineage. A review. *BioEssays*, **13**, 151–156.

Nonomura, N., Nakamura, M., Namiki, M., Kiyohara, H., Mizutani, S., Okuyama, A. & Sonoda, T. (1991). Mixed gonadal dysgenesis: reports and a review of 65 Japanese cases. *Archives of Andrology*, **26**, 15–19.

Rabinovici, J. & Jaffe, R. B. (1990). Development and regulation of growth and differentiated function in human and subhuman primate fetal gonads. *Endocrine Reviews*, **11**, 532–557.

Tho, S. P. T. & Behzadian, A. (1991). Detection and amplification of sequences. *Seminars in Reproductive Endocrinology*, **9**, 46–55.

Tongli, X., Blackburn, W. R. & Gardner, W. A. (1990). Fetal prostate growth and development. *Pediatric Pathology*, **10**, 527–537.

3 Neuroendocrine control of puberty

Puberty is often defined as the period of first becoming capable of sexual reproduction. It is marked by the adolescent growth spurt, maturation of the genital organs, development of secondary sexual characteristics, first ejaculation in boys and onset of menstruation (the menarche) in girls. There are also psychosocial characteristics resulting from the effects of increased gonadal sex steroid production and the resultant maturational changes in the CNS. Puberty is a temporal process involving transition and development. One conspicuous feature of sexual development in the human is the prolonged interval between birth and initiation of the pubertal process. Initiation of spermatogenesis and ovarian cyclicity does not normally occur until early in the second decade of life, and then only after a stage of 'entering' puberty lasting several years has taken place.

A change in gonadal activity as a major component of puberty was apparent more than 2000 years ago as knowledge of the effects of castration during prepubertal development began to accumulate. Similarly, knowledge that these changes in gonadal activity are of a humoral nature is not modern either. The ancients used castration as a form of punishment, and well beyond the Middle Ages castration was used in European cultures to retain the quality of the unbroken male voice and ally it with the lung capacity of the adult. These days, of course, the pleasant counter tenor is the result of voice training not castration. In the early 1900s, transplantation experiments gave new insights in the understanding of the mechanism of puberty. It was observed that transplantation of the ovaries from an immature animal into an adult led to the premature onset of pubertal activity in the transplanted gonad. This observation later led to the important law of puberty which states that the onset of pubertal changes in the gonad is determined by the maturation of a somatic component, rather than by maturation of the gonad itself. The onset of puberty is a brain-driven event via pulsatile release of hypothalamic gonadotrophin releasing hormone (GnRH). It is initiated even in the absence of gonads and puberty can be activated prematurely by exogenous pulses of GnRH.

THE PHASES OF PUBERTY

The complete maturational and functional process of primate puberty can be considered by phases from development *in utero* through pubertal development which is the transitional, often difficult period between childhood and adulthood. Four stages can readily be identified: the developmental or fetal stage, the neonatal stage, the prepubertal juvenile stage and the pubertal adult stage. Generally one stage has to be completed successfully before the next is initiated.

In the fetal stage, gonadal steroidogenesis progressively increases, reaching levels comparable with the onset and first half of puberty before declining around birth. The episodic release of GnRH and gonadotrophins, with intact negative feedback, demonstrates that the hypothalamic–pituitary–gonadal axis is functional at an adult level by the time of birth. Puberty is merely the reactivation or potentiation of the pre-established functions and characteristics. The second, neonatal phase encompasses the period towards the end of gestation and the first few months of postnatal life. During this time hormonal secretion is greater than during the remainder of childhood. Secretion of gonadotrophins and sex steroids increases after the first few days of life and persists for weeks without triggering the physical changes that might be expected with such hormonal stimulation. For the first few months of the neonatal period the human baby has intermittently high circulating levels of oestrogens and androgens, such as testosterone, before these stabilize at low levels. The third juvenile phase starts in the first year and lasts from 9 to 13 years. It begins with a decreased hypothalamic stimulation that results in a decrease of circulating levels of gonadotrophins and persists throughout childhood. The fourth pubertal phase is marked by a resurgence of GnRH, gonadotrophin and sex steroid secretion. This increased secretion is initiated by release of gonadotrophins in pulses coincident with sleep. With maturation, this pulsatile pattern of secretion becomes regularly episodic and advances progressively, involving more of the sleep period until waking time and eventually encompassing the entire 24-hour period. This final phase is defined as adolescence where the body changes progressively into that of the adult male or female.

The first menstruation, or menarche is often considered the obvious signal that adulthood has commenced. In most young women, however, ovulation does not occur until 6 or more months after menarche, and regular ovulatory menstrual cycles are not established in many until several years later. Consequently, fertility is low during these years, a fact acknowledged by Pacific South Sea islanders who were allowed to be reasonably promiscuous during this time. Sexual freedom was considered socially advantageous because it was believed that experience would mature the adolescents into responsible adults. At marriageable age (or if the girl became pregnant) a strict code of behaviour was enforced to ensure a reasonable certainty of paternity. First ejaculation in boys likewise does not necessarily signify fertility as, at this time, seminal plasma typically lacks mature spermatozoa. Prolonged adolescence is characteristic of all primates and is adaptive allowing the opportunity for increased peer socialization, communication and learning.

GONADAL FUNCTION DURING PUBERTY

The testis

Growth of the testis during the first decade of postnatal life is negligible. The adult volume, however, is some 6–12 times that of the prepubertal volume and one of the earliest heralds of male puberty is the initiation of a rapid increase in testicular size between 9.5 and 13.5 years of age. This increase in testicular size results largely from an increase in the diameter and length of the seminiferous tubules due to the appearance and proliferation of Sertoli cells and spermatocytes. It is during this period that the Sertoli cells acquire the apparatus needed to support spermatogenesis in the mature male, including the development of the organelles supporting metabolic and synthetic activities. The testis of the immature human male contains few, if any, recognizable Leydig cells and testosterone secretion is minimal. Leydig cell development and the consequent rise in circulating testosterone concentration (blood levels rise from approximately 0.2 ng/ml to 6 ng/ml) precedes the acceleration in testicular growth.

The initial activation of testosterone secretion occurs nocturnally via pulsatile GnRH and gonadotrophin release. Androgens, testosterone and 5α-dihydrotestosterone in particular, are the final cause of sexual, metabolic and systemic changes accompanying puberty. The most obvious physiological and anatomical effects induced by androgens may be summarized as follows.

- Stimulation of a taller, leaner body growth with muscular development; protein anabolism
- Increase in hair growth

- Maturation of sexual organs (penis, prostate, seminal vesicles) and thickening of the vocal cords; deepening of the voice
- Increase of libido and aggressiveness via CNS interactions
- Increase in sebaceous gland activity and pheromone production.

The ovary

In contrast to the postnatal pattern of testicular growth, the size of the ovary increases in a linear fashion from infancy to adulthood. Moreover, growth and atresia of ovarian follicles occur throughout childhood (Chapter 2). The enlargement of the ovary prior to puberty is the result of an age-related increase in the number and size of antral follicles and in the quantity of medullary stroma. Graafian (preovulatory) follicles, however, are not observed during prepubertal development. In girls between 8 and 10 years of age, circulating oestradiol concentrations begin to increase to values typical of adult women during the early follicular phase of the menstrual cycle. It is at this stage that the first physical signs of puberty, such as breast enlargement and the appearance of labial hair, are observed. Augmentation of follicular growth and steroidogenesis, without concomittant ovulation, continues as early puberty progresses with the further development of the secondary sexual characteristics. The menarche occurs, on average, between 11 and 13 years of age.

Oestrogens, oestrone and oestradiol-17β in particular, cause major sexual, metabolic and systemic changes that accompany puberty. These effects may be summarized as follows.

- Stimulation of lipid and carbohydrate metabolism resulting in fat deposition and a female form
- Induction and maintenance of a proliferative endometrium within the uterus; maintenance of ciliary and secretory elements within the fallopian tubes (oviducts)
- Production of cervical mucus and maintenance of a healthy vaginal mucosa
- Stimulation of breast development, blood clotting rate, bone maturation and strength.

Figure 3.1 illustrates the reproductive organs of the mature male and female.

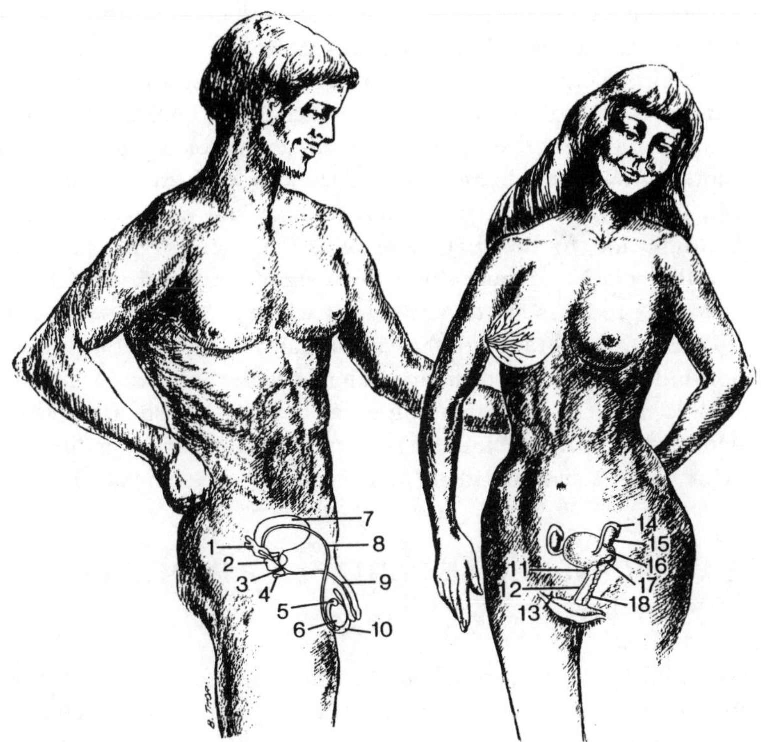

Fig. 3.1 The reproductive organs of the male and female. (1) Seminal vesicle, (2) prostate gland, (3) ejaculatory duct, (4) Cowper's gland, (5) epididymis, (6) testis, (7) urinary bladder, (8) vas deferens, (9) urethra, (10) scrotum, (11) urinary bladder, (12) urethra, (13) clitoris, (14) fallopian tube, (15) ovary, (16) uterus, (17) cervix, (18) vagina.

THE ADRENARCHE

In both sexes, puberty is controlled by two independently sequential processes. The reactivation of the hypothalamic–pituitary–gonadal unit is preceded by increased androgen and oestrogen secretion by the adrenal glands, the adrenarche, which occurs 2 to 3 years before gonadal maturation, the gonarche, in children 6 to 10 years of age. The progressive rise in dehydroepiandrosterone and androstenedione is needed for the adolescent growth spurt around the time of puberty. The postperinatal growth velocity is identical in girls and boys but girls commence their pubertal growth spurt about 2 years earlier than boys, who are later advantaged due to their higher circulating levels of testosterone. The timing of the growth spurt depends largely on skeletal maturation rather than chronological age; girls, compared to boys, are advanced in puberty because of advanced skeletal maturation. It may be that the Y chromosome delays skeletal maturation. Increasing secretion of sex steroids indirectly stimulates linear growth by increasing endogenous GH and somatomedin secretion. In girls, lower levels of testosterone and androstenedione of ovarian origin add to the total pool of adrenal androgens as puberty advances. Biological evidence (although its significance was then obscure) that the ovary produces androgens dates back to the 1930s, when it was observed that ovaries, grafted to the ears of male mice, were able to reverse castration-induced atrophy of their seminal vesicles and prostate gland.

In adolescent girls, important pubertal changes which are androgen dependent include the development of pubic and axillary hair, increased libido, changes in sebaceous gland activity and bothersome acne.

THE NEUROENDOCRINE CONTROL OF PUBERTY: AN OVERVIEW

Progression into puberty varies from individual to individual with hereditary and environmental factors contributing to this variance. Generally speaking, in warmer, sunny climates puberty is attained earlier than it is in colder climates. The effect of sunlight in particular, is discussed in the last section of the chapter (p. 51). The stimulus for the onset of puberty is generated within the CNS and transmitted to the anterior pituitary by means of the intermittent hormonal signal of GnRH from the hypothalamus. Because the neural mechanisms activating changes in the pattern of GnRH release are not fully understood, it is described in 'black box' terminology as a GnRH pulse generator. The hypothalamic GnRH pulse generator, together with the pituitary gonadotrophs, comprise the basic neuroendocrine unit responsible for LH and FSH secretion. The transition into puberty in both boys and girls reflects increases in the frequency and amplitude of the GnRH pulse which is responsible for the activation of the pituitary–gonadal axis found in adolescence. Increased circulatory sex steroids, in turn, influence the secretory activity of the gonadotrophs by their actions on both the GnRH pulse generator and the anterior pituitary gland. Both oestrogens and androgens can potentiate the effects of GnRH on LH and FSH release (Fig. 3.2).

In the absence of feedback by gonadal hormones, the hypothalamic GnRH pulse generator of sexually mature males and females appears to

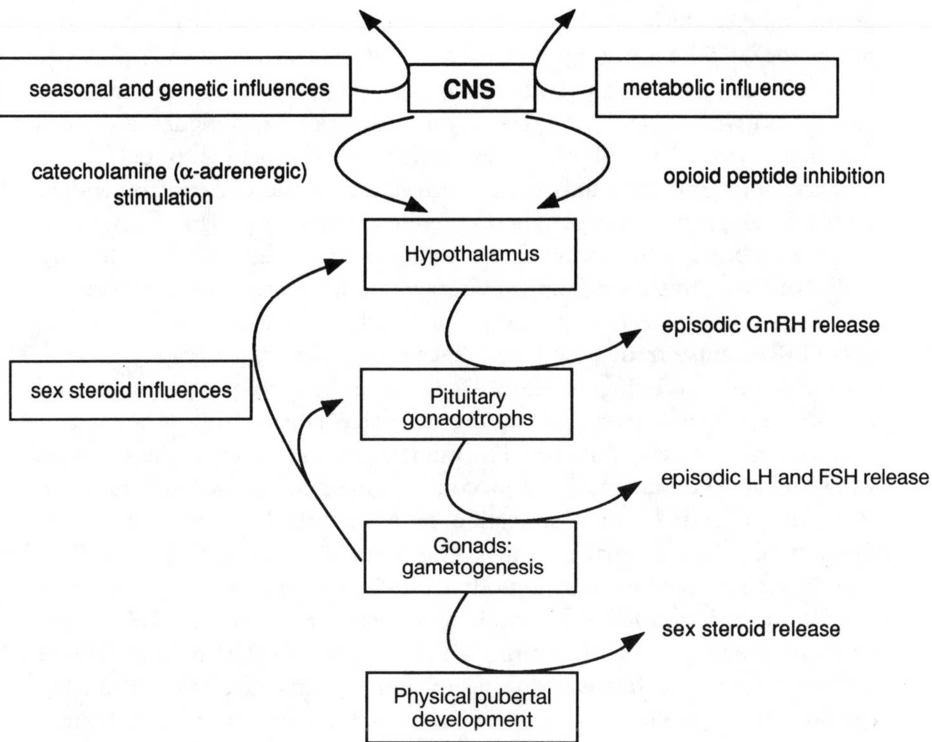

Fig. 3.2 Schematic hypothesis of the hypothalamic–pituitary–gonadal events preceding the onset of puberty. Higher cortical systems are involved in modulating the GnRH pulse generator. The frequency and amplitude of the GnRH pulses activate pituitary gonadotrophin secretion which stimulates gonadal synthesis and release of sex steroids responsible for the physical manifestations of puberty. Sex steroids, in turn, modulate hypothalamic–pituitary activity.

exhibit an operational frequency of approximately 1 cycle/hour. This adult pattern is observed in ovariectomized and postmenopausal women and in hypogonadal men. If the normally episodic release of GnRH becomes continuous this results in a downregulation of FSH and LH receptors and a decreased synthesis and release of FSH and LH. This effect forms the basis for the treatment of sexual precocity in children using GnRH analogues; continuous stimulation with GnRH analogues, for 5 to 7 days, results in a decrease of circulating levels of LH and FSH into the hypogonadotrophic range. Precocious puberty may occur in children as young as 2 years old as a result of premature gonadotrophic stimulation of the gonads. Tina Medina, a Peruvian child with precocious puberty, had a son at 5 years of age. Many cases of central precocious puberty, especially in girls, are idiopathic in origin; other cases have occurred in association with CNS lesions including head trauma, tumours and third ventricle cysts. In children with peripheral precocious puberty, the hypothalamic–pituitary axis is not activated and the cause of secondary sexual development is circulating sex steroids. These can originate endogenously from either gonadal or adrenal tumours or be ingested (Chapter 16).

There are many neurochemical interactions which control GnRH secretion and the final trigger of puberty is uncertain. It is undisputed, however, that GnRH pulse frequency is influenced by genetics, geography, health, nutrition and psychological state. Deprivation of love, psychological and prolonged physical stress, including intensive competitive training and exercise, may all delay puberty. The human is a very social primate species whose social interactions have profound influences on gonadal function (Chapters 6 and 7); the connection between social factors and sexual maturation deserves further study. The immediate regulation of GnRH-secreting neurons within the hypothalamus has been linked to a variety of neurotransmitters, such as catecholamines which stimulate GnRH release, and opioid peptides (β-endorphin) which suppress GnRH release. Opioid neurotransmitters, effective at times of vulnerability, are released during periods of physiological stress such as pain, exercise and starvation (conditions unfavourable for reproduction).

It has been suggested that the genetic time-frame of puberty is ultimately tuned by blood-borne substances which convey information about the growth and nutritional status of the body and direct the activity of the GnRH secretory system. The identity of these signals remains unknown but could include metabolic information relating to carbohydrate or protein utilization and influencing hormonal (for example, insulin, GH) biochemistry. It is obvious that growth to maturity is required before reproduction can take place. The notion that a particular body size or composition may provide the trigger for the initiation of puberty is very attractive because it provides the simple hypothesis that the ability to produce the correct LH pulses is dependent upon a favourable energy balance or level of nutriton. Frisch from Harvard University and her colleagues have pointed out the close correlation between age at menarche and body weight. For example, girls living in the USA experience menarche when body weight reaches approximately 47 kg, and these authors propose that the attainment of a body weight in the critical range causes a change in metabolic rate. This in turn reduces the sensitivity of the hypothalamus to oestrogen, altering the ovarian–hypothalamic feedback and leading to the reawakening of the dormant GnRH pulse generator. Similarly, a critical weight of 55 kg has been suggested as necessary to precede sexual maturation in boys. Frisch later refined her hypothesis to one in which menarche is more closely correlated with the attainment of a particular proportion of body fat. Either hypothesis may explain the well established trend over this century toward an earlier age at menarche in European and north American girls, since the attainment of the threshold body weight or a particular fat composition at a progressively younger age is a result of improvements in the standard of living. In Norway in the 1840s, for example, the average age of puberty was between 16 and 18 years, but this had declined dramatically to between 12 and 14 years by the 1960s. On the other hand, pubertal development and menarche is delayed in competitive athletes and ballet dancers, as it is in geographical areas where nutrition is only marginally adequate (Chapter 18).

Other researchers favour the hypothesis that sexual maturation is not directly related to the proportion of body fat. Rather, puberty is dependent on a stable threshold body weight driven by an adequate overall energy balance. From an evolutionary perspective, such a mechanism makes sense; our nomadic hunter/gatherer forebears would have been vulnerable to loss of body condition at times of food restriction, particularly in the winter months. For a woman to attain maturity and risk pregnancy during times of low caloric intake would not have been adaptive. The finely tuned energy balance required for the maintenance of fertility is discussed in Chapters 13 and 18. It is possible that the universal energy regulator, insulin, may prove to be the final metabolic trigger for puberty. Circulating glucose levels depend on the nutritional status and the integrated endocrine control mechanisms which maintain it within normal limits. Insulin, by increasing glucose utilization for glycogen and fat synthesis, prevents glucose levels from rising too high. Glucagon, catecholamines, corticosteroids and GH, by decreasing glycogen and fat synthesis, prevent circulating glucose levels from falling too low. In the context of the regulation of appetite, the primate hypothalamus has the capacity to monitor circulating insulin levels. A child has a metabolic milieu similar to that associated with fasting in the adult (that is, increased glucose utilization) which may explain the 'sweet tooth' so commonly found in children. The developmental patterns of those hormones with potential metabolic impact, such as the pubertal rise in GH secretion in humans and concomitant increase in circulating growth-promoting peptides, are probably important. Growth is inhibited in children suffering from mental and/or physical trauma, a fact acknowledged by psychiatrists when naming a particular form of dwarfism often seen in institutionalized children, as 'psychosocial short stature'. As mentioned before, adrenarche, the progressive rise in circulating concentrations of adrenal androgens, precedes the increased secretion of gonadal sex steroids by approximately 2 years. There is also the possibility that a putative somatometer responds to the maturation of the adrenal glands.

ADOLESCENT SEXUALITY: BEHAVIOURAL ASPECTS

Puberty represents the culmination of developmental processes over a prolonged period including the sexual differentiation of the brain. Adolescence means transition, a stage between child and adult. For the young person it can often be a difficult period because this transition involves enormous physiological change, in addition to psychological and mental adjustment. At the same time, there are changes in the responsibilities to, and the expectations of, society. It can also be an exciting, dynamic and creative period. During the extended period of adolescence there are critical periods similar to those observed in fetal development. The effects of gonadal steroids are multifocal, influencing the tissues producing them, their targets, the higher CNS centres and the anterior pituitary gland. In the establishment of both morphological and behavioural sexual patterns, hormonal interactions critical in driving mental and physical development are vulnerable to modifying epigenetic influences.

An essential part of sexual maturation is the formation of a personal identity which has a strong relationship to key developmental aspects in the formation of interpersonal relationships and the enhancement of self-esteem. Early relationships may be based on juvenile behaviour patterns where, usually a transient, desynchronization of the physio/psychosexual elements exist. The first coitus is usually a milestone in the continuum of psychosocial development. It is an event mostly remembered as the transition from childhood to adulthood. As summarized in Fig. 3.3, sexual identity has two key modalities. Firstly, gender identity which is defined by the individual and by physiology. Gender bestows a biological identity and is dependent on genetic and hormonal factors. Secondly, sex roles which are defined by society are variable and do not necessarily follow biological imperatives. Sex roles are influenced by peers and reflect subcultural values and practices. For instance, in western society a male is discouraged from crying, while in some other cultures crying is considered a natural release for emotional and physical upsets in both male and female. Role expectations may also be manipulated because sexual behaviour is an especially sensitive area, vulnerable to parental or peer control. A great many rules and expectations over dress, hair style, curfew

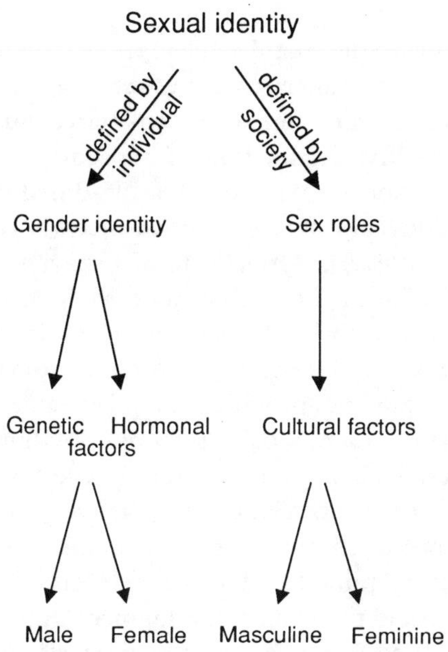

Sexual identity

defined by individual

defined by society

Gender identity

Sex roles

Genetic Hormonal factors

Cultural factors

Male Female Masculine Feminine

Fig. 3.3 Biological and cultural modalities in the development of a sexual identity.

and friends are in effect indirect attempts to regulate sexual behaviour, though this basic issue may never be faced directly and honestly. However, no matter what the environmental situation, each adolescent must still learn to deal with personal relationships and the establishment of an individual feel of the feminine or masculine. Stunting of sexual development is seen by many behaviourists as a contributing factor to the confusion of sexual and aggressive outlets. It should be remembered that libido is androgen dependent in both males and females, that both sexes possess all the steroid hormones, and that the differences are merely quantitative. The human female's potential for sexual arousal and orgasm at all times helps maintain long-term relationships and is an evolutionary device to maximize parental care of the young.

LIGHT, SOLTRIOL AND REPRODUCTION

The correlation between light and reproduction has often been noted in the context of seasonal breeding in many types of mammals. Biological clocks are synchronized with the environmental rhythms and enable the organism to anticipate the regular changes in conditions. The most prominent changes are the alternation between day and night (circadian) and the regular succession of annual seasons (circannual).

There is evidence that modulation of reproduction may involve the steroid hormone soltriol, or vitamin D (1,25-dihydroxycholecalciferol), which mediates a wide range of reproductive effects beyond its well-known regulation of calcium levels. The effects of ultraviolet light on the skin result in the production of the prohormone cholecalciferol (vitamin D_3) which is hydroxylated to the active soltriol in the liver, kidney, placenta and possibly other sites. The steroid hormone soltriol and the pineal hormone melatonin appear to provide a complementary endocrine regulation in the adaptation to the daily/seasonal rhythms of sunlight and darkness. Considerable changes in circulating levels of soltriol have been reported in the human in association with age and sex, during neonatal life and during puberty. In both sexes, a several-fold rise in circulating soltriol is seen during puberty. The peak levels correlate with breast development in girls and pubic hair development in boys rather than with chronological age. The increase in soltriol levels are paralleled by a decrease in melatonin levels, and a nocturnal fall in circulating melatonin has been observed during adolescence.

Humans, unlike most other mammals, reproduce throughout the year. However, several investigations have shown that the age at menarche may be related to seasonal factors. Studies in the USA, where latitude changes in daylength are pronounced, and in Japan and Holland have shown that menarche occurs at the highest rate in spring and summer and less frequently in autumn and winter. There is also a coincident increase in conceptions during the spring equinox. Thus, season may play some crucial role in determining the onset of menstruation within groups of women. Popular folklore has it that sexual activity can be related to the amount of light given by the different phases of the moon. Increased exposure to sunlight may, in part, explain earlier puberty in western cultures where children spend more time out of doors and a sun-bronzed body is considered an object of beauty.

The mechanism of action of soltriol is not understood. If its action is analogous to that of other steroid hormones then it is possible that it entrains a certain organizational biological rhythm during development. This responds, later in life, to activational changes in soltriol levels linking it to effects on the onset of puberty and changes in fertility. Whatever the mechanism, it makes adaptive sense for all mammals, including the human, to adjust their reproduction to optimal seasonal conditions and thus maximize their procreative success.

General references

Bogin, B. (1993). Why must I be a teenager at all? *New Scientist*, **1864** (March), 34–37.

Frisch, R. E. (1988). Fatness and fertility. *Scientific American*, **253**, 70–77.

Grumbach, M. M., Sizonenko, P. C. & Aubert, M. L. (eds.) (1990). *Control of the Onset of Puberty*. Williams & Wilkins, Marylands, MD.

Lee, P. A. (1988). Pubertal neuroendocrine maturation. Early differentiation and stages of development. (Mini review.) *Adolescent & Pediatric Gynecology*, **1**, 3–12.

Ojeda, S. R. (1991). The mystery of mammalian puberty: how much more do we know? *Perspectives in Biology & Medicine*, **34**, 365–386.

Perera, A. D. & Plant, M. (1992). The neurobiology of primate puberty. In *Functional Anatomy of the Neuroendocrine Hypothalamus*, eds. D. J. Chadwick & J. Marsh, pp. 252–267. John Wiley, Chichester, UK.

Pescovitz, O. H. (1990). The endocrinology of the pubertal growth spurt: review. *Acta Paediatrica Scandinavica*, **367** (Suppl.), 119–125.

Pescovitz, O. H. (1990). Precocious puberty. *Pediatrics in Review*, **11**, 229–237.

Roenneberg, T. & Aschoff, J. (1990). Annual rhythm of human reproduction. II. Environmental correlations. *Journal of Biological Rhythms*, **5**, 217–239.

Stumpf, W. E. & Denny, M. E. (1989). Vitamin D (soltriol), light, and reproduction. *American Journal of Obstetrics & Gynecology*, **161**, 1375–1384.

4 Control of the menstrual cycle

In the preceding chapters, the sexual differentiation of the gonads, functional characteristics of the gametes and reproductive activation at puberty were discussed. This chapter describes the mature ovary and control of the menstrual cycle. The 1980s was an exciting decade with major advances in understanding in this area following the discovery of the neuroendrocrine nature of the ovary. The ovary not only produces steroids but also a number of other hormones, regulatory peptides and growth factors. Some of these, like inhibin, have extra ovarian functions, whereas others, like oxytocin and relaxin, exert their function within the ovary. A number of cytokines (immune hormones) and growth factors, like epidermal growth factor (EGF), fibroblast growth factor (FGF) and insulin-like growth factor (IGF), regulate ovarian function from within by modulatory effects on immunity, steroidogenesis and connective tissue remodelling of differentiating cell types. Most importantly, at all times information regarding the ovarian condition is conveyed to the CNS by autonomic nerve fibres with which the ovary is intensely innervated. Such functional inter-relatedness can lead, at times of stress, to confusing symptoms when the causative factors may remain obscure. For example, in patients with nutritional defects, infertility caused by hypothalamic and ovarian disturbances may occur in seemingly healthy persons (Chapter 18).

In functional terms, the menstrual cycle can be viewed as the integration of three fundamental activities controlling fertility. These functions are the control of the ovarian cycle (maturation and release of the oocyte), the uterine cycle (provision of a suitable environment for the blastocyst to implant) and the cervical cycle (selective sperm entry into the female reproductive tract only during the periovulatory period).

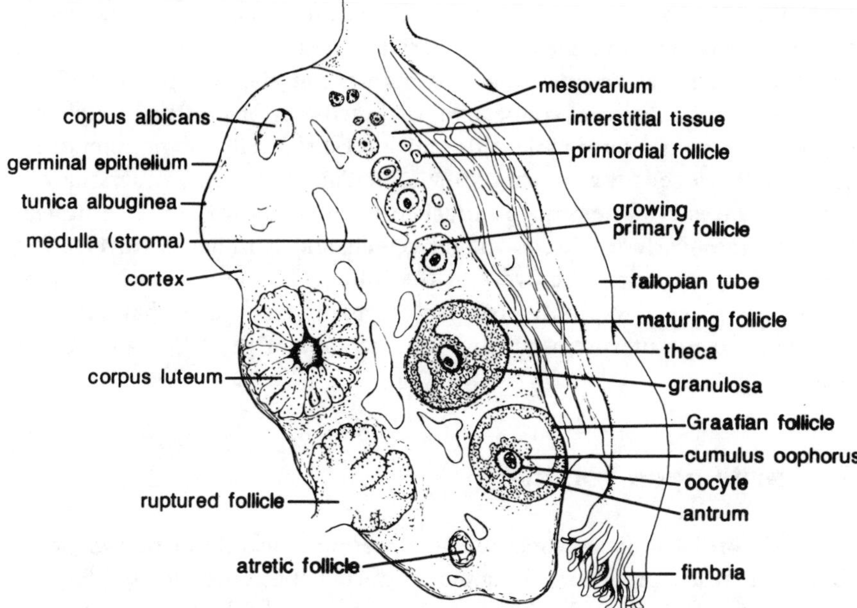

Fig. 4.1 Cross-section of the mature ovary showing events of the ovarian cycle.

GROSS ANATOMY

The ovarian follicles are the principle functional units providing the necessary support system for the oocyte to attain its potential on fertilization, that is, the development of a viable offspring. The follicular somatic cells contribute in many ways to this function, particularly through the early initiation of oocyte growth, the provision of nutritive requirements and the control of the subsequent maturation of the oocyte contained in follicles selected for ovulation and atresia of those that are not. Communication is both by the cumulus cells (the specialized granulosa cells in the innermost follicular layer surrounding the oocyte) which are coupled metabolically by gap junctions through which nutrient and regulatory molecules can be directly delivered to the ooplasm and by changes in the microenvironment within the follicular fluid bathing the oocyte. The hormones of the hypothalamic–pituitary–ovarian axis provide overall cycle integration.

Figure 4.1 is a composite diagram of the mammalian ovary providing a morphological representation of its cyclical physiological functions. The adult ovary is covered with coelomic epithelium, the germinal epithelium, overlying a tough connective membrane, the tunica albuginea. Immediately beneath the tunica albuginea is the ovarian cortex, consisting of the stroma and the follicles at various stages of maturation and degeneration. The cortical stroma consists of connective tissue cells performing the customary support functions, contractile cells and interstitial cells. Some interstitial cells are theca cells that remain after follicular atresia, and others have a different ancestry. The ovary is attached to the body wall by a peritoneal fold, the mesovarium. Nerves, blood vessels and lymphatics traverse the mesovarium and enter the ovary at the hilum. The uterine extremity of the ovary gives rise to the ovarian ligament, the cord that attaches the ovary to the uterus just below the entrance of the fallopian tube or oviduct.

The privileged follicle

The primordial follicles constitute the resting stockpile of non-growing follicles that are progressively depleted during the reproductive lifespan. During the fertile period of the woman, primordial follicles continuously leave the non-growing pool by being converted into primary follicles. This is followed by a period of growth and differentiation that culminates in either ovulation or, more likely, atresia. The follicle which eventually does ovulate is selected during the late luteal phase from a group of small follicles 2–5 mm in diameter. All follicles greater than 5 mm diameter become atretic during the luteal phase and, at the onset of menstruation there is, typically, only a single follicle whose follicular fluid contains measurable amounts of FSH. The mechanism by which a single follicle is selected for ovulation while others become atretic is not fully understood. It seems to be pure chance, not necessarily related to oocyte quality, that a particular follicle happens to be at the stage of its development to profit from a selective advantage over the other candidates. FSH, especially, enhances early follicle cell development and oocyte growth and a final selection takes place in the early follicular phase when a single antral follicle is activated to proceed to maturation. Products of this privileged follicle then exert dominance over other antral follicles ensuring their atresia. The follicles in the growing pool can become atretic at any stage of development; the follicles in the non-growing pool may also spontaneously degenerate. It is estimated that more than 99% of follicles becomes atretic in the human ovary.

Once the primary follicle is close to its maximum size, there is a second maturation phase in which follicle growth consists of rapid mitotic activity

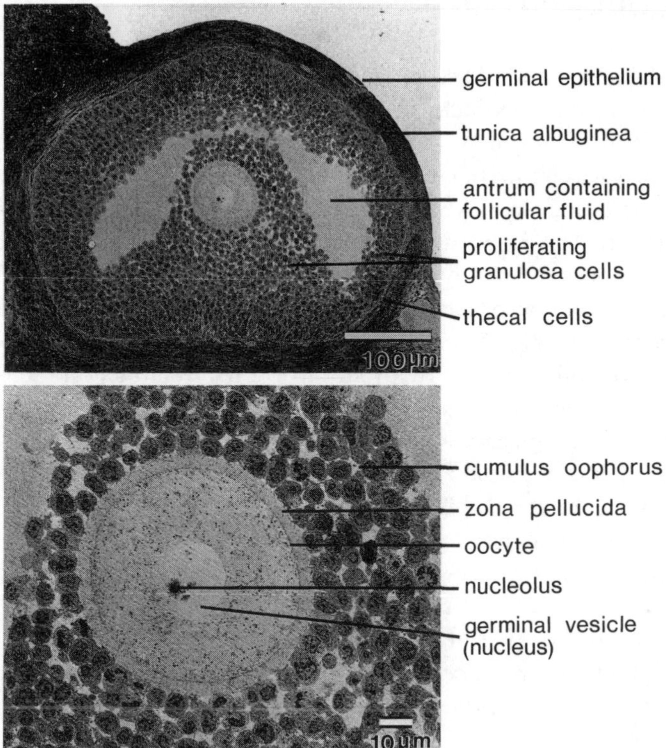

Fig. 4.2 Scanning electron micrograph of an antral follicle in the rat. Photograph Ms Coral Gilkeson, electron microscopist in Biological Sciences, Macquarie University.

of the granulosa and theca cells and the accumulation of follicular fluid. Formation of the antral cavity signals the transformation of the primary follicle into a Graafian follicle in which the oocyte occupies an eccentric position surrounded by the cumulus oophorus (Fig. 4.2). Ovulation consists of mechanical rupture of the follicle with extrusion of the oocyte and adhering cumulus oophorus and, therefore, marks the end of the follicular phase of the menstrual cycle. Following follicle rupture the mural granulosa cells undergo changes referred to as luteinization, and the surrounding theca cells and blood capillaries intermingle to give rise to the corpus luteum. Normally, unless pregnancy occurs, the functional life of the corpus luteum is 14 $\pm$ 2 days, after which it regresses and is replaced by an avascular scar, the corpus albicans.

STEROID HORMONES

The best known and characterized secretory products of the follicle are the steroid hormones which fulfill many functions related to reproduction. They function as hormones in the classical sense and act on a wide variety of target tissues comprising not only the reproductive system but also other systems including the CNS and the musculo-skeletal, cardio-vascular and immune systems, adipose and cutaneous tissues, and the liver. In addition to their endocrine function on structures remote from the ovary, these follicular steroids also act locally within the follicles both as paracrine agents (acting on adjacent cells) and as autocrine agents (acting on or within the cells in which they are produced). For example, granulosa cells influence, in paracrine fashion, the secretory pattern of the theca cells and the meiotic maturation of the oocyte; in autocrine fashion they influence their own mitotic rate and differentiation. Paracrine and autocrine influences are universally important in the control and synchronization of changes in physiological functions.

The three major classes of sex steroids, progestogens, androgens and oestrogens, are found in the ovary, placenta, testis and adrenal cortex. There is a high degree of similarity between the biosynthetic processes in the different steroidogenic tissues so the ovary may act as a model. Follicular steroids are produced from the catabolism of cholesterol derived from several sources.

- Preformed cholesterol taken up from the blood, primarily in the form of circulating lipoproteins; cholesterol associated with low density lipoprotein (LDL) is the major circulating form of steroidogenic cholesterol in humans
- Preformed cholesterol within the ovarian cells obtained either as free cholesterol or liberated from stores in cytoplasmic lipid droplets
- Cholesterol synthesized *de novo* in the ovarian cells from carbon components derived from cellular metabolism of carbohydrate, fat or protein.

The rate-limiting enzymes in cholesterol biosynthesis are under hormonal control, but cholesterol synthesis is also regulated by the intracellular levels of cholesterol through a negative feedback mechanism. If the dietary supply of cholesterol is insufficient, intracellular stores may become depleted resulting in increased *de novo* synthesis for essential steroid production.

The steroid hormones may be classified on the basis of either their chemical structure or their major physiological function. The hormones

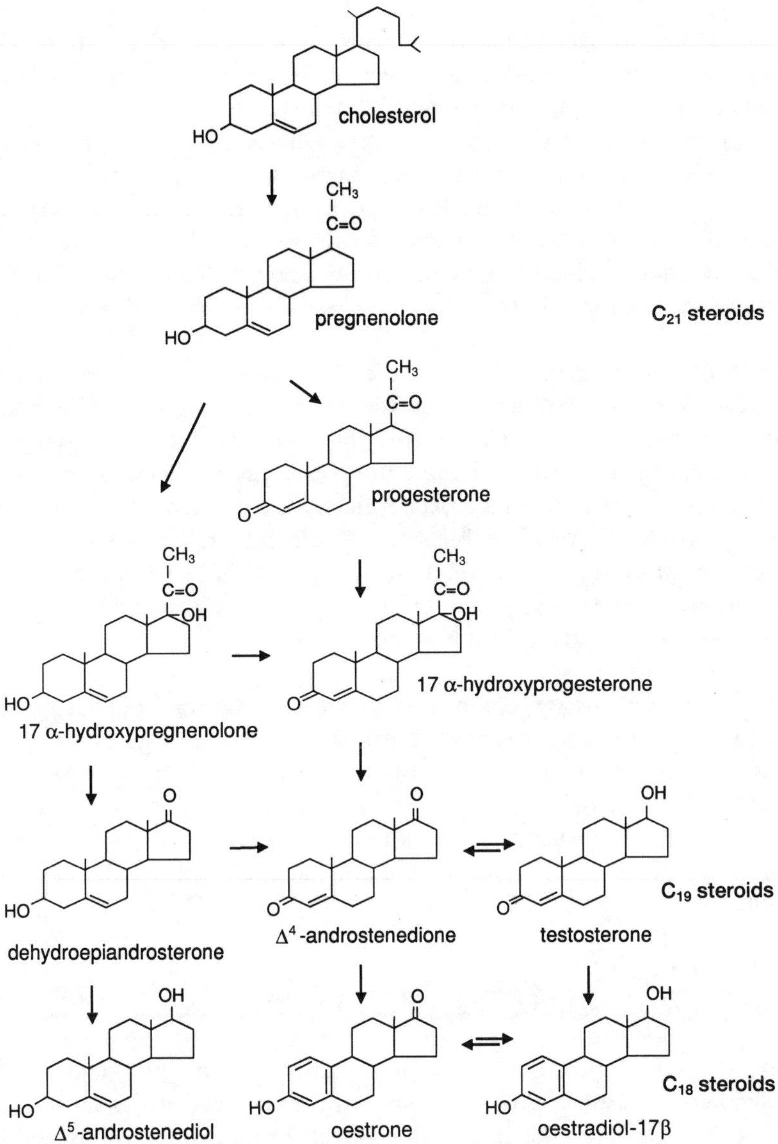

cholesterol

pregnenolone

C_{21} steroids

progesterone

17 α-hydroxypregnenolone

17 α-hydroxyprogesterone

dehydroepiandrosterone

Δ^4-androstenedione

testosterone

C_{19} steroids

Δ^5-androstenediol

oestrone

oestradiol-17β

C_{18} steroids

Fig. 4.3 Principal biosynthetic pathways for steroid production in the ovary.

of main reproductive interest are the sex steroids grouped as progestogens, androgens and oestrogens. Their major physiological actions have been described in the previous chapters. Chemical classification refers to the decreasing number of carbons as cholesterol is metabolized to the major C_{21}, C_{19} and C_{18} classes of steroid hormones (Fig. 4.3).

Progestogens Progestogens (progestins, gestagens) belong to the C_{21} (pregnane) series of compounds. Pregnenolone is a most important progestogen because it is the precursor of all the steroid hormones. Pregnenolone's most abundant C_{21} product is progesterone which is produced by follicles at all growing stages of development to supply the necessary precursor in hormone production and as a hormone in the peri- and postovulatory periods of the menstrual cycle. Other C_{21}-steroids of follicular origin include 17α-hydroxyprogesterone, 20α-dihydroprogesterone and the pregnenediols.

Androgens Androgens belong to the C_{19} (androstane) series of compounds. Biological evidence that the ovary is a significant source of androgens dates back to the 1930s with the observation that ovaries, grafted to the ears of male mice, were able to reverse castration-induced atrophy of the seminal vesicles and prostate glands. Androstenedione and testosterone are the immediate biosynthetic precursors of the oestrogenic steroids oestrone and oestradiol-17β, respectively. The 5α-reduced androgens 5α-dihydrotestosterone (DHT), androsterone and epiandrosterone have also been identified in the ovary.

Oestrogens The oestrogens belong to the C_{18} (oestrane) series of compounds. Oestrone and oestradiol-17β are the most important of the follicular steroids. Oestrone was the first sex steroid to be isolated and identified; in 1929 Doisy and his colleagues crystallized it from human pregnancy urine. Oestradiol-17β is approximately 10 times as potent as oestrone and, on a molar basis, is the most active of all steroids produced by the ovary.

PATHWAYS OF STEROID BIOSYNTHESIS

Figure 4.3 provides a visual representation of the principal ovarian biosynthetic pathways in the synthesis of the major progestogens, androgens and oestrogens. The first step in the conversion of cholesterol to steroids, and the rate-limiting step, is the production of pregnenolone. The enzyme system that catalyses this reaction is a multienzyme complex located on the mitochondrial membrane. Pregnenolone, the intermediate substrate common to all classes of steroid hormones, is converted to progesterone by a microsomal enzyme complex, Δ^5-3β-hydroxysteroid dehydrogenase: Δ^{5-4}-isomerase. Under physiological conditions in mammalian tissues, the two enzymes appear to function as a single entity and this reaction is essentially irreversible. Similar enzyme complexes bring about the conversion of 17α-hydroxypregnenolone and dehydroepiandrosterone (DHEA) to 17α-hydroxyprogesterone and androstenedione, respectively. The rate-limiting step in the biosynthesis of androgens is the microsomal 17α-hydroxylase:$C_{17,20}$-lyase enzyme complex. This reaction can utilize pregnenolone (Δ^5 pathway) or progesterone (Δ^4 pathway) as substrate, resulting in the formation of DHEA or androstenedione, respectively. This enzymic step is subject to hormonal feedback regulation and is, therefore, one of the key points of physiological control over follicular steroid secretion. Androstenedione and testosterone are converted to oestrone and oestradiol-17β, respectively, by a microsomal enzyme complex referred to as 'aromatase' because of the aromatic structures of the end products. 17-Ketone and 17β-hydroxy C_{19} and C_{18} steroids are readily interconverted by a reversible 17β-hydroxysteroid dehydrogenase present in microsomes. Since 17β-hydroxysteroids of both oestrogens and androgens are biologically more potent than the 17-ketones, this reversible reaction may have important implications for biological activity.

It is useful to point out that in most reactions the enzyme-bound intermediate steroids do not exist as free compounds, but rather as transition states in which the substrate remains bound to the enzyme complex and proceeds through the entire metabolic sequence before release of the final product. This final step results in the liberation of the biologically active hormone and it is for this reason that chemical flow charts which depict intermediary metabolism can be misleading. Figures 4.4, 4.5 and 4.6 provide the further metabolic pathways of the major sex steroids oestradiol-17β, testosterone and progesterone. The steroids, as well as their metabolites, are excreted in the urine as sulpho- or glucuro-conjugates. Conjugation increases the polarity of the steroids making them water soluble so that they can be excreted. The conjugation reactions occur in the liver, the kidney and the intestinal mucosa.

The 'two-cell-type:two-gonadotrophin' control of follicular steroidogenesis

Follicular steroidogenesis is a function of the joint action of the two types of follicle somatic cell, the theca and granulosa cells. Steroidogenic profile differences are due to compartmentalization within the follicle (only the theca is vascularized), hormonal receptors and enzyme activities. These factors lead to the creation of substantially different microenvironments for the two cell types. The follicle is under the primary control of

oestradiol-17β ⇄ oestrone → 16-epioestriol

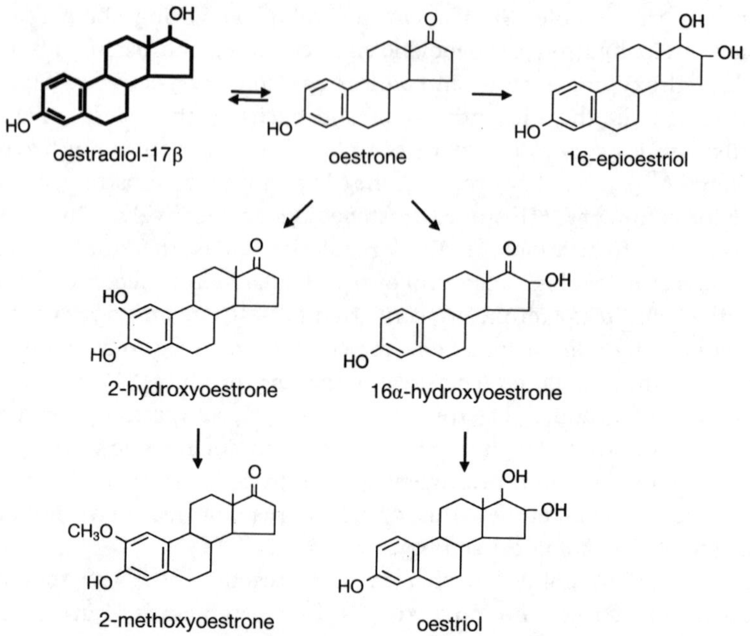

oestrone → 2-hydroxyoestrone → 2-methoxyoestrone

oestrone → 16α-hydroxyoestrone → oestriol

Fig. 4.4 Metabolism of oestrogens.

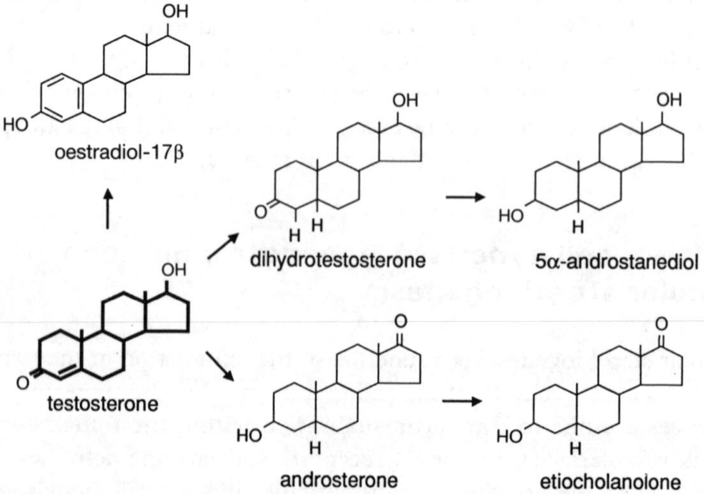

testosterone → oestradiol-17β

testosterone → dihydrotestosterone → 5α-androstanediol

testosterone → androsterone → etiocholanolone

Fig. 4.5 Metabolism of androgens.

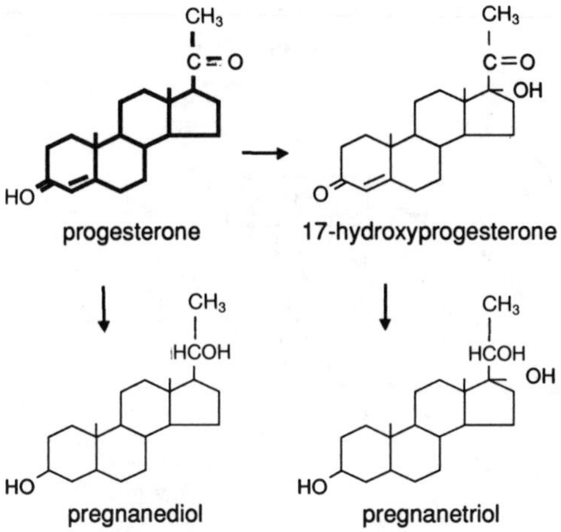

progesterone → 17-hydroxyprogesterone

↓ pregnanediol ↓ pregnanetriol

Fig. 4.6 Metabolism of progesterone.

the pituitary gonadotrophins FSH and LH, but their actions are modulated by many other intraovarian factors. For example, variation in the cellular concentrations of specific receptors dictates the fate of a given follicle including its level of steroid output at a given stage of development. The actions of FSH are restricted to granulosa cells as other ovarian cells lack FSH receptors. The actions of LH are exerted on both follicle cell types as well as on the interstitial cells and corpus luteum.

Granulosa cells have FSH receptors at all stages of development and also acquire LH receptors and responsiveness to LH at later developmental stages. In the theca and interstitial cells the steroidogenic pathways function primarily in the production of androgens. In the granulosa cells the pathways are organized principally for aromatase activity and oestrogen biosynthesis by the large antral and preovulatory follicles, and for the *de novo* synthesis of progesterone and its C_{21} metabolites. Therefore, the earliest steroidogenic response of undifferentiated granulosa cells to FSH is increased activity of the aromatase enzyme complex. Given an adequate supply of aromatizable testosterone and androstenedione (substrate supply is the rate-limiting factor) the granulosa cells are able to increase their rate of synthesis and secretion of oestradiol-17β. Progesterone biosynthesis occurs in granulosa cells, initially in response to FSH stimulation, but this action is later augmented by LH after its receptors have differentiated. In antral and preovulatory follicular stages, intracellular cholesterol is probably derived entirely from *de novo* synthesis, since cholesterol associated with lipoproteins in blood cannot penetrate the avascular granulosa cell layer.

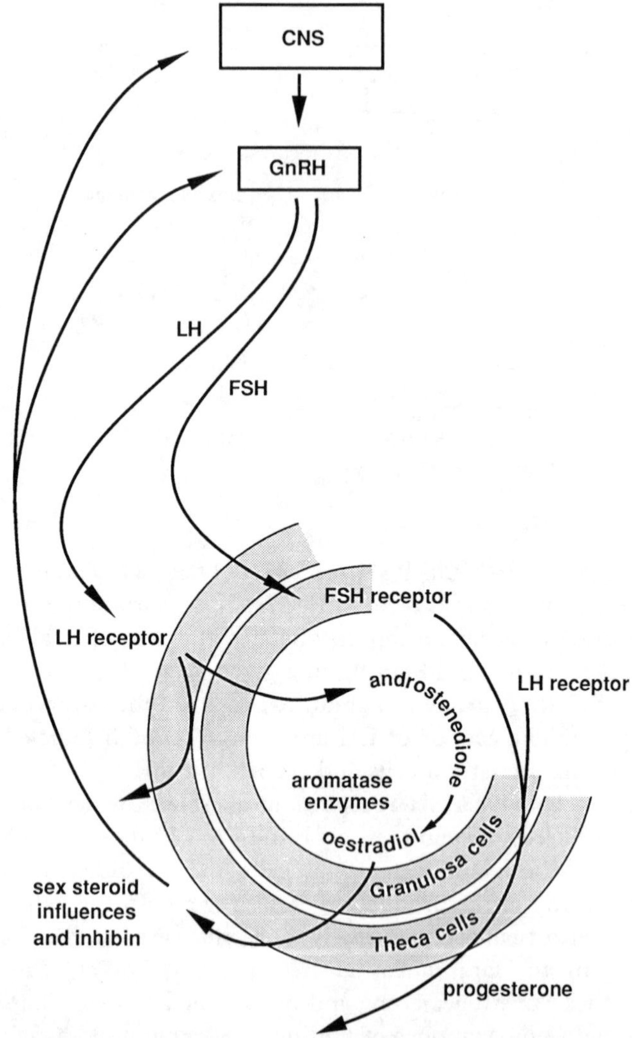

Fig. 4.7 Schematic representation of the 'two-cell-type: two-gonadotrophin' control of steroidogenesis and oestrogen secretion in the ovarian follicle (see text for explanation).

Theca cells at all follicular stages possess specific LH receptors and respond to LH with increased androgen secretion. Androstenedione is the principal aromatizable androgen in the human, together with lesser amounts of testosterone. The theca interna becomes highly vascularized as the follicle matures and blood-borne lipoproteins provide the major source of necessary cholesterol. Figure 4.7 illustrates the basis of the 'two-cell-type: two-gonadotrophin' control of steroidogenesis and oestrogen secretion in the ovarian follicle.

As mentioned above, the privileged follicle very rapidly establishes dominance over the other follicles, probably by suppressing the secretion of FSH. During the mid and late follicular phase of the cycle, the concentration of FSH falls progressively, while the concentration of LH rises. This divergence in gonadotrophin secretion is due to the negative feedback control of FSH by secretion of inhibin from the granulosa of growing follicles and, most importantly, oestradiol of dominant follicle origin. About 90% of the oestradiol secreted by the ovary is derived from the dominant follicle, giving it exclusive regulation of FSH secretion through feedback by oestradiol. The dominant follicle is able to continue development with levels of FSH below the threshold necessary for the smaller follicles. This allows dominant structures (the ovulatory follicle and postovulatory corpus luteum) to retain, by strategic CNS feedback, control of the gonadotrophin fluctuations throughout the menstrual cycle.

MECHANISM OF OVULATION

At midcycle an oestradiol-17β level of 200–500 pg/ml, sustained for 36 to 48 hours, stimulates the LH surge resulting in ovulation. Only follicles that have differentiated to the preovulatory stage are capable of responding to the LH surge that will result in follicular rupture. A significant increase in GnRH pulse frequency and amplitude occurs during the periovulatory period. Although there is general agreement that pulsatile GnRH is required for the basal gonadotrophin secretion, the role of a GnRH surge and the sites of oestrogen positive feedback remain controversial. Oestradiol may modulate gonadotrophin secretion at the anterior pituitary, hypothalamus and other CNS sites by mechanisms which still need to be fully clarified. One suggestion is that oestradiol may modify the gonadotroph response to GnRH through a quantitative change in the pituitary GnRH receptors. As well as a preovulatory increase in oestradiol level, the follicle also produces a slight rise of progesterone which augments the preovulatory gonadotrophin surge (large-amplitude pulses are produced by progesterone feedback effects on the hypothalamic GnRH pulse generator). Subsequently, the corpus luteum produces a marked increase in progesterone during the luteal phase with a small oestrogen peak in the late luteal phase. Oestrogen and progesterone inhibit FSH and LH during the luteal phase, perhaps via an increased inhibitory opioidergic influence on the GnRH pulse generator. Thus, as for oestradiol, progesterone may stimulate or inhibit gonadotrophin secretion with apparent feedback at both the pituitary and hypothalamus; its effect may be correlated with dose or duration of exposure. Levels of both steroids drop 2–3 days prior to menses, allowing the return of FSH and the next cycle of follicular development.

As described in Chapter 2, the primary oocyte remains in meiotic arrest from its formation in fetal life until reactivation about 36 hours before ovulation. In association with the midcycle LH surge there is resumption of meiosis and shortly before ovulation the meiotic division is completed with the release of the first polar body. The midcycle LH surge disrupts the anatomical links between the oocyte and cumulus granulosa; this ensures that, on ovulation, an oocyte of appropriate maturity is released. The timing between oocyte maturation and ovulation is crucial: if the time is extended the released oocyte may be 'aged', less fertilizable and may become vulnerable to chromosomal disturbances and faulty development. The second meiotic division is completed only if the oocyte is fertilized. It is misleading, however, to suppose that the oocyte is inactive and profits from a unidirectional, finely tuned regulation in its own development and maturation (after all, no follicle ever differentiated without its egg). On the contrary, to ensure survival, the oocyte participates in the follicle's development by optimizing its response to hormonal signals. For example, oocyte secretions empower the cumulus cells to synthesize hyaluronic acid bringing about their necessary expansion and separation prior to fertilization.

A model for follicle rupture

The mechanism determining the rupture of the Graafian follicle has been controversial since the study of ovulation began around 1670 with the discovery of the follicle and its ovum. In particular, more recent debate has concerned the cellular composition of the follicle wall and the role of proteolytic enzymes. In most mammals, including the human, smooth muscle and collagen fibres are present in the theca externa and these are primarily responsible for the follicle wall's tensile strength. At ovulation, it is important for the tensile strength of the wall to decrease to the point at which gentle rupture occurs to avoid the intrafollicular pressure exceeding the normal 15 to 20 mmHg. Enzymatic degradation of the follicle wall is probably the best hypothesis explaining follicular rupture and the plasminogen-activator–plasminogen system (or a modification of it) may be operative in situations where controlled proteolysis and tissue degradation are necessary. The plasminogen-activator–plasminogen hypothesis is based on the ability of the granulosa cells to secrete small amounts of a specific protease (plasminogen-activator) to generate a second protease (plasmin) from the proenzyme (plasminogen) normally present in follicular fluid and extracellular oedema fluid. Plasmin, in turn, activates collagenase attached to collagen fibres, and, with additional proteases, the proteolysis of collagen is completed. The net effect is to decrease the tensile strength of the follicle wall. Plasminogen-activator production in granulosa cells can be stimulated by LH, FSH, prostaglandins, progesterone and androgens.

The corpus luteum and its control

Luteinization of the granulosa cells and the formation of the corpus luteum is merely an extension of follicular growth and ovulation. After ovulation many gap junctions reappear between the developing luteal cells, and cellular LH receptor increases stimulate 3β-hydroxysteroid dehydrogenase activity and progesterone biosynthesis. In primates, LH is the hormone responsible for the maintenance of the function and lifespan of the corpus luteum; however, prolactin may influence the number of receptors for low density lipoprotein and thus regulate substrate availability for steroidogenesis. In primates, the corpus luteum is also a source of oestradiol, inhibin, relaxin, neurohormones (for example, oxytocin), growth factors and prostaglandin E_2 and $F_{2\alpha}$ (PGE_2 and $PGF_{2\alpha}$). Like the follicle, the corpus luteum consists of subpopulations of luteal cells (granulosa-lutein/theca-lutein) that differ in function as well as responsiveness to endrocrine and local factors. Likewise, there are many putative para- and autocrine regulators of the corpus luteum and more of these are discovered every week. However, the primary function of the corpus luteum is to secrete progesterone. Progesterone has several biological effects on reproductive target tissues concerned with the preparation for pregnancy and the provision of nourishment to the conceptus.

Luteolysis

In fertile cycles, the functional lifespan of the corpus luteum is prolonged in early pregnancy by chorionic gonadotrophin, an LH-like hormone secreted by the implanting blastocyst. In non-primate species there is uterine involvement in the luteolytic process; for example, in the guinea pig, rat and cow, hysterectomy lengthens the life of the corpus luteum

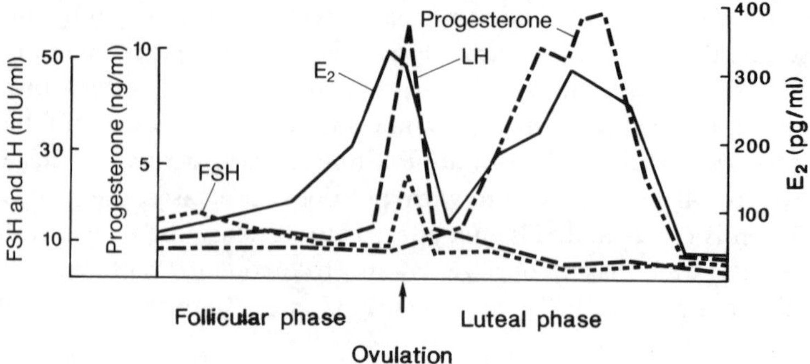

Fig. 4.8 Changes in the levels of circulating hormones during the menstrual cycle, E_2 is oestradiol-17β. Note that the hormone concentrations are drawn to different scales.

to the time span of the normal pregnancy. In these species, luteolysis is attributed to the uterine synthesis and secretion of $PGF_{2\alpha}$ and its subsequent transport to the ovary by a countercurrent mechanism involving the uterine vein and the ovarian artery. Because hysterectomy does not alter ovarian periodicity, the corpus luteum of the menstrual cycle must be independent of uterine control. It has been postulated that in primates a luteolysin is generated within the ovary and/or corpus luteum itself which regulates luteal function and lifespan. Evidence is available to suggest that exogenous $PGF_{2\alpha}$ promotes luteal regression. For example, the human corpus luteum synthesizes $PGF_{2\alpha}$, has receptors for $PGF_{2\alpha}$ and increasing amounts of a circulating metabolite of $PGF_{2\alpha}$ (13,14-dihydro-15-keto-$PGF_{2\alpha}$) have been observed in the late luteal phase. Although the mechanism of luteolysis is still under investigation, it is likely to involve the subprimate mammalian pattern utilizing $PGF_{2\alpha}$ or an analogous compound together with other local influences to create an environment conducive to luteolysis. The mechanism of action of $PGF_{2\alpha}$ is also not clear but it may influence factors such as luteal blood flow, LH receptors and the balance of luteotrophic/luteolytic influences. Figure 4.8 depicts diagrammatically the circulating hormone levels during the menstrual cycle.

MECHANISM OF MENSTRUAL BLEEDING

The endometrium is a dynamic tissue showing a series of cyclical changes which correlate with events in the ovary. During menstrual bleeding the two superficial zones of the endometrium (known collectively as the zona functionalis) are shed. The endometrial cycle is commonly divided into two phases: the proliferative phase (corresponding to the follicular phase of the ovarian cycle) and the secretory phase (equivalent to the luteal phase). Under the influence of the rising oestradiol levels, the endometrium undergoes proliferation of epithelial and stromal cells and glandular dilation. The major effect of oestrogen on the endometrium is that of growth. Oestrogen also increases the number of oestrogen and progesterone receptors on endometrial cells. In the luteal phase, rising progesterone levels cause further secretory changes in the endometrium. These include changes in the glands, which become coiled and secretory, together with increased vascularity and oedema of the stroma. Progesterone also halts the endometrial growth caused by oestrogen and stabilizes the endometrium. The major vaginal effects of rising oestradiol levels in the follicular phase are maturation of the vaginal basal cells and the formation of watery vaginal mucus. Under the influence of luteal progesterone, cervical mucus becomes thick. On corpus luteal involution and the loss of oestradiol and progesterone support, the endometrium sloughs. This results in the local release of prostaglandins, causing vasoconstriction and myometrial contractions. Endometrial production of prostaglandins reaches high concentrations as menstruation progresses and dictates the degree and duration of menstrual bleeding. Menstrual sloughing releases lysosomal enzymes which cause the release of phospholipids from cell membranes. The conversion of phospholipids to prostaglandins is responsible for the local rise in prostaglandin production. This process is enhanced by oxytocin which stimulates prostaglandin production. $PGF_{2\alpha}$ causes myometrial contractions, vasoconstriction and ischaemia. PGE_2 causes vasodilation and inhibits platelet aggregation, augmenting the non-coagulability of menstrual blood.

Dysmenorrhoea (pain associated with the menstrual flow) is one of the most common gynaecological problems, with about 52% of postpubescent girls being affected and 10% missing school regularly (a USA statistic). Dysmenorrhoea relates to prostaglandin-mediated high frequency uterine contractions, which may become dysrhythmic. The symptoms rarely last for longer than 48 hours but an effective treatment, for severe cases, is prostaglandin synthetase inhibitor. Exercise has been shown to benefit women with dysmenorrhoea.

PREMENSTRUAL SYNDROME

Premenstrual syndrome (PMS), first described in 1931, is a complex disorder which appears with a consistent and predictable relationship to menses. The diverse symptoms are often sufficiently severe to interfere with daily activities but resolve promptly at or soon after the onset of the menstrual flow. In extreme cases, because of intractable symptoms, women choose surgical castration and hysterectomy rather than continuing with the disease. The most common symptoms are grouped into physical (breast tenderness, abdominal bloating, oedema of lower extremities, headache, joint and muscle pain, nausea, acne, fatigue, constipation), psychological (emotional lability, irritability, anger, anxiety, depression, lethargy) and behavioural (increased appetite, food cravings and changes in libido) categories. These symptoms are an amalgam of specific endrocrine interactions and personality; any woman suffering from several of the above can become increasingly anxious, restless, irritable and hostile, especially if she is unfamiliar with the psychological effects which hormones can have. An estimated 20–40% of women have some symptoms before menses and up to 10% are real victims where it recurs cycle after cycle. Many treatment strategies, including steroid replacement therapy, tranquillizers, vitamins and other nutritional supplements, have been recommended with little efficacy. In one study, however, 75% improvement was found in the PMS sufferers as reflected by a decreased mean score on a Menstrual Distress Questionnaire when ovarian activity was abolished by treatment with a GnRH agonist. When the GnRH agonist was given in combination with a low dose of oestrogen/progestogen replacement (to counteract the effects of hypooestrogenism) there was still a 60% improvement. In another study adverse PMS symptoms were rectified or ameliorated with careful dietary guidelines and nutritional supplementation.

Many hypotheses, mostly related to inappropriately timed or sudden hormonal withdrawal, have been offered in explanation for PMS. Although the menstrual cycle is endogenously driven, it is subject to many exogenous influences that may alter its basic rhythm. Thus PMS probably represents a group of interrelated symptom complexes with different pathophysiological etiologies. The basic normal physiology described in this chapter allows an appreciation of the many variables involved in the correct timing and integration of menstruation and a small imbalance at any stage may cause discomfort.

One plausible aspect of PMS identifies cyclic changes in endogenous opiates as these have important effects on endocrine secretion, mood and behaviour. In primates and corroborated in humans, a cyclic pattern of β-endorphin is associated with the menstrual cycle: highest concentrations during the luteal phase, less during the late follicular phase and lowest during menstruation and the early follicular phase. During a faulty luteal phase, excessive exposure and then abrupt withdrawal of β-endorphin may trigger psychoneuroendocrine manifestations of PMS. Several authors have observed plasma β-endorphin levels to be lower in the luteal phase in women suffering from PMS when compared to asymptomatic women. Whether the transient decrease of β-endorphin is contributed by or contributes to symptoms of PMS is not clear. Neuroendocrine changes may be primary or secondary to PMS. It is of interest that within the bowel, endogenous opiates are known to decrease muscular propulsive activity and this may account for the bloating and constipation noted by many women during the luteal phase. On the other hand, increased prostaglandin activity, subsequent to a fall of endogenous opiate activity, might explain episodes of loose bowel movements that frequently precede menstruation.

Neuroendocrine regulation depends on the correct balance of activities of neurohormones and neurotransmitters (catecholamines, β-endorphins, for example), resulting in the characteristic pulsatility of GnRH. The symptoms of PMS may be amplified by social, environmental and lifestyle influences because circulating glucocorticosteroids have an effect on ovarian function. For example, granulosa cells possess cortisol receptors which inhibit FSH-stimulated aromatase activity and enhance FSH-stimulated progesterone production associated with increased 3β-hydroxysteroid dehydrogenase activity. The presence of these receptors and a concentration of cortisol-binding globulin in follicular fluid suggests a physiologically significant local regulation of follicular function by glucocorticosteroids. On the CNS level, glucocorticosteroids are known to modulate opioid and GnRH release. Several studies have shown an association of the prevalence and severity of PMS with diet, smoking, consumption of alcohol and caffeine, and lack of exercise. A study of university students aged 18–22 years in Oregon, USA, revealed that foods and beverages that are high in sugar content, taste sweet (artificial sweeteners in 'junk-foods'), contain caffeine (chocolate) and ethanol (beer) were associated with an increased prevalence of premenstrual symptoms. Nutritional observations, of course, are not necessarily alternative to theories of endocrine imbalance as an explanation of PMS.

HISTORICAL ATTITUDES TO MENSTRUATION

Historically the menstrual cycle was seen by man as an unnatural occurrence. Folklore is full of false instructions regarding this biological function. Because blood is associated with injury and death, woman's blood was viewed with fear and superstition. Up to relatively recent times in every culture, menstruation was the subject of strict taboos and served as an excuse for men to dominate women; there are still echoes in our 'modern' age of old misnomers such as 'the curse'. It is amusing to read of some of the bizarre statements made, in all seriousness, as to menstruating women's disastrous effect on quite ordinary things. For example, a woman whilst bleeding was said to have the fantastic power to turn wine sour, seeds sterile, meat rotten and stop dough from rising. Perhaps more sinister were ancient customs forbiding contact with menstruating women who were segregated into special huts or were forced to draw attention to their condition by smearing bright coloured dyes on their faces and calling out 'unclean'. Menstruating women were also forbidden to enter sacred buildings (including Christian churches) and in some areas of commerce they were also forbidden to work during their period; for example, in sugar refineries it was said they could turn the sugar black!

General references

Adashi, E. Y. & Mancuso, S. (eds.) (1990). *Major Advances in Human Female Reproduction.* Raven Press, New York.

Beckman, J. C., Krug, L. M., Penzien, D. B. *et al.* (1992). The relationship of ovarian steroids, headache activity and menstrual distress: a pilot study with female migrainers. *Headache*, **32**, 292–297.

Braden, T. D. & Conn, P. M. (1991). Gonadotrophin-releasing hormone and its actions: a review. *Canadian Journal of Physiology & Pharmacology*, **69**, 445–458.

Brann, D. W. & Mahesh, V. B. (1991). Regulation of gonadotrophin secretion by steroid hormones: a review. *Frontiers in Neuroendocrinology*, **12**, 165–207.

Burger, H. G. (1989). Inhibin, a member of a new peptide family. *Reproduction, Fertility & Development*, **1**, 1–13.

Doisy, E. A., Veler, C. D. & Thayer, S. (1929). Folliculin from urine of pregnant women. *American Journal of Physiology*, **90**, 329–330.

Geisthovel, F., Moretti-Rojas, I., Rojas, F. J. & Asch, R. H. (1990). Insulin-like growth factors and thecal-granulosa-cell function: a review. *Human Reproduction*, **5**, 785–799.

Irianni, F. & Hodgen, G. D. (1992). Mechanism of ovulation. *Reproductive Endocrinology*, **21**, 19–38.

Krasnow, J. (1991). Advances in understanding ovarian physiology. Regulation of steroidogenic enzymes in ovarian follicular differentiation. *Seminars in Reproductive Endocrinology*, **9**, 283–302.

Mortola, J. F., Girton, L. & Fischer, U. (1991). Successful treatment of severe premenstrual syndrome by combined use of gonadotrophin-releasing hormone agonist and estrogen/progestin. *Journal of Clinical Endocrinology & Metabolism*, **72**, 252A–252F.

Piva, F., Bardin, C. W., Forti, G. & Motta, M. (eds.) (1988). *Cell to Cell Communication in Endocrinology*, *Serono Symposia Publication*, Vol. 49, Raven Press, New York.

Seifer, D. B. & Collins, R. L. (1990). Current concepts of β-endorphin physiology in female reproductive dysfunction. *Fertility & Sterility*, **54**, 757–771.

5 The testis and control of spermatogenesis

Both the following account concerning testicular physiology and the control of spermatogenesis and the preceding chapter concerning the menstrual cycle highlight the numerous common mechanisms in gonadal control of egg and sperm production. It was not until modern molecular biological techniques became available that it was possible to reinterpret the older cytological and endocrine findings and reach a better understanding of the testis as a functional unit. The 1980s was a particularly good decade in the advancement of our knowledge of both ovaries and testes. Numerous studies have focussed on auto- and paracrine factors influencing intra- and extratubular elements and the relationship of the Sertoli cell to the configurational changes which occur in the germ cell during spermatogenesis. Many secretory products of Sertoli cell (morphological equivalent of the granulosa) and Leydig cell (morphological equivalent of the theca) origin have been identified, including growth factors, cytokines, regulatory peptides, transport proteins and plasminogen activators. The isolation of the differentiating germ cells in a physiologically distinct compartment by means of a Sertoli–Sertoli junctional barrier, vascular restriction and receptor exclusiveness is homologous to the 'two-cell-type: two-gonadtrophin' control of follicular steroidogenesis in oocyte maturation. In a wider context, environmental influences can interrupt the complex process of spermatogenesis with important consequences for male reproduction, as they do for the female. Beyond spermatogenesis, there is also the potential for environmental stresses to affect sperm motility, both before and after ejaculation. In future a deeper understanding of the whole individual, within an environmental context, will be an important area of focus.

GROSS ANATOMY

Since early time man has domesticated animals through castration and this implies that the testis was one of the first organs to be studied at a

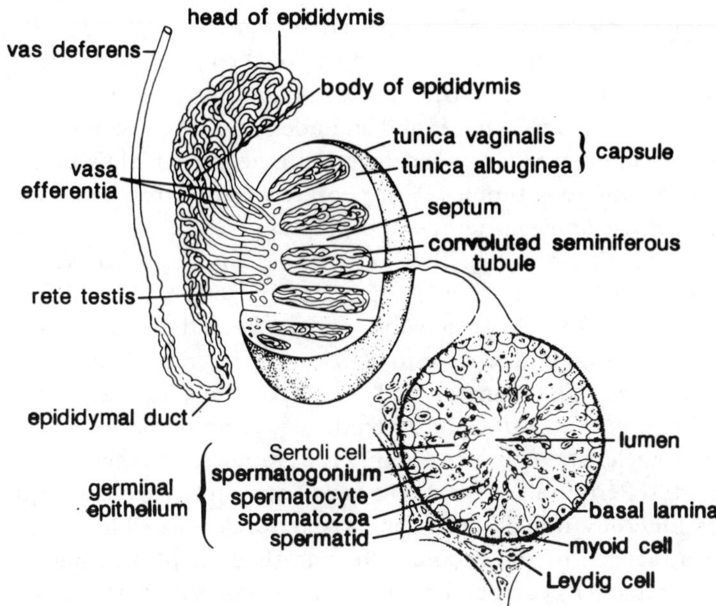

Fig. 5.1 The testis sectioned through the long axis displaying component parts. An enlarged cross-section of a seminiferous tubule shows the germinal epithelium and adjacent interstitial tissue.

physiological level. This awareness is supported by references to testis function in ancient Assyrian, Egyptian, Greek and Roman writings. Later, in 1850, the source of testosterone production, the interstitial or Leydig cell, was described by Franz Leydig. On first impression, man's testicles seem unremarkable; they are about 0.08% of bodyweight, which is small compared with many other mammals such as the rat or common domestic animals where the testes comprise between 0.5 and 1.5% of bodyweight. Sperm length is also not exceptional in the human (approximately 50 μm long). The longest mammalian sperm is from the marsupial honey possum *Tarsipes rostratus*. However, such differences reflect differing reproductive strategies (Chapter 7) and are not important for the purpose of this chapter. The endocrine and biochemical control of spermatogenesis is remarkably similar in all mammalian species.

The testes (Fig. 5.1) are coated with an outer coelomic epithelium, the tunica vaginalis, and lie free within the body cavity attached to their scrotal sac by a membrane, the mesorchium. Each testis contains two basic elements: the seminiferous tubular component composed of germ cells and non-germinal somatic Sertoli cells, and the intertubular or interstitial tissue component composed of connective tissue, blood/lymphatic vessels, nerves and the Leydig cells. These functional components are encased in a tough fibrous capsule, the tunica albuginea. The tunica albuginea contains smooth muscle fibres and an underlying network of blood vessels (tunica vasculosa) which makes contact with the interstitial tissue surrounding the seminiferous tubules. The capsule's contractility is an important factor in the maintenance of an appropriate pressure within the testis. In many mammals, including the human, the closely packed seminiferous tubules are arranged in lobules separated by septae or bands of fibrous tissue (in rodents there are no subdivisions). Each tubule is a single coiled loop that opens at both ends into a series of collecting channels known as the rete testis. The number and length of these tubules vary among individuals and are probably determined during puberty. An estimated average total length of human seminiferous tubules, for both testes, exceeds 100 metres. Peritubular myoid cells become organized around the developing seminiferous tubules during puberty and are responsible for peristalsis-like contractions of the tubules affecting their shape and lumen width. These contractions exert an important stimulatory influence on the Sertoli cells and are the major force for the movement of fluid and propulsion of sperm through the seminiferous tubules. Along with the basal lamina, the peritubular myoid cells also provide the structural underpinning on which the basal compartment of the germinal epithelium rests. The rete testis, situated at one side near the epididymis, is a tubular system which connects both with the straight tubule ends of the seminiferous tubules (tubuli recti) and with the efferent ducts (ductili efferentes). The efferent ducts lead from the rete testis into the epididymis where they unite to form the epididymal duct, a single, much convoluted tube that eventually gives rise to the vas deferens or ductus deferens. The vas deferens, blood vessels, nerves and cremasteric muscle form the spermatic cord suspended between the inguinal canal and the testis.

THE SPERMATOGENIC CYCLE

The germinal tissue occurs in the seminiferous tubules in association with the Sertoli cells. The permanent germinal epithelium results in the simultaneous presence of all stages of spermatogenesis within the seminiferous tubule. Immature spermatogonia are located at the base of the epithelium, while the spermatocytes and spermatids occur at successively higher levels. The spermatozoa border the lumen of the tubule into which they are released, flagellum first and head orientated towards the Sertoli

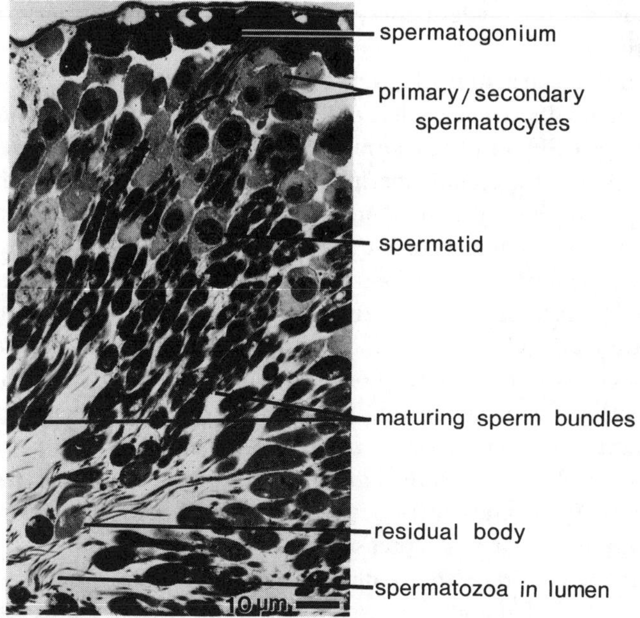

Fig. 5.2 Scanning electron micrograph of the seminiferous tubule in the rat illustrating stratification of cell types, from base to lumen. Photograph by Ms Dianne Hughes, Electronmicroscope Unit, School of Biological Sciences, Macquarie University.

cell. The proliferation of spermatogonia at the base of the epithelium and their displacement toward the luminal surface results in a distinct stratification of germ cells as they develop within the seminiferous tubules (Figs. 5.1 and 5.2).

Germ cells at different phases of maturation provide a well-defined succession of cellular stages, repeated indefinitely. Two successive appearances of the same cellular stage in a given segment of the seminiferous tubule is defined as the cycle of the seminiferous epithelium or, more simply, the spermatogenic cycle. The number of stages constituting the cycle is constant for a given species; for example, six segments lasting 64 days in the human and 14 segments lasting 48–53 days in the rat. The timing of each stage of spermatogenesis does not seem to depend on endocrine factors but is genetically determined. If the germ cells do not progress through the spermatogenic process at the predetermined rate they subsequently degenerate. It is important to know the timing of the various stages of spermatogenesis because, in a practical sense, this allows comparisons to be made between putative drugs for therapeutic or contraceptive use.

Although the spermatogenic cycle is genetically controlled, the number of spermatogonial stem cells that enter the cycle is endocrine controlled, specifically by FSH and high intratesticular levels of testosterone. These hormones, therefore, dictate sperm density and fertility in healthy men. Since the endocrine system is influenced by environmental stressors, intermittent stress may reflect intermittent minor fluctuations in seminal sperm concentration. In fertile men, the sperm concentration varies between 60–80 million active sperm per ml of semen and one ejaculate is approximately 2.5–3.5 ml in volume. Volume is not generally a good semen parameter because men with the poorest samples in terms of sperm density often have the highest volumes. The cut-off point between fertile and infertile sperm numbers is taken as 20 million active sperm per ml, but sperm density alone is insufficient to assess the quality of a particular semen sample. Parameters such as sperm motility, viability and morphology are especially important (Chapter 1).

It is of interest to note that a defect in the secretory pattern of GnRH associated with psychological stress, nutrition or weight is a less common cause of infertility in men than it is in women where it is frequent and can account for approximately 40–50% of the cases of infertility. A possible explanation for stress-induced subfertility may directly implicate Leydig and/or Sertoli cell influences on spermatogenesis through paracrine factors other than testosterone production. It has been suggested that, under stress, locally elevated levels of opioids may act as inhibitors of sperm motility. The Leydig cells produce β-endorphin which inhibits Sertoli cell function by antagonizing the actions of FSH. Since both Leydig and Sertoli cells have receptors for β-endorphin, which is also present in semen, germ cell viability may be influenced without the disruption of the hypothalamic–pituitary–testicular axis.

SERTOLI CELLS AND THE SERTOLI CELL BARRIER

The Sertoli cell was originally described in the human testis by the Italian histologist Enrico Sertoli in 1865. Sertoli cells develop from the supporting cells of the fetal testis and are arranged in the seminiferous tubule circumferentially with their basal membrane against the basal lamina and their cytoplasm extending towards the lumen. The germ cells mature in association with a permanent population of Sertoli cells. Following puberty, the Sertoli cells do not divide and their number remains relatively stable with only a gradual decline until old-age. Just before puberty, however, there is a burst of mitotic activity the extent of which may be determined by contemporaneous health and nutritional parameters (Chapter 3). Formation of the inter-Sertoli cell junctions coincides with the cessation of Sertoli proliferation. Circumstantial evidence suggests that inhibin plays an important role in the feedback regulation of peripheral concentrations of FSH during puberty when Sertoli and granulosa cells actively divide. Inhibin, therefore, may be an important hormone in the regulation of the length of the seminiferous tubules in adult men and of the number of follicles in adult women.

The Sertoli cells occupy a pivotal position in the seminiferous epithelium as they are the only cells to extend from the outer wall of the tubule to the lumen and possess anatomical and functional subdivisions within their cytoplasm. Their shape is very distinctive: likened to the trunk of a tree with the many cytoplasmic ramifications between the surrounding germ layer being the branches of the tree. The Sertoli cell shape is possibly the most complex yet described in any epithelium because it must continually alter its shape to accommodate the structural transformation and movement of the germ cells (Fig. 5.3). The Sertoli cell cytoplasm forms two permanent, basal and adluminal compartments and one transient, intermediate compartment within the bounds of the seminiferous epithelium. Compartmentalization is by means of an ingenious set of attachments. Sertoli cells are attached to the basal lamina, to each other and to germ cells by desmosomes, gap junctions and tight junctions to form the Sertoli cell barrier. Originally referred to as the blood–testis barrier, this barrier is more appropriately named the Sertoli cell barrier since it is exclusively formed by Sertoli cells (compartments both form and function *in vitro* in cultures where only Sertoli cells are present). The barrier is absent at birth but becomes effective at the same time as meiosis and maturation of primary spermatocytes. Cells in the basal compartment include spermatogonia and spermatocytes which have relatively free access to substances diffusing from the lymph surrounding the seminiferous tubule. Indirectly, through lymph, basal compartment cells also have access to the products delivered from the vascular system. Numerous tight (occluding) junctions between Sertoli cells at their baso-lateral surfaces demarcate the basal from the adluminal compartment. A third compartment, known as the intermediate compartment, is formed during the transit of leptotene cells from the basal to the adluminal compartment and involves successive formation and breakdown of tight junctions and ensures that the integrity of the Sertoli cell barrier is maintained during transit of cells from the basal to the adluminal compartment. The adluminal compartment sequesters the

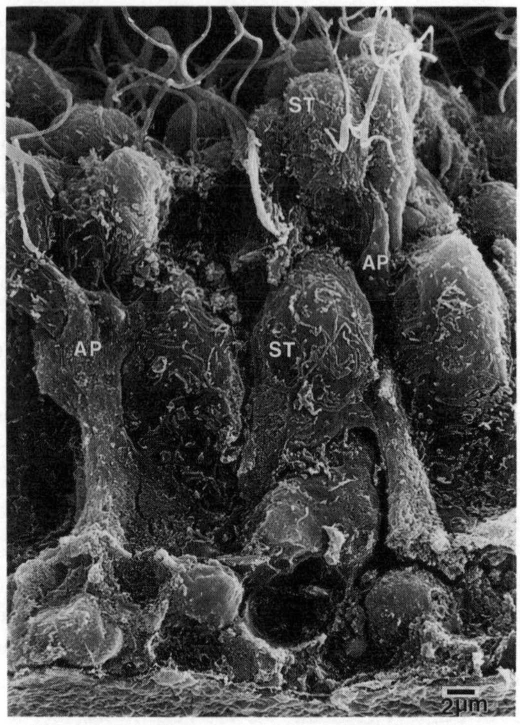

Fig. 5.3 Scanning electron micrograph of Sertoli cells exposed from their basolateral trunks to the apical bough-like processes (AP). The apical profile of the maturing spermatids (ST) is embraced by the Sertoli cell cytoplasmic processes. Reproduced from Hamasaki & Murakami (1986). *Journal of Electron Microscopy*, **35**, 132–143, with permission. I am grateful to Dr Masao Hamasaki, from the Kurume University School of Medicine, Fukuoka for supplying an original copy of this micrograph.

more differentiated germ cells into a unique physiological environment. The Sertoli cell barrier allows many substances to enter the adluminal compartment but the regulation of the levels of these substances and the secretion of materials into it by the Sertoli and germ cells creates a special microclimate in this compartment. Substances, including toxicants and microbes, must either open the barrier or pass through the Sertoli cell to affect adluminal germ cells directly. The Sertoli cell barrier also functions as a barrier preventing the immune system from recognizing autoantigens developed on the surface of the spermatocytes and spermatids.

LEYDIG CELLS AND THE INTERSTITIAL TISSUE

The seminiferous tubules are not penetrated by blood vessels or nerve fibres. These are confined to the spaces between the tubules comprising a part of the interstitial tissue. The interstitial tissue also contains the Leydig cells, many macrophages (which may account for about 25% of the cells in the interstitium) and a complex network of lymph vessels. The proportion of each component may differ markedly in individuals according to age and physical health. The Leydig cells are surrounded by seminiferous tubules reflecting the functional relationship between the two compartments. The macrophages play an important role in scavenging senescent and damaged cells and cellular debris.

TESTOSTERONE BIOSYNTHESIS, TRANSPORT AND METABOLISM

Testosterone is the principal androgen in the male, approximately 95% of it is produced by the Leydig cells with the remainder coming from the adrenal glands. In a normal adult male, the endrocrine effects of testosterone are felt in virtually every tissue as described in Chapter 3. In terms of its effect on fertility, however, the primary role of testosterone is paracrine where it regulates spermatogenesis.

Testosterone, a C_{19} steroid, is synthesized from cholesterol through a pathway summarized in Fig. 4.3 (p. 59). In humans, this cholesterol is mostly synthesized *de novo* in the Leydig cell from the acetate precursor and is stored in cytoplasmic lipid droplets. The cholesterol is then converted in the mitochondria to pregnenolone, which is in turn transported to the microsomes for further processing. Studies of incubated human microsomes suggest that the Δ^5 pathway predominates in our species. Most of the testosterone produced is released immediately into the general circulation where it is bound by the glycoprotein sex hormone-binding globulin (SHBG). Approximately 1–2% is unbound and available to target tissues and a small portion is weakly bound to albumin. Changes in the concentrations of SHBG will, therefore, influence the total testosterone carrying capacity as well as the proportion of circulating testosterone that can enter target cells. A normal production rate of testosterone in the adult is about 6 mg per day. Figure 5.4 outlines diagrammatically the metabolism of testosterone and the approximate weightings attributed to each pathway. Testosterone is catabolized in the liver to the 17-ketosteroids, 5α-androsterone and 5β-etiocholanolone, which are excreted in

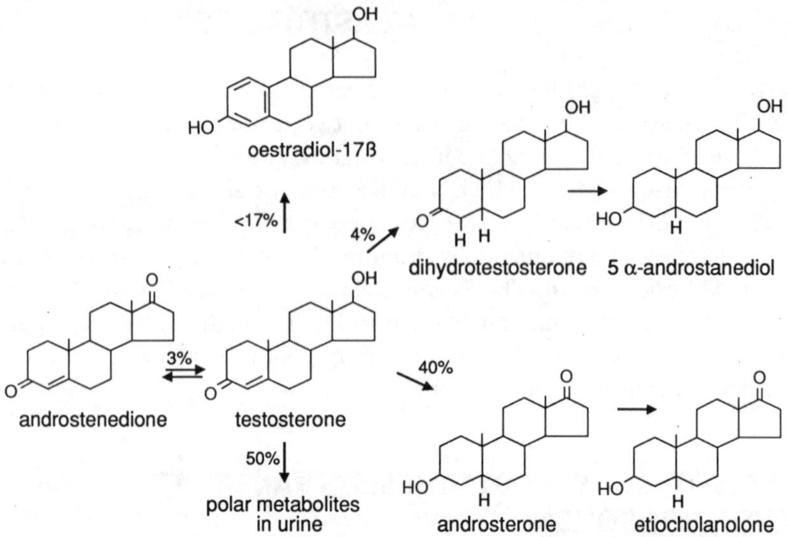

Fig. 5.4 Major metabolic pathways for testosterone and their relative importance.

the urine. Testosterone is also metabolized to a series of hydroxylated polar compounds, some of which can be conjugated to glucuronic acid, and excreted. A small amount of testosterone is excreted unmetabolized. Testosterone is also bioconverted in target tissues to the physiologically active steroids 5α-dihydrotestosterone (DHT) and oestradiol-17*β*.

In certain target tissues such as the prostate, seminal vesicles and pubic skin, testosterone is irreversibly metabolized to DHT which then becomes the androgen of primary physiological significance in those tissues. Chapter 2 described ambiguous external genitalia in individuals who lacked sufficient 5α-reductase activity. DHT binds with two to three times higher affinity to the androgen receptor and is up to ten times more potent than testosterone. Thus local 5α-reduction in the target tissues possessing this enzyme amplifies the potency of testosterone.

Approximately 25% of the oestradiol produced in men is secreted directly by the testes, the remainder is produced by the adrenal glands and from testosterone metabolism in fat cells and the liver. Oestrogen stimulates the synthesis of several hepatic proteins including SHBG and may be important in mediating some aspects of male sexual behaviour. Increased oestrogen secretion by, for example, a testicular tumour may suppress gonadotrophin release and testosterone synthesis resulting in hypogonadism and feminization. Interestingly, obesity in men can lead to abnormal sex steroid metabolism manifested as a reduced testosterone level and accompanied by a subnormal gonadotrophin level. The mechanism(s) involved is not clear but it has been suggested that hyperoestrogenaemia, due to increased aromatization of testosterone in the adipose tissue, may play a role by affecting gonadotrophin regulation.

SPERMATOGENESIS

Virtually all available information regarding the endrocrine control of steroidogenesis and spermatogenesis (the process in which spermatogonia form spermatozoa) is derived from studies in the laboratory rat. These processes appear to be highly conserved in all mammals so it is likely that the rat can serve as a suitable model for the human. Spermatogenesis is successfully completed under specialized conditions requiring the creation of the unique environment within the seminiferous tubule. The production of free testicular spermatozoa commences when sufficiently high circulating levels of FSH and LH are attained. Maintenance of spermatogenesis depends upon the joint actions of the two types of somatic cells: the Leydig cells which have LH receptors and Sertoli cells which are the target cells for FSH. The effect of LH on spermatogenesis is indirect because its function is to stimulate Leydig cell synthesis of testosterone which must be present in high concentrations in Sertoli cells. Both testosterone and FSH are required by the Sertoli cells to support all phases of spermatogenesis. Testosterone is transported from the interstitium into the seminiferous tubules bound to androgen-binding protein (ABP), a Sertoli cell-specific protein. ABP transports testosterone within the Sertoli cell, maintains high concentrations of the steroid in the seminiferous tubules and also acts as the carrier transporting testosterone from the testis to the epididymis. Hence ABP serves the same function in the reproductive tract as SHBG serves in the systemic circulation.

FSH intiates the proliferation of the spermatogonia and testosterone is necessary for spermiogenesis. However, the interactions are complicated since all aspects of Sertoli cell function seem to depend on adequate stimulation by FSH which is emerging as the central metabolic control point for testicular function. In addition to controlling ABP, FSH also controls the nutritional interaction between Sertoli and germinal cells; for example, Sertoli cells synthesize and secrete iron (transferrin), copper and other transport proteins whereby metals, nutrients and vitamins can be transported to the germ cells. FSH also controls the production of oestradiol by aromatization of testosterone. There is evidence that oestradiol may be implicated in a simple closed-loop endrocrine regulation by diffusing back to the Leydig cell to inhibit further steroidogenesis. Peripherally the oestrogens stimulate SHBG synthesis by the liver. Circulating testosterone exerts negative feedback control over LH secretion, and inhibin, secreted by Sertoli cells, exerts negative feedback control over FSH secretion. However, there is some debate as to whether inhibin is the only substance responsible for FSH control and a multiple feedback control of FSH secretion may be involved.

Spermatogenesis may be divided into three phases based upon functional characteristics.

The proliferative phase

There are three types of spermatogonia: stem cell, proliferative and differentiating. Stem cell spermatogonia are relatively resistant to insult and often survive when other germ cell types have been depleted. Their infrequent divisions give them resistance to noxious substances that adversely affect the transforming cell types. Proliferative spermatogonia undergo, under FSH control, renewal by successive mitotic division. This renewal is thought to occur by the stem cells dividing sporadically to provide pairs of spermatogonia. These pairs then engage in a series of synchronous divisions leading to the formation of chains of spermatogonia joined by intercellular bridges. Each chain is the progeny of a single stem cell. On approaching its final size, mitosis ceases in each chain but it continues to differentiate synchronously into mature spermatogonia.

The meiotic phase

At the end of the differentiation phase, the most mature spermatogonia go through two successive meiotic divisions to form primary and secondary spermatocytes. This results in a halving of the genetic material with the production of four haploid spermatids. In addition to synchronizing development, the cytoplasmic bridges between the spermatids now provide a device by which each haploid cell, by sharing cytoplasm with its neighbours, can be supplied with all the products of a complete diploid genome. Prophase of the first meiotic division, in which genetic recombination occurs, is exceptionally long lasting (about 3 weeks in man and rat). Cytogenetic errors during the mitotic and meiotic phases can be responsible for the birth of a child with chromosomal aberrations such as Down's syndrome, XXY, XYY constitutions and sex reversal. The etiology of sex reversal, leading to testes in apparently XX individuals, is most often due to the presence of *SRY* on the paternally derived X chromosome. This abnormality arises during meiosis with the transfer onto the X chromosome of a specific region of the Y chromosome that includes *SRY*. An estimated 20% of male infertility has been attributed to gamete impairment due to faulty mitosis and/or meiosis.

The spermiogenic phase

Spermiogenesis is the final phase of differentiation of spermatids into spermatozoa just prior to their release from the seminiferous epithelium. This process occurs without cell division and is one of the most phenomenal sequences of transformations in the body. During spermiogenesis the round, relatively undifferentiated haploid spermatids undergo morphological, biochemical and physiological changes that result in the formation of a highly polarized, motile cell having a morphology that is distinctive for each mammalian species, for example flattened and rounded in humans, hook-shaped in rats. The mechanisms driving spermiogenesis are still poorly understood despite the recent increased insight into cellular and molecular mechanisms operating in mammalian cells.

During spermiogenesis several events occur simultaneously, but for convenience the changes can be divided into a series of developmental steps to focus on one cytological process at a time. The process is illustrated diagrammatically in Fig. 5.5.

Formation of the acrosome The acrosome arises from the Golgi apparatus which is situated in a perinuclear position in immature spermatids. The Golgi is involved in producing secretory products to form the acrosomal vesicle. This vesicle is deposited at the pole of the nucleus where it spreads out and, with the nearby Golgi apparatus contributing more and more material, eventually forms the acrosomal cap. With the movement of the spermatid nucleus to the cell surface, the acrosomal region is welded to the plasma cell membrane surface. This imparts polarity to the cell, dividing the head or nuclear anterior region of the cell from the tail or flagellar caudal region; the junction between becomes the neck or midpiece. Concomitantly, the cell begins to take on an elongate shape as the cytoplasm is stretched out along the flagellum. As the nucleus progressively elongates, the Golgi apparatus moves away from the acrosome and migrates to the caudal region of the cell. The acrosome is lysosomal in nature and contains the enzymes hyaluronidase and acrosin. Between the

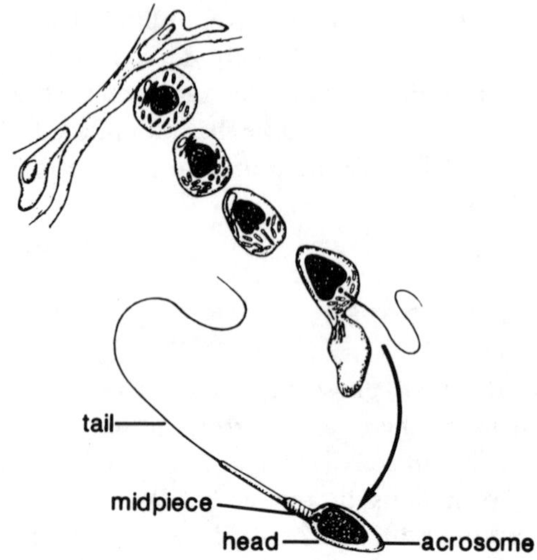

Fig. 5.5 The progressive stages of differentiation of a round spermatid in spermiogenesis.

inner acrosomal membrane and the nuclear membrane is situated the perforatorium which enables the mechanical and chemical penetration of the egg vestments; it carries lysin, a protease distinct from the acrosomal proteases.

Nuclear changes Associated with the changed position of the nucleus is a progressive condensation of chromatin (heterochromatin) to form the head shape characteristic of the species. Nuclear DNA condensation is caused by changes in histones and other specific basic proteins associated with DNA and unique to the testis. The degree of condensation is particularly variable in human spermiogenesis which may be a reflection of environmental differences. The nuclear volume decreases substantially as a result of the tight packing of DNA and the elimination of fluid. Through changes in size and shape, sperm heads become hydrodynamically streamlined and so well equipped to move in fluid environments.

Development of the flagellum The sperm tail arises from the two centrioles which migrate towards the periphery at the luminal side of the spermatid. The distal centriole differentiates into the axial filament or axoneme (the central core of the flagellum) and the proximal centriole comes into contact with a differentiating zone of the nucleus to form the implantation fossa of the flagellum. Connection of the flagellum to the nucleus and aggregation of mitochondria around the axial filament in the midpiece forms a spiral sheath which provides the least resistance to bending.

Elimination of cytoplasm The spermatid is reduced in volume to approximately 25% of its original size before sperm release and there are at least three phases in the process of making spermatids smaller and more streamlined. First, water and cytoplasm are eliminated in the condensation and elongation process of the spermatid nucleus. Secondly, some cytoplasm is eliminated just before sperm release by structures called tubulobulbar complexes and, finally, on release, a cytoplasmic package called the residual body is left behind to be engulfed by the Sertoli cell. Apart from cytoplasm, the residual body contains packed RNA and organelles which were earlier used by the spermatid but are no longer necessary for sperm survival. A small amount of cytoplasm, the cytoplasmic droplet, remains around the neck of the spermatid.

Spermatid migration By opening the occluding inter-Sertoli cell junctions, plasminogen activators of Sertoli cell origin play a crucial role in facilitating the release of mature spermatids and the migration of stem cells from the basal compartment to the adluminal compartment of the seminiferous tubule. Plasminogen activators also participate in remodelling the seminiferous epithelium.

TEMPERATURE AND SPERMATOGENESIS

Spermatogenesis is susceptible to errors if testicular temperature is raised above normal levels for any length of time; it typically proceeds to completion in an environment a few degrees (4–7 °C) lower than that of all other internal organs of the body. The lower temperature is effected by the development of two structures.

The scrotum The scrotum is in a cooler environment and the skin is richly furnished with sweat glands. Whole-body heating, as in repeated sauna bathing, marathon running and tight clothing may increase the danger of mutational damage in men (Chapter 15).

The pampiniform plexus The pampiniform plexus of testicular blood vessels is an interconnecting network of multiple small veins which ramify over and among the coils of the artery leading into the testis. Close association of the extended artery and the venous plexus constitute a highly efficient countercurrent heat exchanger in which the arterial blood is precooled to the temperature of the scrotal skin before it reaches the testis and the venous blood is warmed to body temperature before its return to the abdomen.

Luminal fluid production begins at puberty when the first spermatozoa are released from the Sertoli cells and carried away to eventually reach the rete testis and the efferent ducts. Normal fluid production, stimulated by FSH, is dependent on the lower temperature found in the scrotum.

MATURATION OF SPERMATOZOA

In all mammals, including man, the spermatozoa that leave the testis are infertile. To be able to meet, recognize and fertilize an oocyte, the postgonadal sperm must undergo several successive biochemical transformations during its transit in the male (epididymal maturation) and the female (capacitation) genital tracts. In the male genital tract the principal post-testicular sperm modifications occur in the epididymis. Under testosterone stimulation, modifying factors which cause a complete reorganization of the sperm plasma membrane are secreted by the epididymal epithelium. During their 2-week epididymal journey, the sperm change in surface charge, surface sulphydryl groups, lipid composition, ATPase activity and membrane antigens. In addition, there is a shift from oxidative metabolism to glycolysis, with fructose instead of glucose as the main substrate. The main purpose of these changes is to give the spermatozoa flagellar motility and the ability to fertilize ova.

A second wave of membrane modifications (coating with several seminal plasma proteins and glycoproteins), takes place at ejaculation. These modifications are not essential for fertilization but act as a protective mechanism for the ejaculated sperm.

IMMUNOLOGY OF SPERMATOGENESIS

Spermatids and spermatozoa are antigenic. Although testis-specific autoantigens first appear in late pachytene spermatocytes, the bulk differentiate during the establishment of sperm surface domains later in spermatogenesis. The absence of a routine autoimmune response to spermatozoa in normal men is a result of anatomical and physiological factors including the Sertoli cell barrier, antigen-mediated mechanisms, seminal immunosuppressive factors and suppressor T-lymphocyte activity. Sperm are segregated behind the Sertoli cell barrier which effectively isolates germ cell components in the seminiferous tubules away from the testicular blood supply as well as preventing the passage of immunoglobulins and lymphocytes from the blood supply into the tubules. This barrier is not, however, always complete and some soluble sperm antigens can leak from the seminiferous compartment into the interstitium where antigen-mediated mechanisms help maintain immunological unresponsiveness. Lymphocytes migrating through the interstitium are targets of several testicular inhibitory substances. These provide a protective mechanism preventing an autoimmune attack on the germ cells. Inhibition of immune system activation may be the operative mechanism behind prolonged transplant survival in the interstitial tissue. Immune modulators which decrease the antigenicity of sperm have also been detected in the seminal plasma. The absence of antisperm antibodies in normal women is also of interest. The specific factors preventing development of antisperm antibodies in sexually active healthy women are unclear, but the sperm itself may be responsible for the suppression of female cellular immune responses.

The presence of antisperm antibodies is clinically relevant because it may cause a reduction in normal fertility which may be due to interference with sperm transport and/or gamete interaction. Common effects of antisperm antibodies include reduced cervical mucus penetration, increased phagocytosis of sperm, impaired interaction with the zona pellucida, lack of fertilization and an increased risk of spontaneous abortion.

ACCESSORY SEX GLANDS: THE FORMATION AND FUNCTIONS OF SEMINAL PLASMA

Differentiation of the accessory sex glands in the fetus, their extensive growth and secretory activity initiated at puberty and their function in the adult are completely androgen dependent. All accessory glands possess 5α-reductase activity and its product DHT is the principal mediator of androgen activity. Secretions of the prostate, seminal vesicles and bulbourethral glands, together with a small volume of epididymal and rete testis fluid, make up the seminal plasma. Secretion from the accessory sex glands is enhanced during sexual excitement; seminal plasma volumes in different ejaculates from the same individual vary depending on the level and duration of precoital male sexual excitement and on fluctuations in circulating testosterone.

The principal substances in seminal plasma are water, inorganic ions (making it isotonic with blood), high concentrations of fructose (for energy requirements) and proteins (for coagulability of semen), all of which assist sperm survival and transport in the female. Other substances act as buffers against the quite acid (pH 5 or less) vaginal secretions and are antibacterial and immunosuppressive in both the male and female genital tracts. High concentrations of prostaglandins are also present and assist sperm transport by modifying the contraction of smooth muscle in the female tract and aiding female orgasm (Chapter 8).

COMMON MODIFIERS OF STEROID HORMONE SECRETION AND FERTILITY

There is little variation between species in the amount of testosterone secreted per unit bodyweight. In mammals, such as the human, comparatively small testicular size is usually compensated for by increased testosterone release into the bloodstream. The social environment, however, has a profound effect on testosterone production. For example, proximity of a desired female causes a transient rise in testosterone levels. This response may be mediated by LH stimulation or, initially, by an increased bloodflow through the testes effectively 'flushing out' hormone stores. Desire for the female is enhanced by exposure of the male to the female's sex pheromone (Chapter 6). Sperm density is also increased by a positive sexual environment.

Many factors including anatomical, physiological, endogenous and exogenous variables can be causally related to in- or subfertility. Common factors include sexually-transmitted disease (STD), primary testicular failure, cryptorchidism, illness, high fever, neurological and immunological disorders, alcoholism and excessive drug use. For example, there is an association between cigarette smoking, testicular dysfunction and lesions in sperm (Chapter 19). Physiological variation in the spermogram (distribution of sperm characteristics) is also climate dependent. A French study reported that ejaculate volume and sperm number were at a minimum in September and at a maximum in March and November, whereas sperm motility and morphology were better in August and September and less satisfactory in March and April. The common denominator linking many of the above disparate syndromes may be stress-induced endrocrine changes.

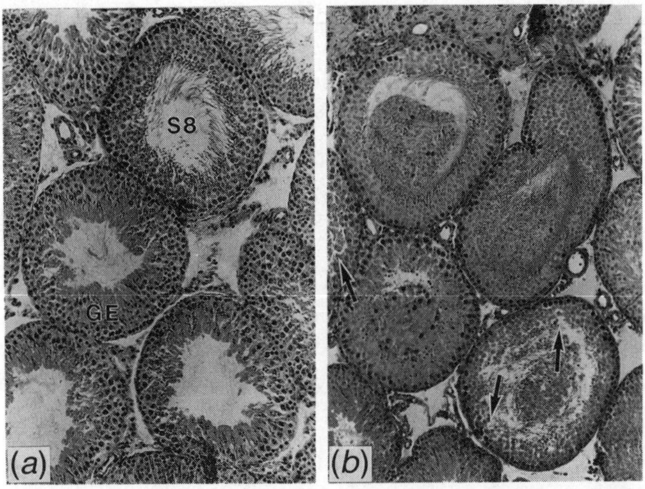

Fig. 5.6 Light micrograph of (*a*) the control rat testis with intact germinal epithelium (GE) and seminiferous tubules in stage VIII (S8) of spermatogenesis, compared with (*b*) the caffeine-exposed rat testis with cellular debris in the seminiferous tubular lumen and a partially degenerating germ layer (arrows). Reproduced from Pollard & Smallshaw (1988). *Journal of Developmental Physiology*, **10**, 271–281, with permission.

Chronic stress is consistently associated with lower testosterone secretion and is usually associated with reduced levels of LH, but a direct effect on the testis may also be operative. As many endocrine aspects of the physiology of stress are discussed in Chapters 13 and 18, only one possible factor in the etiology of testis dysfunction will be considered here. As described above, a good blood supply to the testis is essential to deliver the nutritional, endrocrine and other products necessary for spermatogenesis to proceed optimally. Any condition which reduces testicular blood supply may, therefore, have the potential to disrupt the unique germinal epithelial environment, directly by affecting the Sertoli cells, or indirectly via Leydig cell function, or both. Figure 5.6 illustrates a patch of testis with partially degenerating germinal epithelium in a rat consuming caffeine equivalent to a human caffeine intake of 15–18 cups of brewed coffee daily over a period of 5 weeks. The cellular debris found in the lumen of the rat testis included spermatocytes, spermatids and sperm fragments. It is possible that tubular integrity deteriorated as a result of deprivation of oxygen or other supplies following ischaemia produced by repeated vasoconstriction of the testes. Caffeine is a strong sympathetic nervous stimulant (one reason we enjoy its effect) causing the release of catecholamines, and adrenaline is a known vasoconstrictor of various abdominal organs including the testes. Human findings, likewise, have demonstrated that cytotoxic agents directly affect the germinal epithelium and sperm differentiation.

The presence of possible maternal toxicants, such as caffeine in the food supply, and the effects these may have on fetal and neonatal development is an often discussed source of concern. However, much less publicity is given to paternally-mediated drug effects on the offspring. An understanding of testis function and increased awareness of the risks of germinal epithelial degeneration should increase men's care of their special cargo. After all, the elaborate interrelationships of the Sertoli cells with each other and with germ cells are unsurpassed in complexity by any organ in the body and, therefore, its unique structural complexity, co-ordination and cell production needs special protection.

General references

Chapelle, A. de la, Hästbacka, J., Korhonen, T. & Mäenpää, J. (1990). The etiology of XX sex reversal: review. *Reproduction, Nutrition & Development*, **Suppl. 1**, 39–49.

Cropp, C. S. & Schlaff, W. D. (1990). Antisperm antibodies. *Archivum Immunologiae et Therapiae Experimentalis*, **38**, 31–46.

De Braekeleer, M. & Dao, T. N. (1991). Cytogenetic studies in male infertility: a review. *Human Reproduction*, **6**, 245–250.

Drife, J. O. (1987). The effects of drugs on sperm: review article. *Drugs*, **33**, 610–622.

Leydig, F. (1850). Zur Anatomie der Mannlichen Geschlechtsorgane und Analdrusen der Saugethiere. *Zeitschrift für Wissenschräftlige Zoology*, **2**, 1.

Maddocks, S., Parvinen, M., Söder, O., Punnonen, J. & Pöllänen, P. (1990). Regulation of the testis. *Journal of Reproductive Immunology*, **18**, 33–50.

Olshan, A. F., Teschke, K. & Baird, P. A. (1991). Paternal occupation and congenital anomalies in offspring. *American Journal of Industrial Medicine*, **20**, 447–475.

Piva, F., Bardin, C. W., Forti, G. & Motta, M. (eds.) (1988). *Cell to Cell Communication in Endocrinology. Serono Symposia Publication*, Vol. 49. Raven Press, New York.

Pollard, I. & Smallshaw, J. (1988). Male mediated caffeine effects over two generations of rats. *Journal of Developmental Physiology*, **10**, 271–281.

Sertoli, E. (1865). De l'esistenza di Particulari Cellulse Ramificate nei Canalicoli Seminiferi dell 'Testicolo Umano. *Morgagni*, **7**, 31–40.

Setchell, B. P. (1986). The movement of fluids and substances in the testis. *Australian Journal of Biological Sciences*, **39**, 193–207.

Sharpe, R. M. (1990). Intratesticular control of steroidogenesis: review. *Clinical Endrocrinology*, **33**, 787–807.
Skinner, M. K. (1991). Cell-cell interactions in the testis. *Endocrine Reviews*, **12**, 45–77.

6 Sexual behaviour and pheromones

In the preceding chapters, processes with a greater degree of autonomy, such as growth, maturation and ageing, were examined. This chapter and the next examine sexual communication and the effects of procreational biology on the foundation of social structure. To effectively synchronize social interactions an animal needs both the ability to send information to others and the capacity to receive (and correctly interpret) signals from others. Both aspects of communication must evolve together and involve co-ordinated nervous and hormonal interactions. Reproductive behaviour has its origin in genetic coding, but its final expression is vulnerable to modifying epigenetic influences and past history. The extent to which our behaviour follows the dictates of our genes or is the expression of conscious inventions of our thinking brains is not known, but it is the subject of much controversy. Many behavioural patterns, for example, feeding, sleeping and breathing, are essential and recur throughout our lifespan. Other behavioural patterns, not essential for individual survival, are brief and may occur rarely or even only once in a lifetime but still have great significance. Such patterns may include parturition, lactation or even copulation. The first coitus is usually a milestone in the psychosocial development of an individual as it represents a critical, dynamic conjunction of two bodies. Copulation co-ordinates, through a complex series of neuroendocrine events, the transition between two major life stages, prereproduction and reproduction. The co-ordination of the behavioural aspects again represents the interaction of another complex series of neuroendocrine events. Gamete production can proceed in total social isolation, but the production of viable young (ignoring IVF technology) requires social exchanges. Because we are a socialized species, copulation has a greater significance than just the transmission of the genotype and serves many other functions (Fig. 6.1). On the positive side, sexual pleasure is particularly important, so much so that evolutionary pressures have engineered the human female's escape from dependence on ovulation for

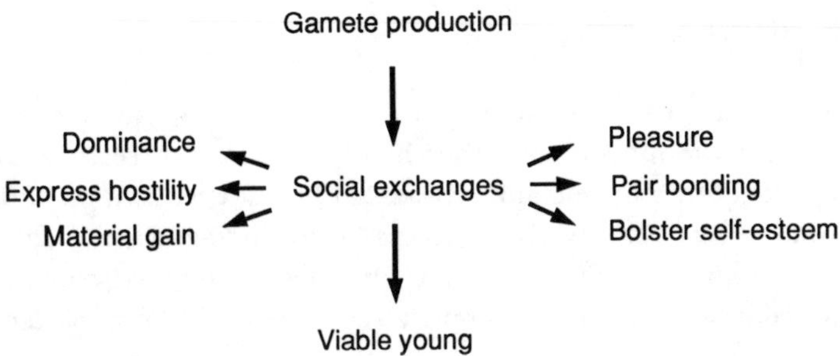

Fig. 6.1 The social functions of copulation.

sexual satisfaction. Continued sexual receptivity favours pair-bonding, essential for the maintenance of long-term relationships and the raising of offspring. Self-esteem reinforces the sufficiency and security of the human being and for that reason helps in raising the offspring. On the negative side, sexual behaviour may be used to establish dominance in relationships, express hostility or be used for material gain (it is said that prostitution is the oldest profession in human society). The sociobiological functions of sexuality go far beyond reproduction and pair-bonding. They influence the shaping and linking of groups of individuals into communities and communities into societies, that is, in anthropological terms, the humanization of biological relationships. Procreational sex in becoming recreational sex also became the glue of society.

The critical factor in the transition of small communities into complex societies was the unparalleled development of the human cerebral cortex. The cortex has taken over much of the control of sexual behaviour in humans from the rigid hormonal control of sexual behaviour seen in other mammals, including most primates. This switch from hormonal to cerebral control opened up vast new possibilities because there are parts of the cortex that are able not only to interpret sensory cues but which are also capable of judgment, spontaneity and free will. The disadvantage of a flexible, open system, however, is that environmental influences (including the pressures of modern society) can have adverse effects on sexual behaviour; for example, personal interests may conflict with the pressures that society can exert. The cortex, as developed in humans, is uniquely effective in shaping the environment, both social and physical. Nevertheless, despite the evolutionary trend toward the selection for increased individual freedom, behaviour, particularly sexual behaviour, is still to a large extent under subconscious genetic control.

THE HUMAN MALE

The human male has to some extent deviated from a strict stereotype of sexual behaviour but, on the other hand, he does not seem to be very different from other male mammals in that he is dependent on androgens for normal sexual activity. It is necessary to distinguish the copulatory behaviour from the behaviour that precedes it, that is, courtship. In humans, courtship involves complex sequences of motor patterns and multisensory stimulation, especially for the sexually inexperienced. There is also a useful distinction to be made between seeking sexual contact (sexual motivation or libido) and being able to complete the copulatory act (performance/potency or ejaculation). Among humans, the importance of this distribution is clearly reflected in the high demand for the treatment of male erectile dysfunction.

Well-controlled studies on the mechanisms underlying sexual behaviour are difficult to obtain and are, therefore, largely lacking. Anecdotal accounts exist, for example the anonymous data of an island-bound ornithologist who weighed the content of his electric razor each day and found it significantly heavier on the days preceding and during his occasional visits to his mainland girlfriend. The extra beard weight was attributed to a rise in testosterone in anticipation of, and during, sexual activity. In normal men, however, the relationship between circulating testosterone and sexual behaviour is indirect because, as long as the testosterone exceeds the critical threshold level, higher concentrations do not necessarily result in a greater frequency or intensity of sexual behaviour. Individual differences in sexual behaviour are influenced by a particular partner, experience and learning, as well as differences in the level of circulating androgen. Behavioural accounts of the effects of testosterone replacement and withdrawal on sexual activity in hypogonadal men demonstrate that sexual decline (as measured by frequency of fantasies, sexual intercourse and self-rated sexual interest) usually becomes apparent 2–4 weeks after the withdrawal of testosterone therapy, whilst restoration effects are seen within 1–2 weeks of androgen replacement. Replacement experiments have also demonstrated that visually erotic experiences, for example films, are effective in androgen-depleted men but sexual interest, erections in response to fantasies and intercourse are more androgen dependent. There are, however, many individual differences and perhaps, not surprisingly, there is a more marked male sexual response if the female partner takes the initiative.

The biological functions of courtship may be summarized as follows.

(a) Advertising a readiness to mate
(b) Assessing the competence of a member of the opposite sex as a potential partner and parent; this aspect of courtship frequently involves demonstrations of physical prowess and competitiveness among men, as often seen in games of 'upmanship', which are heightened during courting
(c) Reducing fear of aggression in the partners, for example by the presentation of gifts or by mimicking infantile behaviour as seen in the use of 'baby talk'.

THE HUMAN FEMALE

Mate choice is important for both sexes but for the female it is essential that she chooses well as she makes a considerable investment in time, energy and food reserves in raising her offspring. In lower mammals, female reproductive behaviour requires circulating oestrogens. The behavioural effects of oestrogens may also be amplified by progesterone, for example in the rodent, lordosis (the immobile posture of the receptive female) will be performed only if she is in the correct endocrine state. In female primates, and particularly humans, the situation is much more complex. Female sexuality as defined by Frank Beach (from the University of San Francisco) can be subdivided into three principal components: attractiveness to the male, receptivity of the female to that male and proceptivity or sexual approaches initiated by the female.

As stated before, in the human female and some other primates, such as the rhesus monkey, androgens (testosterone and/or androstenedione) are necessary for normal proceptive behaviour; however, as in the male, little stringent research has been done in this area. Investigations are difficult because the many factors involved in sexuality cannot be controlled so the results are open to conflicting interpretations. Also, to squeeze the whole repertoire of female sexual behaviour within the confines of Beach's three categories seems unduly restricting. Some studies have shown that many women tend to show a peak of sexual activity around the middle of the cycle when androgen and oestrogen levels are relatively high, but again many do not. Other reports show a peak either just before or just after menstruation (this peak also coincides with high androgen levels but may also be due to appreciation of being infertile). What is of interest, however, is that physicians report a dampened libido to be a major symptom in women taking oral contraceptives when compared with controls. In order to separate the hormonal component of sexual behaviour from all other cues, hormonal measurements need to be linked with sexual interest, independent of male partners. Because a woman is influenced by many social factors there is often no discernible cyclical pattern. In many societies, a woman cannot take the initiative in sexual interaction as she is more obligately linked to the male's initiative. It may be that midcycle peaks of sexual interactions may stem from an oestrogen-induced increase in sexual attractiveness of the female to her mate ('she's hot'). Conversely, a decline in sexual activity in the luteal phase could be due to a progesterone-induced reduction of the woman's attractiveness to the male rather than her own sexual interest. Exogenous administration of androgens enhances sexual desire and response in women (in addition to an enhanced libido, androgens also increase the sensitivity of the clitoris; in contrast the normal sexual responsiveness of the vagina, that is its lubrication, is oestrogen dependent).

MATE CHOICE: CONSCIOUS OR UNCONSCIOUS?

We think that our choice of mate is conscious for very obvious reasons: we are compatible, we have lots of interests in common and furthermore our senses are pleased when we look at, talk to or touch our loved one. Many studies indicate, however, that our bodies have not relinquished us as completely as we would like to believe. There are many subliminal cues forcing us to choose, without conscious control, one of several possibilities. That is, the thinking brain can be manipulated by unthinking 'gut' feelings. Variations in levels of sexual activity between individuals is genetically determined and driven through hormones but modified by the environment. In social species particularly, the environment is of paramount importance; for example, the reproductive organs atrophy more rapidly in socially isolated or unisexually reared male rats than in heterosexually reared males. Exposure of the male to the female is sufficient to have major endrocrine effects which enhance the secretion of LH and testosterone. Free-ranging rhesus macaques are seasonal breeders, but the rise in blood testosterone can be accelerated by exposing the males to females brought into oestrus early by exogenous hormones. The testosterone levels of such males rose to levels normal for the breeding season. This appropriate influence of the females on the males reflects the social communication of endocrine states and acts as a means of synchronizing the reproductive condition of the sexes. In addition to the effects of females on the male's gonadal state, the male's social interactions and status, relative to other males, also affects gonadal secretions. For example in rhesus monkeys, social defeat by other males lowers the testosterone levels in the blood. The company of females brings it back to normal if the male is dominant to the female (this response is related to stress as discussed in Chapter 13).

TYPES OF COMMUNICATION

Despite the bewildering complexity of communication systems, there are only a few pathways by which signals from one individual are received by another. Communication between individuals can involve just one sensory modality or a combination depending on such factors as species, habitat and context. These sensory modalities are:

- Visual signals, particularly in humans, which involve expression of the face and posture of the body are important
- Tactile signals are important but usually occur together with other modalities; tactile stimulation is particularly important in induced ovulators such as cats
- Auditory signals are not so important in humans
- Chemical signals are very important in the human but the perception is largely subliminal.

PHEROMONES AND MAMMALIAN REPRODUCTION

Perception of chemical cues can produce a variety of reproductive endocrine and behavioural responses. These responses are under the regulation of pheromones (from the Greek *pherein* 'to transfer' and *hormon* 'to excite'), which are volatile hormones affecting the behaviour and physiological states of other individuals. In mammals, pheromones are implicated in communication signals relating to aggression, defence, territory marking, individual identifications, social interactions such as infant–parent relations, recognition of sexual condition, sex attraction and the evocation of sexual behaviour. Under natural conditions, pheromones rarely operate in the absence of other sensory signals and never out of an environmental context. An assemblage of physical, social and environmental factors is utilized, especially by humans, to co-ordinate reproduction and ensure its success. It is probably in this area of sexual choice that we humans exaggerate our intellectual independence by denying our biological heritage. We are unwilling to accept that our decisions may be moulded, or even driven, by physical phenomena beyond our conscious control.

Pheromones are subdivided into those substances that directly affect the CNS of the recipient organism leading to a rapid behavioural response, and those substances whose effects stimulate changes in the recipient's nervous and endocrine systems resulting in physiological and behavioural effects over a longer time course.

Signalling pheromones Signalling pheromones are those that elicit a more or less immediate change in behaviour, for example adoption of a mating stance. Veterinarians have exploited this response by using male sex pheromone sprays to elicit the characteristic mating stance in receptive oestrus sows, for instance, which can then be identified as fertile and suitable for artificial insemination. Mammals release an especially wide variety of odours as a result of the action of skin glands. The skin is a major site of steroid metabolism and many hormones, including oestrogens, androgens and pituitary hormones, affect skin gland function. Sex steroids serve to regulate skin gland size and the level of secretion (for example the action of oestrogen in the lubrication of the vaginal epithelium and in the pubertal growth of the secretory tissue of the mammary glands). Urine is a rich source of internal glandular secretions and metabolic breakdown products that can also have signalling value (as demonstrated by familiar dog and cat behaviours). The bulk of pheromone research has been done on laboratory mammals such as rodents and has led to the accumulation of evidence for a similar involvement of chemosignals in the primates. Some of the best publicized research has shown that, under oestrogen stimulation the vagina of the female rhesus monkey is the source of odour cues (a complex mixture of volatile aliphatic acids) that are attractive to males and stimulate their sexual activity and this has led to the identification of the same short-chain fatty acids in human vaginal secretions.

Priming pheromones The priming pheromones trigger a chain of physiological events involving the endocrine system. Their effects on behaviour are delayed and not always visible. The priming pheromones co-ordinate physiological/behavioural events ensuring adaptation and successful reproduction. They are evoked by particular issues pertaining to social conditions, for example density and dominance, and are also influenced by non-social environmental factors such as day length and nutrition. The primer pheromone in mice that influences puberty can have significant consequences on population biology. When puberty is delayed, the generation time is lengthened and the growth rate for the population will decrease; when puberty is accelerated, the generation time is shortened and the growth rate and population size will increase.

Production, release and reception of pheromones

Pheromones are produced by two types of skin glands: apocrine glands, which are coiled tubular structures opening into hair follicles, and holocrine glands, which make up sebaceous tissue. Apocrine glands secrete aqueous sweat and odourous lipids and are controlled by steroid hormones. Pheromones are also produced by the action of microorganisms on the secretions of skin glands. Not all body odours, of course, serve as pheromones. The most important property of a compound used as a pheromone is an appropriate volatility, and the limiting lower and upper molecular weight of a pheromone is estimated to be 50 and 300, respectively.

Identification of pheromones in mammals has revealed a variety of compounds that are responsible for odour. Pheromones commonly consist of more than one component, and functional groups recorded include acids, phenols, amines, alkanes and alkenes. Commonly, pheromones are odourous lipids derived from steroids. The sex steroids testosterone and oestrogen are ideal precursors for pheromones so that there is a close co-ordination of physiological and behavioural functions. Hair associated with skin glands is important for the dispersal of odour (less odour is detected from shaved as compared to unshaved armpits). Skin glands which produce odour occur with associated hair tufts in the breast, axilla, genital and anal regions. Volatile 16-androstene steroids have been isolated from axillary sweat and semen in humans. Of particular significance is androstenol, a metabolic product of testosterone.

Detection of olfactory stimuli can occur through two systems. The accessory olfactory system consists of chemoreceptor neurons in the vomeronasal organ and their neural pathway through the olfactory bulb and forebrain, whereas the main olfactory system has a more generalized function as a chemical analyzer of environmental odours with no predetermined function. The accessory olfactory system is involved in pheromone detection and in the integration of this sensory information with reproductive behaviour mediated by the hypothalamus and limbic system. In this role, the vomeronasal organ ties external chemical signals to neurotransmitters in the CNS and thus to endocrine changes in the circulation. It has long been acknowledged that the olfactory bulbs have an important role in the neural regulation of reproductive behaviour. For example, damage to the olfactory system produces severe male copulatory dysfunctions in some species (in hamsters it eliminates copulation altogether, while in rats copulation takes longer and is clumsier). Administration of exogenous gonadotrophins or testosterone does not alter these effects, although prior sexual experience may reduce the degree of behavioural deficit.

It has been suggested that in the adult human the vomeronasal system is vestigial having a functional significance only in infant–maternal recognition. However, recent research has revealed that the vomeronasal organ in adults is identifiable, functional and responds to the putative human pheromones in a gender-specific fashion. There is the additional possibility that there are internal pheromone receptors that allow the absorption of pheromones from the vagina or mouth into the blood stream where they can exert an influence on the endrocrine system.

The same pheromone can have both signalling and priming effects and so can cause immediate and long-term behavioural and/or physiological responses.

PHEROMONES IN THE RODENT

Much research has been done in assessing the behavioural and physiological effects of pheromones that affect social behaviour in various mammalian species. Work on rodents can be used to provide an overview of the types of interaction important in social species, including the human. In rodents, the onset of puberty is controlled by pheromones: exposing juvenile females to soiled bedding induces puberty in them about 10 days more rapidly than in females exposed to clean bedding. The puberty-priming pheromone comes from male urine. The level of pheromone in the urine is androgen dependent and urine from juvenile or castrated mice has no effect whereas testosterone replacement restores the pheromonal effect. The chemical identification of the puberty-accelerating pheromone is not yet definite although there is a suggestion that it could be one or both of two amines, isobutylamine and isoamylamine. Conversely, the onset of puberty occurs later in mice reared in groups of females than when they are reared singly, since socializing among the females results in suppression of sexual maturation. The puberty-delaying pheromone comes from female urine and its production is dependent on adrenal gland activity. Grouped females that are adrenalectomized do not produce the puberty-delaying pheromone whereas pheromonal activity is restored to the urine when the adrenalectomized females are given replacement glucocorticosteroids.

Once puberty has been reached, chemosignals continue to play a role in regulating ovarian function. For example, when female rats or mice are grouped together, the oestrous cycle is longer and oestrus becomes synchronized by pheromonal communication. Conversely, an oestrus-synchronizing effect by the male due to shortening of the oestrous cycle has been described in mice and is termed the Whitten effect. The introduction of a male into a group of females induces an early synchronous wave of oestrus 3 days later. The Whitten effect also operates in seasonally mating species such as sheep and goats and in enhancing reflex ovulation in voles and cats. The adaptive significance of the Whitten effect is that an optimum number of females may be impregnated when conditions for mating are good. In contrast, when a pregnant mouse is exposed prior to implantation (4 days postcoitus) to pheromonal stimulation from a strange male, the unfamiliar pheromone terminates the pregnancy and the female then promptly returns to oestrus (the Bruce effect). Males of a different genetic strain from that of the sire induce the greatest block to pregnancy. The adaptive significance of the Bruce effect, named after Hilda Bruce who discovered it in 1959, is in facilitating outbreeding.

Although much remains to be learned about the ways in which social interactions modify neuroendocrine mechanisms, the overall endocrine shifts underlying the Bruce and Whitten effects are partially understood. The synchronizing action of the Whitten effect is in fact a shortening of the oestrus cycle where the pheromonal cues from the male mediate LH release in response to increased GnRH stimulation and so bring on oestrus. In the pregnant animal, pheromonal stimuli from an alien male decrease prolactin release preventing corpora lutea formation; the consequent decrease in progesterone secretion results in the failure of implantation. It has been suggested that the pregnancy block may also result from the stimulation of gonadotrophin release (and reciprocal prolactin suppression), unifying the Bruce and Whitten effects. Primer pheromones probably have a common endocrine effect, resulting in changes in prolactin secretion from the anterior pituitary, since exposure of female mice to males results in a rapid decrease in serum prolactin. Adrenal gland involvement also may be implicated since in high density populations, where pregnancy-block occurs naturally it is the adrenal function that is affected. Efficient sex attractants and primer pheromones which cause earlier puberty, early seasonal breeding and more regular oestrous cycles would all result in evolutionary advantage. For example, in herding females during the non-breeding season, mutual depression of ovarian activity may then facilitate subsequent synchronization of breeding and the block to pregnancy reduces inbreeding.

POSSIBLE ORIGINS OF A PERSONAL PHEROMONE: THE INVOLVEMENT OF THE MAJOR HISTOCOMPATIBILITY COMPLEX (MHC)

There has been an extended series investigating the genetics and individual variables underlying some of the odours of mammals, particularly in mice. Genotypic influences on reproductive functions have been associated with the major histocompatibility complex (MHC) which is also involved in immunological functions and disease resistance.[1] MHC class I and class II genes encode membrane glycoproteins; class I antigens are expressed ubiquitously throughout the body and are the main basis for graft rejection; class II antigens occur mostly on bone marrow-derived cells, including B-lymphocytes, activated T-cells and macrophages. The class I and II glycoproteins also show a remarkable polymorphism in human (and mouse) populations with multiple alleles of class I and II genes occurring throughout human populations. By using different strains of mice alleleic for MHC glycoproteins, it is possible to analyse the nature, production and perception of a unique mixture of odorants acting as pheromones. For the laboratory mouse there is strong evidence that MHC influences reproduction through preferential selection of males based on the maintenance of heterozygosity of genes in the MHC, which affect fitness.

Three main recognition theories which involve the MHC have been proposed. Animals with different MHC genes are host to different bacterial floras. These, in turn, would excrete diverse chemical wastes thereby creating truly personal fragrances. In this scenario the vaginal fatty acids are not glandular secretions but are instead the product of microbial metabolism on the vaginal substrate secretions. This hypothesis is supported by the observation that addition of penicillin to incubations of vaginal lavages prevents the production of volatile fatty acids.

Another possibility stems from the observations that considerable anatomical and physiological variation is associated with MHC variants. These differences could influence the body's output of odorous compounds such that each individual exudes a distinctive mixture of odours.

The third, simplest proposal involves the release into the circulation of MHC products which are then excreted into the urine where they break into smaller volatile fragments. In a series of elegant experiments, mice

[1] The major histocompatibility complex (MHC) of genes gets its name from the fact that its products govern the success or failure of organ transplants. People have different combinations of genes in the complex and unless the donor's MHC resembles the recipient's the immune system will reject the graft.

were exposed to a radiation dose sufficient to destroy their bone marrow and stop the production of bone marrow-derived cells. Replacement bone marrow was from a donor with a different MHC. After approximately 2 weeks, glycoproteins characteristic of the donor appeared in the blood of the recipient while the recipient's own MHC went into a steady decline. At the same time the mouse took on the scent of the donor's MHC in place of its former fragrance.

Normally the mouse's concept of self derives not from its own smell, but from the scent it experiences during its first 3 weeks of life. The young can be tricked if fostered by parents that are genetically different but are perceived as related (for a human parallel see the kibbutz experience related on p. 107). Typically, mothers and infants of all species are sensitive to each other; for example, ewes as soon as 2 hours after having given birth will not accept or nurse an alien lamb.

Signalling pheromones in humans

Humans can probably respond to pheromonal cues; they certainly possess a level of olfactory awareness as they go to great lengths to manufacture perfumes and deodorants. Further, humans possess a functional accessory olfactory system and have three regions with odour-producing glands and hair tufts. Due to our upright posture, the breast-axilla region probably provides a better opportunity for the release of pheromones than does the genital and anal regions used so extensively by other mammals. On the social level, there is also little doubt that humans both emit and perceive distinctive, personal odours. In the biblical story, Jacob fooled his elderly blind father by donning the clothes, and with them the smell, of his brother Esau. It is generally accepted, for example, that 'a mother recognises the smell of her baby and her baby returns the compliment' (recognition, however, involves both olfactory and facial characteristics; bonding on the other hand, is more complex, utilizing many variables, see Chapter 11).

Sex attractants

Odorous compounds, identified as short-chain fatty acids, have been isolated from human vaginal secretions. They are similar to the substances found in vaginal secretions of rhesus monkeys and termed copulins which effectively stimulate mounting and ejaculatory behaviour in male rhesus monkeys. Copulins consist of a mixture of acetic, propionic, isobutyric, butyric and isovaleric acids. These acids vary in concentration during the human menstrual cycle, being highest prior to and at ovulation. Copulins are said to be sexually attractive to men (perhaps that is why Napoleon sent the message to Josephine '*ne lave pas, je reviens*'). The important biological factor is that ovulation should occur when mating is most likely to have a fertile outcome. Whether human vaginal secretions stimulate coitus is questionable since there are many other factors involved, pheromones being only part of a mixed stimulus. In an unpublished trial of human copulins, 20% of couples whose patterns of sexual behaviour were cyclic, showed increased sexual intercourse following exposure to the copulin mixture as compared with exposure to control substances.

Mate recognition

Gender recognition in humans is in most cases visual, but tests have shown that sexual and individual discriminations can be made on the basis of olfactory cues. Interestingly, it has been demonstrated that humans can discriminate between mice of different MHC strains on the basis of their urine or faecal odours, demonstrating the discriminating power of the human 'sniffing' mechanism. An analogous mechanism to that observed in mice, here operating for humans, would certainly have an adaptive significance since optimal choice of mates, facilitating an appropriate balance between extreme outbreeding and incestuous mating, could be reached by discerning close kin. The biological reason for discouraging breeding between close relatives is the greatly increased chance of homozygosity of recessive genes coding for deleterious characters. Neonatal mortality, for instance, is correlated with the degree of cousin marriages as is the incidence of major congenital malformations and other deleterious health effects. Major postnatal morbidity in the progeny of first cousins is elevated by 1.3–4.1% with prenatal mortality in excess of 1.0–6.4% when compared to non-consanguineous offspring. To-date, little information is available on adult morbidity or mortality rates, and hence the contribution of late-acting recessive genes in consanguineous progeny can only be guessed. The failure to initiate a pregnancy when closely related MHC antigens are shared is also a contributing factor in partner incompatibility.

There are systems operating in human societies discouraging inbreeding. Particularly impressive are the kinship laws of the Australian aboriginals. Practically all Australian tribes practiced exogamy where, to prevent consanguineous unions, marriages were required to take place between specified different subclasses. Other primates, such as the chimpanzee and the Japanese macaque, are also similar in this respect in that sexual relationships between siblings and between mother and son are absent or very rare. A more recent example demonstrating the strong distaste for incest is demonstrated by the occupants of Israeli kibbutzim. In a total of 125 marriages among second-generation inhabitants of three long-established kibbutzim, there was not one instance of marriage between members of the same peer group and only 3% between members of different peer groups in the same kibbutz. There was also little evidence of premarital sexual relations between adult members of the same peer group. It is presumed that this behaviour is based on negative imprinting by pheromones during childhood, thus the closer the relationship the stronger the biological taboo. Major exceptions against outlawing inbreeding in human societies are between royals: families such as those of the Peruvian Incas, the Pharaohs of ancient Egypt or 19th century European monarchs. In these cases brother–sister or cousin matings were common and served to keep power and privilege within a narrow group. In some modern societies (African/Asian Hindus and Moslems, in particular) consanguinity is encouraged for a number of social reasons. Family ties offer advantages in the maintenance of property and are more convenient from a cultural/religious compatibility point of view. How important these customs are in the maintenance of deleterious recessive alleles in the population gene pool is not clear. An indication may be gained, however, by looking at lethal recessive genes responsible for diseases such as Tay–Sachs or infantile amaurotic idiocy. Tay-Sachs disease is about 100 times more prevalent in Ashkenazy Jews living in the USA than it is among the non-Jewish population, with an approximate frequency of Jewish heterozygous carriers of 1 in 45 compared to 1 in 350 in the general population.

Outbreeding works because of the preference of the female for mating with strange males. Males are less selective and in the final analysis it is the female who (if her social setting permits it) makes the choice of mate (Chapter 7). A parallel distaste for close kin can be seen in the behaviour of many non-primate groups where the young males are ousted from the family group and required to disperse into new territory.

Female strategies, such as continuous sexual receptivity and cryptic ovulation, may have evolved to obscure the human female's current reproductive value and confuse males in an anticuckoldry strategy and thus facilitate monogamous pair-bonding through the mechanism of permanent sexual attractiveness. Since the human enjoys non-reproductive sex, it has been suggested that the female orgasm evolved as a lure to keep the male nearby to look after his genetic investment. Alternatively, the female may enjoy sex so much she cannot be trusted to be left alone! Cryptic ovulation and continuous female receptivity promote intense sperm competition.

Analyses of infidelity among women demonstrates that double-mating behaviour is significantly associated with a high probability of conception (Chapter 7). The 'mittelschmertz' (abdominal pain at the time of ovulation) could also permit the female's selection of outbreeding sperm without alerting her primary mate.

Priming pheromones in humans

Instances where priming pheromones may act can be suggested for humans as in other mammals.

Acceleration of puberty onset It has been suggested that the well established decrease in the average age of females at puberty since the end of the 19th century may be due, at least in part, to the increased associations with males. A similar mechanism to that described for mice is implicated here, however, other explanations such as the attainment of a stable threshold body weight at a progressively younger age due to improvements in standards of nutrition and other living conditions may also apply (Chapter 3).

Menstrual synchrony in female company Martha McClintock (from the University of Chicago), in 1971, first reported that menstrual cycles of girls attending an all female school became synchronized as the school year progressed. At the beginning of the year menstrual cycles were about 8 days apart but after 4 months were only 3–4 days out of synchrony in girls sharing the same living quarters. McClintock's study was replicated in experiments in which substances collected from cotton pads worn under the armpits of donor women, when applied to the upper lips of female recipients, caused menstrual onsets which were synchronized with the onset of the donor. Controls receiving ethanol placebos showed no sign of synchrony. Synchronization of grouped females is effective only in the absence of males; the phenomenon was not replicated in other studies where roommates were living on the campus of a co-educational school.

Regulation of the menstrual cycle in male company Evidence also indicates that the menstrual cycle is susceptible to male pheromones. For example, donor extracts collected from cotton pads worn under the armpits of men, when applied to the upper lips of female recipients, reduced menstrual cycle variability whereas a placebo had no effect. Further, the incidence of ovulatory menstrual cycles is significantly higher in sexually active women. According to one study, 92% of women ovulated if they spent two or more nights with a man during a 40-day period compared to a 56% ovulation rate in those who spent one or no nights with a man during the same study period. While the higher incidence of ovulation may result from greater exposure to a male pheromone, it is also possible that ovulation dictates sexual activity. Note, however, the parallel with the *Whitten effect* in mice where the stimulation of male pheromone evokes a high frequency LH pulse in the female and thus begins the sequence of events leading to ovulation.

OVERVIEW

Pheromones affect the behavioural/physiological states of other individuals synchronizing their social life. It seems that the neuroendocrine axes most sensitive to odours involve GnRH, FSH, LH, adrenocorticotrophic hormone (ACTH) and prolactin. The modern human has developed habits which reduce the need for the use of olfaction in normal situations (probably because of substandard hygiene in the past, we have learnt to associate body odour with uncleanliness and have developed methods, other than washing, of masking it). Perhaps for this reason comparatively little attention was paid to the communicative function of body odour in the human prior to the scientific interest generated by advances in the study of other animal groups. It has been demonstrated that the human nose is superior to the most sophisticated modern equipment in detecting odours and, with training, can be made to compare favourably with those of other mammals. One complicating factor is that the reaction to odours can, and often does, take place at the subconscious level. Information gathered mainly by psychologists, psychiatrists and psychoanalysts demonstrates that smell plays an important role in the sexual development of children and may be an important factor in the shaping of heterosexual responsiveness. At the stage of development when dominance-competition between father and son appears, there is an accompanying marked awareness of the sexual odour of adults, with a distaste for that of the same-sex parent. Females, too, show marked preference for alien males, and if the female has been reared by her mother alone she is deprived of the early experience of male odour from her father.

Possible proof of the importance of the axillary organ in the human comes from ritualized greeting ceremonies. Trobriand Islanders, for instance, used to show friendship for a departing visitor by the traditional ritual of wiping sweat from the visitor's armpit, smelling the fingers and then rubbing the sweat on their own chests. Presumably the host is demonstrating that absence of olfactory antipathy (he 'smells' okay to me!) the presence of which may result in hostility and rejection of the individual. Kisses could also owe their origin to the sense of smell as well as to touch. In some cultures, for example Eskimo, American Indian, Polynesian and Japanese, kissing is not practised in public but nose greetings (smell inhalation?) in various forms are used instead.

In normal circumstances there is no need for us to rely on the sense of smell to identify another person. But this can be done in exceptional circumstances with training, for example in the behaviour of blind people. There are also interesting examples like the blind woman working in a laundromat who would identify the owners of the clothes by their smell; Helen Keller could neither see nor hear, but she could tell people apart by their personal fragrance. With training it is also possible to determine the occupation of another person as the sebaceous glands of the scalp and other parts of the skin surface readily adsorb environmental odours. Different races have characteristic odours, as do white people with different hair colours. To quote Huysmans' *Cropuis Parisiens*, published in 1880, odours are 'saucy and sometimes tiring in brunettes and black-haired women, sharp and fierce in redheads, it is heady and pervasive in blonds like the nose of some flowery wines'.

The production and perception of odours also changes in relation to the physiological state of the individual. For example, changes in the odour of a woman's sweat in relation to the menstrual cycle or the appearance of a characteristic odour in relation to the sexual development of a young person is superimposed on the characteristic individual odour. Progesterone changes the quality of odours emitted in females so that each stage of the human menstrual cycle has a distinctive odour. As mentioned before, the product of microbial action on bodily secretions changes as the hormonally dependent substrate varies. Some pheromones such as the sex attractants act as 'turn ons', whilst others, perhaps related to fear or stress pheromones, may elicit hostile reactions. These odours can be readily recognized because odorous steroids are effectively transferred to objects handled by the subject. For example, suitably trained dogs can discriminate between pregnant and non-pregnant women.

Not only odour production but the ability to perceive it fluctuates with the menstrual cycle. Women generally have a greater olfactory sensitivity to most mammalian odours than do men since this ability is oestrogen dependent. It is therefore cyclical and the olfactory threshold, particularly for some musky odours, peaks around the time of ovulation and reaches a minimum during menstruation. It seems ironical that women decorate themselves with male pheromones (female sex attractants) which could elicit danger signals and competitiveness (aggression) in men. The musk steroid androstenol (5α,16-androsten-3α-ol), for example, is a component of the male pheromone present in human male sweat, semen and urine and may provide the physiological basis for the age old use of animal musks, from deer and civet cats, in perfumes used by women. Since the musks are the male sex pheromones of these animals they naturally attract the female (men however, often voice objections to the use of perfume in their women).

To conclude, a quote from *The Joy of Sex* by Dr Alex Comfort:[1]

> Cassolette – French for perfume box. The natural perfume of a clean women: her greatest sexual asset after her beauty (some would say greater than that). It comes from the whole of her – hair, skin, breasts, armpits, genitals, and the clothing she has worn: its note depends on her hair colour, but no two women are the same. Men have a natural perfume too, which women are aware of, but while a man can be infatuated with a woman's personal perfume, women on the whole tend to notice if a man smells right or wrong.

General references

Alexander, G. M. & Sherwin, B. B. (1991). The association between testosterone, sexual arousal, and selective attention for erotic stimuli in men. *Hormones & Behaviour*, **25**, 367–381.

Balthazart, J. (ed.) (1990). *Behavioural Activation in Males and Females – Social Interaction and Reproductive Endocrinology*. Comparative Physiology Series, Karger, Basel.

Bittles, A. H., Mason, W. M., Greene, J. & Rao, N. A. (1991). Reproductive behavior and health in consanguineous marriages. *Science*, **252**, 789–794.

Cowley, J. J. & Brooksbank, B. W. (1991). Human exposure to putative pheromones and changes in aspects of social behaviour. *Journal of Steroid Biochemistry & Molecular Biology*, **39**, 647–659.

Drickamer, L. C. (1989). Pheromones: behavioral and biochemical aspects. *Advances in Comparative Environmental Physiology*, **3**, 269–348.

Etgen, A. M., Ungar, S. & Petitti, N. (1992). Estradiol and progesterone modulation of norepinephrine neurotransmission: implications for the regulation of female reproductive behavior. Review article. *Journal of Neuroendocrinology*, **4**, 255–271.

Lerner, S. P. & Finch, C. E. (1991). The major histocompatibility complex and reproductive functions. *Endocrine Reviews*, **12**, 78–90.

[1] Copyright holders Mitchell Beazley International Ltd., reproduced with permission.

McClintock, M. K. (1971). Menstrual synchrony and suppression. *Nature*, **229**, 244–245.

Monti-Bloch, L. & Grosser, B. I. (1991). Effect of putative pheromones on the electrical activity of the human vomeronasal organ and olfactory epithelium. *Journal of Steroid Biochemistry & Molecular Biology*, **39**, 573–582.

Ryan, M. F. (1990). Sexual selection, sensory systems and sensory exploitation. *Oxford Surveys in Evolutionary Biology*, 7, 157–195.

van Vugt, D. A. (1990). Influences of the visual and olfactory systems on reproduction. *Seminars in Reproductive Endocrinology*, **8**, 1–14.

7 Sociobiology and reproductive success

Sociobiology is the study of the biological basis of social behaviour in animals, including human beings. Viewing social behaviour from a biological standpoint offers insights into behaviours such as sexism, nepotism, altruism, parenting and conflict. Since social behavioural patterns are the products of the same evolutionary pressures which drive morphological and physiological traits, they also are designed to be adaptive and confer reproductive success upon the individual or the species as a whole. Sociobiology looks at the underlying human pattern beyond the influences of culture in the context of a purpose in animal societies. Looked at from an anthropologist's viewpoint, behaviour can be infinitely variable and flexible; however, from a biologist's point of view, *Homo sapiens* shows, in broad terms, behavioural consistency. Philosophical contemplations about 'man's true nature' go back to antiquity. Aristotle, for example, felt that society was the product of the nature of human beings, whilst Plato maintained that human beings were the products of society. The contribution made by Wilson from Harvard was crucial to the definition of sociobiology as a science because previously sociobiology was a loose amalgam of evolutionary theory, ecology and animal behaviour. The combination of different aspects of mating, parental or fighting strategies has led to a different, more comprehensive understanding of their adaptive significance. A more recent synthesis has given rise to the subdiscipline socioendocrinology which deals with the synergy among hormones, physiological regulation and social context. Pheromones as mediators of socio-endocrine action are prominent (Chapter 6), but there are many other agents which link social environment with structure/function and behaviour.

In historical terms, the great strength of sociobiology was that its foundation was grounded in evolutionary thinking and it considered both genetic and epigenetic influences, influences which cannot be clearly separated, especially where these impinge on social structure. Sociobiology looks from the viewpoint of the genes, rather than that of the individual, a view popularized in 1976 by Richard Dawkin's book *The Selfish Gene.*

Over time, as some organisms are more successful in perpetuating their genes, natural selection results in a change in the nature of the genes and the bodies in which they reside. As described in Chapter 6, fitness, mediated by natural selection, is not limited simply to factors relating to the structure and function of a body but also includes complex social behaviours such as courtship, mating, care of offspring, association with friends and fighting. Further, the social environment created by these activities influences both hormones and behaviour to affect the outcome of reproductive success. Generally individuals that function better socially are better perpetuators of their genes and are favoured by natural selection. However, genes which enable their bodies to benefit others, but at some net cost to themselves, will leave fewer copies in the next generation compared to the selfish genes which did not. This fundamental selfishness has profound implications for human behaviour, as the predominant strategy of each gene is to make the best deal it can to ensure that copies of itself will occur in the future. This also implies 'to reproduce well', since a large number of feeble, inadequately cared-for offspring is also a losing strategy compared with a smaller but stronger number who are likely to succeed. In the sense of DNA as the central entity in the evolutionary struggle for survival, everything we do has consequences for reproduction. The question is not 'what's in it for me?' but 'what's in it for my genes?'. In this scenario, parental behaviour occurs not out of unselfish love for the offspring but because parental love represents a part of the strategy whereby genes replicate themselves. They do this by making the parent's body behave in such a way as to enhance the success of the copies of the genes called children, and thus giving the genes potential immortality. Genes coding for contraceptive behaviour are selected against; nevertheless, humans, with their reasoning brains, can, if they wish, choose to ignore aspects of their reproductive strategies which were advantageous in the past but are incompatible with future survival. Biological equilibrators will eventually stabilize the relationship between birth and death rates; if technology favours genetic longevity, fewer genes need to be replaced for optimal social survival and environmental rejuvenation.

STRATEGIES OF REPRODUCTION

An interesting example of a fitness-enhancing reproductive strategy is demonstrated by Huntington's chorea. Huntington's chorea is a hereditary disease produced by a single autosomal dominant gene and fatally impairs the nervous system. The disease arises very rarely by mutation, the gene persists in the population because carriers, before they reveal symptoms of Huntington's chorea, are often sexually promiscuous. The gene is deleterious to the individual, to the society and to the species yet modifies behaviour to increase fertility. More commonly, however, defects in genes adversely affect fertility or are dependent on environmental circumstance. Reproductive modification, for instance, occurs at the maternal/fetal level where healthy, viable genes can reduce fertility because of genetic incompatibilities. The rhesus or RH red cell antigen and other types of antigen systems described in Chapter 9 are such examples. In order to survive, individuals have to behave in co-operation with their genes which reflect successful strategies that have evolved for replicating themselves.

Biological basis of sexism

The female sex produces a relatively small number of eggs (large sex cells) and the male sex produces large numbers of sperm (small sex cells). The exploitation of women by men, as cynically pointed out by Richard Dawkins, probably began very long ago when the smaller, more active sperm began to take advantage of the rich food reserves present in the larger, less active eggs. The biological difference between men and women is at the basis of all sociobiologists' arguments of behavioural differences between the sexes. Since sperm are cheap and eggs are expensive, different reproductive strategies have evolved. Men are aggressive sexual advertizers; for them the maximum fitness goes to individuals with fewer inhibitions, 'play fast and loose', 'love 'em and leave 'em' is a winning combination. Men, in reality or fantasy, seek many different partners (female prostitutes outnumber male prostitutes; girlie magazines are much more popular than their beefcake counterparts). Women, on the other hand, are much more discriminating since their choice of a sexual partner involves a greater reproductive investment. Studies from several different cultures show that a smaller proportion of women think sex is all right with someone known only for a few hours when compared with the proportion of men of that opinion. Society has attitudinized reproductive differences to create the double standard where males are expected to be sexually less discriminating, more aggressive and more available than females. They are also more intolerant of infidelity by their mates than are their partners. Regrettably, the sexual double standard is moved from its genetic context and used to deny equality of opportunity and encourage the exploitation of women.

Sperm competition

Gamete differences set the stage for sperm competition. Sperm from different individuals compete for the same egg as do different ejaculates and different sperm within the same ejaculate. Sperm might succeed not only by leaving the testis first or swimming more rapidly but also by leaving more often. Where more than one male mates with a single female within a short period of time a fertilization contest is promoted. In sheep, lions, rats and humans, several males may mate with a single female during oestrus (ovulation) and this can result in multiple paternity (as commonly observed in rodent litters). Associated with multiple mating is sperm longevity, which in mammals can be protracted; bats, for example, use sperm stored from autumn matings in spring, and in humans fertilization 4–6 days after coitus is possible.

It is a common perception that only human beings engage in rape, but rape is common amongst many groups of animals and is often practised by sexually excluded young males. Rape as a form of sperm competition has also been important in the evolution of reproductive behaviour in other groups as well as mammals. One of the most amazing cases exemplifying the strategy of rape is seen in the bedbug (*Xylocoris maculipennis*). Insemination in some varieties of bedbugs, and their relatives, is homosexual rape. Sperm are injected through the body wall of another male from whence they migrate to the seminal vesicles. Here the sperm mingle with that of the victim and are found in his subsequent ejaculates. The parasitization of the rival's genitals multiplies a male's ability to deliver sperm. The rapist's sperm have countered the phagocytic defences mounted by the victim and, in addition, have evolved enlarged motor organelles enabling them to move fast and overcome any mechanical resistance. This ruthless survival strategy is a variation of the gene's ultraselfish drive to manipulate the organism it inhabits to replicate and ensure its continuation in the next generation.

EVIDENCE OF STRATEGIES FAVOURING SPERM COMPETITION

In species where multiple paternity in a litter is the norm, male sperm competition is severe. Selection pressure favours a variety of physiological devices that increases the chances of outcompeting the sperm from other males; one such device is an increased sperm number in each ejaculate. In those genera in which females are likely to mate with more than one male during oestrus, the males have larger testis in relation to their body size than in those genera where only one male usually gains access to the female. Individuals in such multimating genera also have a higher sperm production capacity per unit weight of testis. Chimpanzees, baboons and macaques, all with multiple mating systems, have ratios of tubules to testis connective tissue that range from 2.2–2.8:1; while men, gibbons and langurs, which are not typically multi-male mating species, have ratios of 0.9–1.3:1. In addition to larger ejaculate volume, the daily rate of sperm production is also increased: in the rhesus macaque it is about 23×10^6 sperm per g of testicular tissue, while the corresponding value for man is only 4.4×10^6 sperm per g of testis.

In rodents, multiple ejaculations are important not only in pregnancy initiation (the luteal phase is induced by mating) but also in the context of sperm competition. Males that achieve multiple ejaculations with a female gain a reproductive advantage over males attaining a single ejaculation. The explanation lies in the sperm concentration achieved in the vagina rather than in the time elapsed between ejaculations. Where oestrus is prolonged, an effective strategy to ensure paternity is to deter the subsequent displacement of sperm by a second male. In rodents, for instance, mating plugs (where the semen coagulates on exposure to air) effectively acts as a sperm barrier, and prolonged copulation in dogs achieves the same result.

Sperm competition in humans

Sperm competition does occur in humans and has presumably been a selective force in the evolution of certain human characteristics, in particular the common (monogamous, serially monogamous, polygamous, promiscuous, or a mixture of these) sexual inclinations of males and females. It may be that the basic male mating strategy consists of a high investment pair-bond with one or more females but also opportunistic mating with other females. This, of course, represents the infrastructure required for the double standard. In fairness to men, however, this double standard could not have been selected if women had always denied them the opportunity for its realization. Since human sperm is fertile for 4–6 days in the reproductive tract of the female, any circumstance that places ejaculate from two or more males in the vagina within this period promotes strong sperm competition for the fertilization of the ovum. Such circumstances are rape, courtship, communal sex, prostitution and faculative polyandry.

Rape

Rape is a cultural phenomenon of most societies. According to an 1988 study in the USA, more than 60 000 rapes are reported each year and it is estimated that this represents only 10% of actual cases. Rape has occurred throughout history in all cultures but is at its highest during wars. For example, in the India–Pakistan war over Bangladesh lasting 9 months, between 200 000 and 400 000 Hindu women were raped by Pakistani soldiers.

Rape was probably biologically more important in hunter–gatherer societies when evolutionary options were still available but ultra-aggressive genes survived to drive *Homo sapiens*. Now rape is the product of a sinister cultural pattern coinciding with biological selection. The stress-induced release of GnRH in rape trauma may stimulate an early ovulation which gives the rapist's sperm a good chance of achieving fertilization. Evidence suggests that an inordinate proportion of rapes result in pregnancy.

Courtship and communal sex

In cultures that permit adolescent females to choose their primary mates, the time between puberty and marriage may represent a brief period of relatively dynamic sexual activity and concomitant potential for sperm competition. Adolescent promiscuity (under the guise of courtship by both sexes) is well documented and generally tolerated in many modern cultures; cross-cultural studies reveal that 45% of indexed societies approve of premarital sex or tolerate it without punishment. These attitudes are not different in principle from many traditional cultures where, for instance, the Trobrian Island females were free to choose their lovers as an adolescent privilege but after marriage female adultery was an offence which could be punished by death. A mercenary, but biologically effective, female strategy is the use of multi-male mating as a primary mate acquisition strategy. By mating repeatedly with several youths she sets up competition between the different sperm until impregnated and then names as the father the favoured phenotype, or best potential provider from among them. If there is a correlation between the sperm's ability to reach the egg and the overall quality of the genes it carries, then the female's offspring have genetic advantages which increase the chance of survival of her genes.

Prostitution

Prostitution, the oldest profession, is indiscriminate sexual activity for profit and has always existed in human society. For example, the use by males of female prostitutes was routine and almost universal in ancient Greece and Rome. In the USA in 1981 there was an estimated 250 000–350 000 full and part-time prostitutes and similar figures can be obtained for anywhere in the world; in Ethiopian cities, a 1974 survey uncovered that about 25% of the adult female population were prostitutes. The practice of female prostitution has potential for creating intense sperm competition but this competition can be potential rather than actual since many modern prostitutes, working in affluent surroundings, reduce fertility by abortion and contraception. Unprotected prostitution, on the other hand, has been successfully exploited by parasitic genes coding for sexually-transmitted diseases such as AIDS. In some of the world's worst afflicted areas, 30–60% and up to 80% of all female prostitutes working in brothels are HIV positive. From these high-risk centres, the disease can easily spread to the general population. In 1993, 14 million people, worldwide, were estimated to be HIV positive. By taking advantage of human sexual behaviour, the HIV virus has enhanced its fitness and, because the disease typically affects people in their child-bearing years, it is now becoming a main determinant of infant mortality in places of most rapid transmission (Chapter 17).

Facultative polyandry

The typical human female has one principal mate from whom she and her children receive care, protection and material resources. This arrangement is biologically adaptive for women, and the male benefits by having routine sexual access to her and thus a consistent competitive advantage over that of any facultative or extra-pair matings. Sporadic female extramarital sex occurs in 73% of indexed societies and is common in 57%. For example, over 50% of respondents in the *Cosmopolitan* survey in 1981 indicated that they had had extramarital liaisons. In the long term, facultative polyandry is rare because on balance a monogamous relationship is adaptive for a woman and her offspring. However, it may also be advantageous to encourage sperm competition. A 1989 analysis of infidelity among British women demonstrated that double-matings (female mating with a second male while still containing sperm from her regular mate) show a significant association with a good probability of conception because matings took place between days 6–15 of the menstrual cycle. A subconscious desire to outbreed may be implied by the accompanying contraceptive practice;

averaged over the entire cycle, contraception was not practised in 25% of extra-pair matings compared with 15% of intra-pair matings.

The potential female benefits of facultative polyandry may be summarized as follows.

(a) The acquisition of valuable genes. In a significant number of societies the first marriage is arranged by the parents often on the basis of wealth or power not biology. Pheromones can subconsciously assist in mate choice which increases fitness (Chapter 6).
(b) The son's effect. This assumes that a male child fathered by a particularly charming man who is successful in seduction may inherit that paternal charm and compete successfully in seduction when his turn comes. 'Charming' sperm may also be exceptional in egg contests.
(c) The maintenance of genetic diversity. This is valuable in a changing, dynamic environment; since half of the genetic complement of the female goes into the next generation, any behaviour pattern which increases her genes' chances of survival is a good strategy.
(d) A fertility backup against male sub- or infertility. It is estimated that up to 20% of couples in developed countries (with the percentage much higher in developing countries) are infertile. Of these, 30–40% have an exclusive male factor (Chapter 1). Fertile females bonded to males with fertility problems can gain reproductively with extra-pair mating.
(e) Acquisition of material resources. The evolutionary origin of gift presentation was probably the presentation of food in courtship. However, the demand of material resources in exchange for sexual favours may be seen as another form of prostitution. Enhancement of social status is a related non-material form of acquisition.
(f) Protection of self and offspring in the absence of the regular mate. This behaviour may have been particularly important at times of social upheaval such as war, where women could monogamously bond with officer-grade men in order to avoid rape by the common soldiers.
(g) Sexual pleasure and comfort for the highly sexed human female.

In the final analysis, sperm competition is kept alive by female choice. The potential cost of seeking facultative mates is the ever present risk of being found out and that male jealousy may result in desertion or severe punishment for female adultery. Risks may, however, be minimized and benefits maximized by concentrating double matings around the times of peak fertility.

Evidence for sperm competition in the human

At first sight it may seem that evolution has given males an overwhelming reproductive advantage over females. However, whenever a child is produced it is still the product of just one male and one female. The child is their ticket to evolutionary success and on balance the two sexes are equally successful or not. There is, of course, a difference in favour of the male in the number of copies of genes that each individual can pass on; King Ismail of Morocco fathered 1056 children. Among females there is less difference between the most and least fit reproductively (12 offspring or so is a realistic maximum), but by promoting monogamy many societies have narrowed the reproductive gap between the sexes.

The main evidence that sperm competition is important in the human comes from three sources, which can be summarized as follows.

(a) The fertilization of separate ova in a single female by different males and the subsequent plural birth of half-siblings provide strong evidence of human sperm competition. A most conspicuous case was reported in Germany where twins were fathered by a black US soldier and a Caucasian German businessman, neither twin being the offspring of the woman's husband. Analysis of blood groups in Britain indicated general levels of paternal discrepancy (offspring sired by fathers other than their putative fathers) of between 5 and 13% with up to 30% in certain communities that are suffering socioeconomic depression.

(b) The concentration of prostaglandins, which stimulate contractions of the uterine myometrium, is about 100 times greater in semen than in other reproductive tissues. Prostaglandins are also effective in promoting the female orgasm, prompting a theory that multiple orgasms in the female may have been selected for by the male to facilitate sperm transport and fertilization and not by the female to increase her pleasure. Sperm competitiveness could be associated with increased levels of prostaglandins in semen as they are in significantly lower levels in 40% of infertile men whose infertility cannot be otherwise explained.

(c) Cryptic or concealed ovulation and continuous female sexual receptivity promote intense sperm competition. These female strategies (described in Chapter 6) obscure the female's current reproductive value, counter and confuse the male's promiscuousness and facilitate monogamous pair-bonding through fear of losing a permanent sexually attractive mate. Non-reproductive sex and the female orgasm may also have evolved as a lure to keep the male nearby to look after his genetic investment. Abdominal pain at the time of ovulation can signal a good term for the female to permit sperm competition without alerting the primary mate.

THE BIOLOGICAL BASIS OF FEMALE CHOICE

Since females are selected to behave in ways that maximize their fitness, a major component of their mate-selection strategy is 'what's in it for me' (see p. 114). From a female's point of view there are three major concerns in choosing a mate. These are: coupling with the best possible mate whose genes will maximize her own genetic fitness; gaining access to the best environmental resources, for example the best combination of food and defence against predators; and acquiring a male whose behaviour will contribute directly to her reproductive success, that is, helping with the care of her offspring.

The connection between resources and reproduction is strong in human societies. For example in Muslim society (where men can have up to four wives) and other societies where polygyny is common, a man acquires wives as he acquires resources, that is, material wealth. Among Australian aboriginals the women tended to aggregate themselves in collectives of co-wives around middle-aged men at the peak of their social capacity. Such men have the greatest capacity to provide resources and shoulder the burdens of child-rearing. It should be pointed out that it was the women who took the active role in establishing these polygynous collectives which provided optimum conditions for the rearing of the younger generation in the harsh Australian environment.

THE BIOLOGICAL BASIS OF PARENTING

Since the *raison d'être* of genes is self-propagation, most of us provide our children with what they need. Love of parent for child is an evolutionary strategy ensuring that parents will invest in the child in a manner that maximizes each parent's gene survival. The child is equally motivated to love its parents because such a strategy enhances it own evolutionary success, that is, parental and filial love is selfish. According to Trivers, of Harvard University, this sets the stage for parent–offspring conflicts. Children want to exact more parental investment than it would be in the best interests of the parents to give; parents are seeking to give the most they can to their offspring at least cost to themselves. Such a conflict implies that the accurate communication of offspring need is evolutionarily unstable since offspring will be selected to demand extra resources. An alternative and attractive sociobiological theory suggests that parental conflict is not essential since the allocation of parental investment is made using accurate information about the condition of the young.

Parental behaviour is typically influenced by genetic relatedness to the children. Many psychological studies show that children in families containing at least one step-parent are more likely to be abused or neglected than are children living with both parents. Humans have an open social programme which may have evolved because they lived in small related bands of people where an orphaned child would typically be adopted by another adult. In human society there is also a role for the menopausal grandmother who may continue the nurture of 25% of her genes through her grandchildren (Chapter 12). Primate studies demonstrate that genes gain fitness by infanticide. Infanticide by an unrelated male langur monkey, for instance, increases the probability that his offspring will be present in the next generation. A langur troop is composed of a single male and several adult females with their offspring, and infanticide occurs when a single male, or group of males, attempts to take over an existing troop of females. Removal of the offspring provides a more immediate access to receptive females because, on cessation of lactation, they move to oestrus.

Intentional childlessness (contraception) is the most desirable hope for alleviating our pressing over-population problem, but for it to be effective it must prevail against several billion years of evolution. One fear is for the long-term genetic consequence because selection will favour those people who wish to have children, and any genetically influenced tendency for lesser reproduction will eventually disappear leaving a population of ever more eager breeders. If our somatic investment cannot subvert the selfish gene, harsher methods than free choice may then have to be employed to effect equal distribution of resources amongst the breeding population, disallowing some gametes from taking a disproportionate share of the available scarce resources. There seems to be, however, a direct association between high child mortality and the wish for many children, strengthening the socioeconomic solution to population problems.

Women often have expressed the impression that men are less parental than they are. A lower paternal, as compared to maternal, instinct may reflect the male/female difference in parental investment: sperm are cheap, so wild oats can be scattered far and wide. In may also reflect the difference between faith and knowledge; men have faith that the children are genetically theirs, whereas females know with certainty that their genes have been passed on.

THE BIOLOGICAL BASIS OF ALTRUISM AND AGGRESSION

The British geneticist J. B. S. Haldane speculated on the possibility that a trait may be selected that confers an advantage for the group but at some net cost to the individual. He used the term altruism for such a trait and defined an altruistic act as one that decreases the personal fitness of the individual performing the act but is beneficial to the population as a whole. For example, selflessly sounding an alarm to protect the group from a predator and, as a result, attracting attention to oneself. In evolution, true altruism should never occur because genes producing altruism should be less fit than genes that produce selfishness, unless the act is linked to a special gene selection. Altruistic behaviour can be selected if the probability is high that the beneficiaries of the altruistic act also have the same genes as the altruist, that is, kin selection. For example, by protecting the offspring the parent invests in its own genetic representation in the next generation. If altruism is evolutionarily acceptable as long as related genes are implicated, the boundary between altruism and self-interest becomes blurred. Nepotism is a favoured human trait and kinship (bloodlines) is a basic organizing principle in all human cultures. *Homo sapiens* maximizes fitness-enhancing behaviour when he treats family differently from strangers, the familiar differently from the strange. We fight most strongly for what we believe in and what we believe in is most likely to be closely related to our home and our family, that is, our basic way of life. It may be that selection for altruistic behaviour has also bred suspicion and intolerance of dissimilarity. On the positive side, sharing of food, possessions and caring for the sick probably evolved from kin selection in hunter–gatherer societies. As social creatures we profit by linking reciprocity with fitness and reciprocal altruism can be adaptive, even between individuals and species which are totally unrelated. The primary requirement is that the giver will be the getter at some later time and if this is so genes for such behaviour would be spread by natural selection. An extension of reciprocity is caregiving behaviour. Caregiving to non-relatives among humans is associated with intelligence and culture and is transmitted by language.

William Durham's contribution is important in viewing warlike behaviour in the light of sociobiology. Durham postulated that war was an evolutionary asset; that simply by existing one group can seriously threaten the resources of another. If groups compete for the same resources then, just as in direct competition between individuals, each group can threaten the fitness of the other. The fitness of a population could be increased by war provided that the cost of waging it is less than the benefits received. Because warfare has been ubiquitous throughout human history and forms the basis of economic and political institutions, it must fulfill the fitness selection criterion. Armed conflict between bands and tribes may well have exerted a fatal influence on our evolution; plunder and rape are effective in resource acquisition and gene dispersal. Fighting may also have selected, in part, for large brains, efficient use of weapons, ability to communicate complex strategies and deceit, as well as conformity and obedience to authority. If our genes cannot advise us, we must learn to listen to our reason and acknowledge that traditional motives of power, territory, prestige and ideology are no longer adaptive. The co-operative mix of all peoples' unique contributions can create, if we allow it, a coherent, intelligent and innovative entity effective in aggression control and in transcending our destructive genetic potential.

General references

Barash, D. P. (1982). *Sociobiology and Behaviour*, 2nd edn. Elsevier, New York.

Bellis, M. A. & Baker, R. R. (1990). Do females promote sperm competition? Data for humans. *Animal Behaviour*, **40**, 997–999.

Betzig, L., Mulder, M. B. & Turke, P. (eds.) (1988). *Human Reproductive Behaviour: a Darwinian Perspective*. Cambridge University Press, Cambridge, UK.

Durham, W. H. (1976). Resource competition and human aggression. Part I: a review of primitive war. *Quarterly Review of Biology*, **51**, 385–415.

Scott, J. P. (1989). *The Evolution of Social Systems*. Gordon & Breach, New York.

Smith, R. L. (ed.) (1984). *Sperm Competition and the Evolution of Animal Mating Systems*. Academic Press, Orlando, FL.

Smuts, B. B., Cheney, D. L., Seyfarth, R. M., Wrangham, R. W. & Struhsaker, T. T. (eds.) (1986). *Primate Societies*. University of Chicago Press, Chicago, IL.

Trivers, R. L. (1974). Parent–offspring conflict. *American Zoologist*, **14**, 249–264.

van den Berghe, P. L. (1988). The family and the biological base of human sociality. In *Biosocial Perspective on the Family*, ed. E. E. Filsinger, pp. 39–60. Sage, Newbury Park, CA.

Wilson, E. O. (1975). *Sociobiology: The New Synthesis*. Harvard University Press, Cambridge, MA.

Ziegler, T. E. & Bercovitch, F. B. (eds.) (1990). *Socioendocrinology of Primate Reproduction*. Wiley-Liss, New York.

8 Fertilization and the initiation of development

PRECONCEPTUAL CARE

In a fundamental sense, healthy pregnancies begin well before conception. It has been repeatedly demonstrated that if the parents' health is good, or consciously improved before conception, many subsequent problems can be prevented. Given a minimum standard of living, preconceptual care is easy and can also be pleasant; exercise, fresh air and a healthy diet significantly reduce the social and health risks to the future child. Hereditary and environmental factors should be discussed by a couple when planning to 'have a baby' because both share the privilege and responsibility of being a parent. There is a close relationship between birthweight and survival: below a certain weight the lower the birthweight the greater is the chance of an infant being still-born or dying within the first few days after birth. In Chapter 11 the factors associated with low birthweight and the population most likely to be at risk are examined and the conclusion is that public healthcare and intensive education are the most effective preventitive measures. Governments should take the necessary measures required and give special attention to the education of young people so that they can exercise a responsible attitude toward their children and improve the lives of the present and future generations.

Nutrition is very important. Despite this importance the general public receives little education in this area, but, even worse, many doctors and paramedics are not sufficiently encouraged to master the scientific basis of nutrition. If the general public were more aware of or trained in nutrition, many disease states might be prevented or certainly more readily corrected. Small changes made, either to the food itself or the way it is prepared, could maximize the nutritional value of the food that is consumed. Nutrition is the study of 41 or so nutrients needed to build health and their relation to each other. These nutrients consist of 10 essential amino acids, supplied by such proteins as milk, eggs and meats; a fatty acid (linoleic acid) found in oils and fresh green vegetables; and some 15 vitamins and 15 minerals. When all of these nutrients are obtained in amounts sufficient to meet individual needs, the diet is said to be adequate. Nutrition for a pregnant woman differs little from nutrition for any person under physiological stress. Pregnancy is a natural physiological state and during the course of evolution adaptive mechanisms have developed to protect the fetus from the harmful effects of adverse environmental factors including an erratic supply of food (Chapter 9). Protection against malnutrition is achieved by mobilization of the tissue reserves of the mother whenever dietary supplies of nutrients are inadequate. Moderate degrees of maternal malnutrition are not necessarily harmful for the outcome of a pregnancy provided that the mother was previously well nourished. There is, however, little doubt that prolonged undernutrition or specific deficiencies before and during pregnancy increases the danger to both mother and infant. For instance, there is an established association between living in areas where there is iodine deficiency (endemic goitre) and mental retardation. Hypothyroidism (prevalent in Nepal and Bangladesh) can be simply and cheaply rectified by the provision of iodized salt.

Genetic abnormalities, however, are more often linked with paternal than with maternal DNA damage. Sperm cells are particularly vulnerable to oxidative damage because DNA repair is stopped during mitosis. Smoking, for instance, involves a high intake of oxidizing compounds such as nitrogen oxide from tobacco smoke. Ascorbic acid (vitamin C) is a powerful anti-oxidant sequestered in high concentration in seminal fluid providing the dietary intake is good. A daily dietary intake of 250 mg (approximately equivalent to two servings of fruit and three of vegetables) of ascorbic acid provides protection, as estimated by the reduced accumulation of oxidized products in the seminal fluid. An ascorbic acid dietary supplement may be beneficial in enhancing male fertility when it is suspected that the diet is poor or deficient. Several studies have associated low ascorbic acid with increased sperm precursors, abnormal morphology (see Fig. 1.1, p. 8), agglutination and decreased sperm count and motility. Ascorbic acid deficiency does not appear to affect the libido so that no physical sign can act as a warning of the condition. Fortunately, the forces of natural selection ensure that the vast majority of human conceptuses with major chromosomal anomalies do not survive to term. A worrisome but preventable association is between men who smoke and increased incidence of childhood cancer, such as leukaemia and lymphoma. Whether this is the result of a nutritional deprivation and/or a drug-specific function is not known.

Related preconception precautions, other than nutritional ones, concern the use of drugs. Nicotine and ethanol, in particular, but drugs of any kind (including regular use of medication such as aspirin, antacids and laxatives) are best avoided during the preconception period. Drugs which are prescribed for a medical problem, for example diabetes or epilepsy, are in a different category and need close supervision. It is also advisable for women using steroidal contraceptives to change to a barrier method of contraception for 3 months before trying to conceive. This simple precaution allows the hormonal effects of the oral contraceptive to be eliminated from the body, the menstrual cycle to become re-established and the mineral and vitamin status to be improved. The social consumption of stimulants and prolonged use of steroidal contraceptives, by increasing the overall demand for nutrients, may cause subtle nutritional deficiencies in seemingly well-fed adults. Sadly the full expression of the hereditary potential of children is rarely achieved, and sometimes the damage due to neglect is irreversible bringing lifelong tragedy (Chapter 9). The importance of preconceptual care has long been known: ancient wisdom has emphasized parental responsibility toward the next generation. For example, instructions as to acceptable preconception behaviour can be found in the Old Testament[1] 'And the angel of the Lord appeared unto the woman, and said unto her, Behold now, thou art barren, and bearest not: but thou shalt conceive, and bear a son. Now therefore beware, I pray thee, and drink not wine nor strong drink, and eat not any unclean thing: For lo, thou shalt conceive, and bear a son; . . . And the woman bare a son, and called his name Samson . . .'. Samson's mother, the wife of Minoah, was rewarded by giving birth to a son of extraordinary strength, the implication being that parents have considerable power to determine the type of child they will produce.

SPERM MIGRATION

At ovulation the ovum has already undergone the first meiotic division with extrusion of the first polar body and is arrested at the metaphase stage of the second meiotic division until activation occurs. Activation is generally induced by penetration of the sperm through the vitelline membrane, which marks the beginning of fertilization which ends at syngamy, the union of the two sets of haploid chromosomes to form a new diploid fertilized ovum or zygote. After ejaculation, the sperm have to traverse the cervix, the uterus and the uterotubal junction before reaching the ovum in the oviduct. Each of these barriers substantially reduces the

[1] Judges, Chapter 13, v 3–5; 24. *The Holy Bible*, Authorised King James Version.

sperm population so that, theoretically, only the fittest ones have the opportunity to fertilize the ovum. Barrier negotiation is one point where natural selection can operate to ensure the survival of optimal genetic combinations from generation to generation.

Sperm can pass rapidly through the reproductive tract, the first are found in the oviducts within 5 minutes of vaginal insemination. Sperm migration through midcycle cervical mucus, however, is slow (at an estimated rate of only 2.0 to 3.0 mm/min) so sperm motility alone would be unable to permit passage of sperm to the oviducts within 5 min. The vaginal and uterine contractions following increased emotional and sexual excitement, particularly orgasm, propel the small proportion of sperm that is not trapped in the coagulum of the cervical mucus into the uterus. Human semen contains large quantities of prostaglandins which may further stimulate the uterine myometrium and so ensure the rapid propulsion of sperm from the internal cervical os to the uterotubal junction. It should be pointed out, however, that in species like the human where intercourse may occur before, after or at the same time as ovulation, the first arrivals do not necessarily have the best chance to fertilize the ovum. Seminal prostaglandin production is dependent on testosterone levels. Prostaglandin is a powerful smooth muscle stimulant but it also enhances sperm motility and zona pellucida penetration in addition to being an immune suppressor. Since sperm are antigenic and the vagina is not an immunologically privileged site, an immune suppressant helping to neutralize macrophage activity is beneficial. However, on the negative side, prostaglandin as an immune suppressant may play a facilitating role in the spread of AIDS and other sexually-transmitted diseases since it is delivered together with the infecting pathogen in the semen.

Cervical mucus is a hydrogel composed of two main elements, cervical mucin and soluble components. The mucin, a fibrillar system of glycoproteins, is linked either directly by disulphide bonds or through cross-linking polypeptides and is capable of changing its consistency according to the balance between the secretion of oestradiol and progesterone. At midcycle under oestrogen domination, cervical mucus is composed of macromolecular fibrils arranged as micelles (parallel chains) with enough spaces between them to permit the passage of spermatozoa. At other times, cervical mucus lacks the micellar structure, is thicker and is an effective barrier to the entry of sperm and bacteria into the uterus. Sperm are not uniformly distributed throughout the mucus but tend to be in the vicinity of the mucosal surface, with many lodged in the cervical crypts from which they are subsequently released to continue their upward journey through the reproductive tract. Only motile sperm are lodged in these crypts where they are protected from phagocytosis. The cervical crypts also act as reservoirs, permitting a continued release of viable sperm over a period of hours and so staggering their ascent to the uterus. Motile sperm have been found in the mucus up to 3 days after coitus while dead sperm are eliminated from the reproductive tract either by phagocytosis or by the movement of the cervical mucus to the vagina. Antisperm antibodies on sperm or in cervical mucus reduce the sperm's capacity to penetrate and survive in cervical mucus and such weakened cells are also removed.

The uterotubal junction forms a further barrier to sperm ascent and it is thought that 85 hours may be the upper limit for maintenance of sperm motility in the human. Consequently, when natural family planning methods are used, a minimum of 4 days abstinence should be allowed to compensate for sperm survival.

There are two main types of cells in the oviductal epithelium, ciliated and secretory. The ciliated cells are much more numerous in the fimbriae and in the ampulla where they facilitate ovum 'pick-up' at the time of ovulation. The secretory cells predominate in the isthmus. Both types of cells undergo changes in structure and function under the influence of the ovarian hormones secreted during the menstrual cycle. Sperm movement in the oviduct is a combination of sperm motility, fluid flow and muscular activity.

OVUM 'PICK-UP' AND SURVIVAL

Ageing of the ovum after ovulation and before fertilization progressively diminishes the number of young born and increases the percentage of abnormally developing embryos. Given the possibility of desynchronization of insemination and ovulation in the human, there is a greater chance of an aged ovum being fertilized. The fertilizable life of the human ovum is approximately 24 hours, but it remains fertilizable for longer than it is capable of producing a normal embryo. Consequently there are many conceptuses which, once fertilized, fail to continue their development.

At ovulation, the ovum is surrounded by two distinct layers of follicle (cumulus) cells. Closely attached to the zona pellucida are the coronal cells which are arranged radially in a layer two to three cells thick. Surrounding the corona radiata are more cells, the cumulus oophorus which is a sticky gelatinous matrix composed largely of hyaluronic acid and containing many cumulus cells (Fig. 8.1). The presence of a well-developed cumulus oophorus, which persists around the ovum during and for some time after

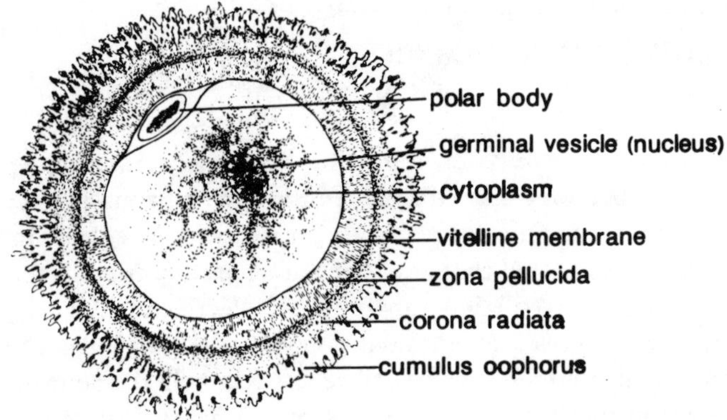

Fig. 8.1 Mature human oocyte with the polar body and cellular investments.

fertilization, is important for normal ovum 'pick-up' and for the early stages of ovum transport. Ovum 'pick-up' is accomplished by the rhythmic contractions of the fimbriated folds, ciliary activity on the fimbriae and the participation of the oviduct.

In most species, the transport time remains the same whether the ova are fertilized or not. In the human the ovum remains in the oviduct for 3 days before it is delivered into the uterus. Secretions of the oviductal epithelial cells are important in maintaining the growth and well-being of the fertilized ovum. Human ova are, however, capable of fertilization and complete development in the uterus without exposure to oviductal secretions as seen by successful human pregnancies following intrauterine deposition of early embryos fertilized *in vitro*. In the human, oviductal secretions may be important for sperm transport and maintenance of sperm viability. The programming of ovum or zygote transport through the oviduct is dependent on the correct hormonal balance of oestradiol and progesterone. An emergency postcoital contraceptive treatment based on the administration of synthetic oestrogens, such as diethylstilboestrol or ethinyl oestradiol, reliably prevents pregnancy provided that treatment is started within 72 hours of unprotected coitus when transport of the ovum is occurring (Chapter 15). Once the developing embryo enters the uterus, it is transported to the site of implantation primarily through muscular activity of the uterine wall. In the simplex uterus of the human, implantation occurs most often on the surface of the midportion of the posterior wall.

SPERM CAPACITATION AND THE ACROSOME REACTION

Whether isolated *in vitro* or following ejaculation, mature epididymal sperm are incapable of immediately binding to the zona pellucida. They must reside for some time in the female genital tract, or culture medium, to undergo the maturational changes which are a prerequisite for fertilization. The acquisition of fertilizing capabilities collectively called capacitation involves, at least in part, the gradual removal of extracellular glycoprotein coating material from the sperm surface, especially in the acrosome region, and changes in motility. The exposure of the plasma membrane permits access to sites necessary for specific interaction with female receptors. Before capacitation, sperm move in a linear fashion which changes to a sinusoidal vigorous movement sometime before they initiate the acrosome reaction. This change in activity facilitates sperm penetration of the cumulus mass and zona pellucida.

The acrosome contains a large array of hydrolytic enzymes, including hyaluronidase and acrosin (a trypsin-like protease) which are released during the acrosome reaction. The acrosome reaction has at least two functions: it renders sperm capable of penetrating through the zona pellucida and facilitates fusion with the vitelline membrane. The natural inducer of the acrosome reaction is the zona pellucida which mediates egg–sperm recognition, sperm penetration and the block to polyspermy. Strong sperm binding to the zona is a species-specific interaction between the plasma membrane of the sperm head and the sperm receptor in the zona pellucida; sperm will not bind to the zona of fertilized or artificially activated ova. The binding of sperm to the zona has been studied most extensively in the mouse where the zona pellucida glycoproteins ZP_2 and ZP_3 have sperm-receptor activities. Functionally homologous elements in the zonae of human eggs similarly initiate the acrosome reaction. Spermatozoa must complete the acrosome reaction at the zona surface before penetrating the zona. This penetration is helped by vigorous sperm tail movements causing a side-to-side and a backward-and-forward slicing of the sperm head. Thus the zona is dissolved through a combination of both mechanical and enzymatic means allowing a passage through to the perivitelline space.

Spermatozoa could never penetrate a thick zona if they were immotile or weakly motile despite the release of acrosomal enzymes on the zona surface. Biologically selected barriers, such as the need to penetrate the zona which serves to exclude abnormal gametes from reproducing, can now be circumvented by micromanipulation in conjunction with IVF technology (Chapter 1). Micromanipulation allows manipulation of the zona (zona drilling) so as to facilitate sperm passage, insertion of sperm under the zona (subzonal sperm insertion) or injection of sperm into the ooplasm (sperm microinjection). In zona drilling the oocyte is manipulated so that a small hole is created through which the oocyte surface protrudes. These oocytes can subsequently be fertilized with sperm from infertile men suffering from, for example, immotile sperm or Kartagener's syndrome. Sperm from patients with Kartagener's syndrome are said to be normal apart from the loss of motility and if manipulated so that they come to lie adjacent to the plasma membrane of the drilled oocyte they will fertilize it. In cases of intransigent infertility, some clinics, prior to micromanipulation, subject the infertile sperm to chemical treatment with methylxanthine substances to promote a possible transient increase in the sperm's biochemical activity. IVF technology for the treatment of male infertility is readily available in western societies but whether artificial fertilization with abnormal sperm is desirable has yet to be determined. The fertility of the children of such unions is not known as they have not yet reached puberty.

FUSION AND OVUM ACTIVATION

Having successfully penetrated the zona pellucida, the sperm gains entrance to the perivitelline space and access to the vitelline membrane. Upon fusion of sperm with the membrane, the metabolically quiescent egg 'awakens' initiating a series of morphological and biochemical events that lead to differentiation and the formation of a new individual. This awakening is referred to as activation and is characterized by completion of meiosis with the release of the second polar body and the exocytosis of the cortical granules. The cortical granules' contents which are released (the zona reaction) from the egg cortex alter the physical and chemical characteristics of the zona pellucida in such a way that the zona becomes 'refractory' to further sperm penetration. In the mouse, for example, the zona reaction is due to a structural modification and consequent inactivation of zona glycoprotein ZP_3 which is primarily responsible for the firm attachment of mouse sperm to the zona. Once the ZP_3 is hydrolysed, the zona can no longer hold sperm and, as a consequence, they are no longer able to pass through the zona. The cortical granule-mediated zona reaction is not the only mechanism to block polyspermy as the egg plasma membrane also reacts with cortical molecules to reject excess sperm. This block to polyspermy, at the level of egg plasma membrane, is called the vitelline block or egg plasma membrane block. The functional consequences of the zona reaction are that free-swimming sperm will no longer bind to the fertilized egg zona and sperm which have commenced penetration will progress no further following the zona reaction.

In polyspermic eggs, only one of the nuclei of the supernumerary sperm fuses with the egg pronucleus, the other nuclei remain haploid and take part in early development. The resulting embryo is a mosaic of diploid and haploid nuclei and as a result of this mosaicism death occurs early in development. Aneuploidy (unbalanced chromosome complement), although it is the major cause of prenatal death and abortion, is more serious because if only one or two chromosomes are lost or added to the normal complement the conceptus is abnormal but still viable. Much research has been devoted to understanding human aneuploidy which causes severe congenital defects such as Down's syndrome (trisomy of chromosome 21). The probability of spontaneous demise of the conceptus is highest in early pregnancy with nearly half of all embryonic loss associated with chromosome abnormalities. Several types of *de novo* chromosome abnormalities such as trisomy (three copies of one autosome), monosomy (one copy of one autosome), triploidy (one extra sperm) or polyploidy (several extra sperm) are chromosome mutations originating either in parental gametes or from a failure of normal fertilization and/or cleavage. Such defects are not generally hereditary but failure of the normal mechanisms for establishing euploidy is increased under stress conditions; for example following exposure to certain drugs or environmental pollutants around the time of ovulation, during spermiogenesis and on ageing.

On completion of meiosis the resulting haploid nucleus transforms into the female pronucleus. Following incorporation of the sperm nucleus into the female cytoplasm there is rapid disintegration of the existing nuclear envelope and decondensation of the tightly packed sperm chromatin. Chromatin decondenses by losing protamines which are replaced by somatic histones derived from the oocyte, and a new nuclear membrane is then reassembled to form the male pronucleus. In sperm, DNA damage before fertilization can be repaired during the pronuclear stage. On the other hand, the reprocessing of the male chromatin also makes it particularly susceptible to modification by the insertion of extragenetic material due to accidental viral infection or deliberate gene therapy. The union of sperm- and egg-derived genomes (syngamy) can be considered as the end of fertilization and the beginning of embryonic development.

EARLY EMBRYOGENESIS

Early embryonic development in mammals is under the control of the maternal genome which regulates the sequential activation and utilization of components from the oocyte. Following ovulation, development is controlled at two levels. An endogenous programme, initiated by oocyte maturation, regulates the 'housekeeping' functions of the zygote, while sperm penetration initiates a further endogenous programme of embryogenesis. During growth of the oocyte, maternal RNA is transcribed and stored and is subsequently translated into proteins essential for the initial cleavage divisions, blastocyst formation and early stages of implantation.

Rapid progress in the study of embryogenesis was made possible by the creation of transgenic animals. By direct microinjection of genetic material into an organism's genome, genetically novel animals can carry integrated foreign genes which are correctly expressed and inherited by subsequent generations. These transgenic animals provide convenient access to the mammalian genome for a wide variety of genetic, biochemical and physiological studies. The transgenic mouse, for example, has supplied fundamental information on gene expression and regulation. Subsequent studies have resulted in transgenic rabbits, sheep and pigs. These have provided numerous insights into the biochemical and physiological aspects of haematopoiesis, immune system development and tumour genesis among many other subjects. As a consequence, gene therapies, involving transplantations of haematopoietic stem cells in humans and gene augmentation in livestock, are realistic possibilities. The major goals of animal husbandry are increased fecundity, feed-use efficiency, food production and disease resistance. For centuries these characteristics have gradually been improved by slow breeding programmes designed to combine beneficient traits of individual animals. Now the animals can be radically changed in just one generation; for example, transgenic pigs expressing the human growth hormone gene grow fast and produce animals with less fat.

Anatomical aspects of implantation and early development

Cleavages in the mammalian embryo are among the slowest in the animal kingdom. They are about 12 to 24 hours apart and are always asynchronous. Asynchronous division from the 2-cell stage creates a 3-cell embryo with the earlier dividing blastomere of the 2-cell embryo, in turn, having descendents that divide earlier than do descendents of the other blastomere. Through early cleavage stages, the blastomeres form a loose arrangement of cells with plenty of space between them. Following late cleavage stages (8-cells in the mouse) the blastomeres begin a process called compaction. During compaction individual blastomeres appear to lose their identity as they merge to maximize their contact and form a compact ball of cells stabilized by differentiating tight junctions between the outer cells sealing off the inside sphere. Gap junctions between the inner cells allow for direct communication. Communication by surface molecule interactions and local autocrine and paracrine signals is fundamental to the co-ordination of development. The importance of growth factors in embryogenesis has been repeatedly demonstrated and many, such as epidermal growth factor and insulin-like growth factor, have been identified as essential in early embryonic growth. Some growth factors are secreted by particular tissues of the embryo and others originate from maternal tissues. The compacted embryo is now called a morula and is composed of two cell types: outer- and inner-placed cells.

The next morphogenetic event in development is the formation of the blastocyst cavity or blastocoele. Fluid secreted into the morula creates the blastocoele which partitions the morula into two compartments; positioned on one side is an inner coherent mass of cells, the inner cell mass, and surrounding the blastocoele and inner cell mass is the outer epithelium (Fig. 8.2*a*). The inner and outer cells of the blastocyst differ from one another morphologically, functionally and biochemically. The predominant outer cell type constitutes the trophoblast (or trophectoderm) layer. This group of cells will produce the tissue of the chorion or outer portion of the placenta. The inner cell mass will generate the embryo proper. Once the trophoblast is formed, 'hatching' commences, the blastocyst escapes from the zona pellucida and the embryo begins to implant in the endometrium.

Implantation results in the functional juxtaposition of the embryonic and maternal blood systems which enables the fetus to survive within the uterus. Figure 8.2 illustrates progressive stages of implantation and early embryo development with its associated extraembryonic structures. All mammalian embryos penetrate the uterine epithelium, but the extent and process of penetration varies from species to species. In the human, penetration is highly invasive (haemochorial type of placentation) where the syncytiotrophoblastic processes (chorionic villi) make direct contact with maternal blood and the cytotrophoblast mingles with maternal tissue. Both contact points pose an immune challenge (see p. 140).

Under the inductive influence of the inner cell mass, nuclear division of

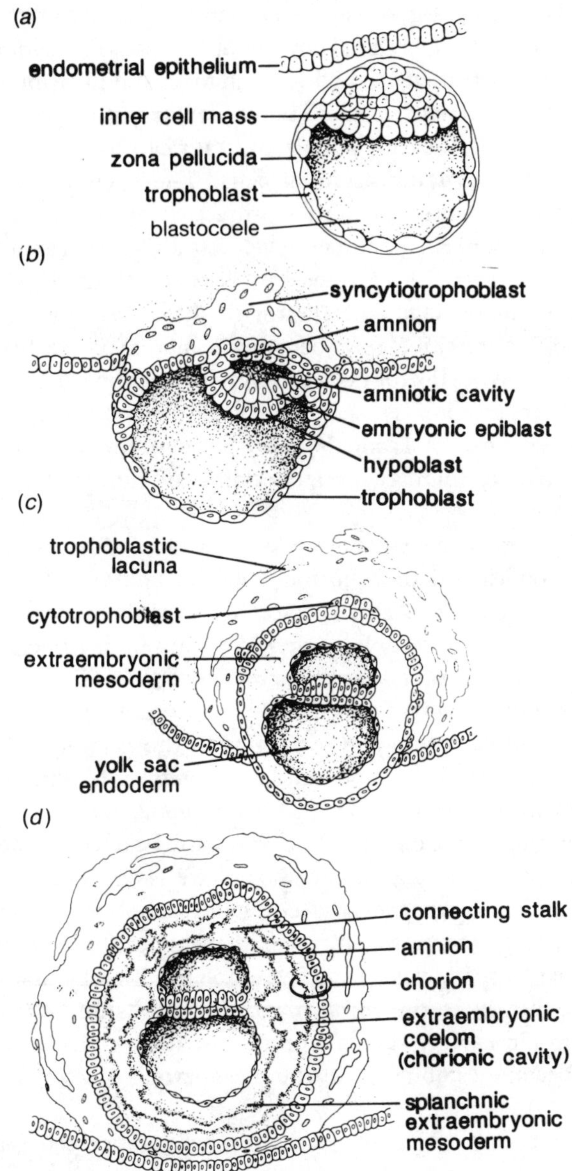

Fig. 8.2 Stages of human implantation and the initial stages of development of the embryo and its associated extraembryonic structures. (*a*) The blastocyst immediately before implantation. (*b*) Development of the blastocyst during implantation. (*c*) The formation of the extraembryonic mesoderm and embryonic ectoderm and endoderm. (*d*) The formation of the extraembryonic coelom and the completion of implantation.

the overlying trophoblast cells occurs in the absence of cytokinesis to form a syncytium known as the syncytiotrophoblast. The syncytium is reactive and capable of proteolysis, which enhances the invasion of the blastocyst into the endometrial stroma. The endometrium responds very rapidly to the invading blastocyst's signals by transforming the surrounding endometrial cells into large decidual cells. The decidua, well supplied with blood vessels, forms a maternal protective barrier against the extremely invasive trophoblast. Lacunae develop within the trophoblast which first contain endometrial secretions but later, at about 13 days of gestation, become interconnected and filled with sluggishly moving maternal blood. Histiotrophic nutrition now becomes a matter of breakdown of maternal blood by the syncytiotrophoblast lining the lacunae. The molecular basis of human implantation is still poorly understood because of a lack of adequate experimental models; however, with the developments in assisted reproductive technology understanding of human fertility control will increase.

The two most conspicuous extraembryonic tissues derived from the cytotrophoblast are the chorion and the amnion. The amniotic cavity initially develops as a small space between the inner cell mass and the adjacent cellular cytotrophoblast and expansion is associated with changes in growth and shape of the embryo (Fig. 8.2*b*). The chorion is a highly modified composite structure that forms from the fusion of two cell layers. Following implantation and formation of the amniotic cavity, the original trophoblast capsule becomes underlaid with an additional layer of cells, the extraembryonic mesoderm, which soon splits into two to form the extraembryonic coelom dividing the extraembryonic mesoderm into an outer somatic layer and inner splanchnic layer (Fig. 8.2*c*,*d*). The outer layer, together with the adjacent cytotrophoblast, transforms into the chorion which integrates with the maternal decidua to create the placenta. The inner layer gives rise to the blood vessels that carry nutrients from the mother to the embryo and the extraembryonic coelom becomes the chorionic cavity. Haemotrophic nutrition commences as fetal blood vessels become established in the mesodermal cores of the chorionic villi (Chapter 9).

The first segregation of cells within the inner cell mass involves the formation of the hypoblast (primitive endoderm) layer. These hypoblast cells line the blastocoel cavity where they give rise to the yolk sac which, in the mammal, is used as a blood cell factory. The overlying inner cell mass tissue is the epiblast, which is destined to form the embryo proper (Fig. 8.2*b*). Gastrulation begins with the formation of a localized thickening in the embryonic epiblast, the primitive streak, through which the endodermal and mesodermal cells migrate. The newly formed mesoderm then migrates laterally and anteriorally to contribute to the formation of the anterior–posterior body axis. Development of the somites from the mesodermal central axis leads to segmentation of the body plan. Once the basic body plan is drafted, further differentiation and organogenesis can proceed. At this stage, the first phase of embryonic development in the human is completed, 25 days after fertilization.

Twinning

Early blastomeres are totipotent, that is, an isolated blastomere can give rise to an entire embryo, a property that is lost as differentiation proceeds. Human twins are classified into two major groups: identical or monozygotic and fraternal or dizygotic. Fraternal twins are the result of two separate fertilizations when two oocytes are shed in a single ovulation; identical twins are formed from a commom embryo whose cells dissociated from one another. Identical twins (roughly 0.25% of human births) are produced by the separation of early blastomeres or by the separation of the inner cell mass into two regions within the same blastocyst. About 33% of identical twins have complete and separate placentae, indicating that separation occurred before the formation of the trophoblast tissue. The remaining identical twins share a common placenta, suggesting that the split occurred within the inner cell mass after the trophoblast had formed. If the split is incomplete, that is, the two new embryonic axes fail to separate in their entirety, the outcome is conjoined twins. The degree of union may be slight or extensive, and the twins may be joined at any part of their bodies. Most conjoined twins do not survive after birth and frequently suffer from major heart malformations. Ever since medical science made the separation of conjoined twins a possibility, there have been concerns about the ethics involved; sometimes one of the twins is sacrificed for the sake of the other. The original 'Siamese' twins, Eng and Chang Bunker, were born in 1811, lived for 63 years and had 22 children between them. Their wives lived in separate houses and the twins spent alternate weeks with each of them. The famous Russian sisters Masha and Dasha Krivoshlyapova (still alive in 1992) were isolated from their family at birth and kept for 20 years in an institution for the mentally handicapped and the aged on the outskirts of Moscow. In 1990 a dicephalic girl with two well-formed heads was born in Chirpan (south-east Bulgaria), an area described as highly polluted with radioactivity after the Russian nuclear accident at Chernobyl in 1986. The majority (about 70%) of conjoined twins are female. The incidence of spontaneous dizygotic twinning varies among different populations as genetic and environmental (dietary) factors can affect basic gonadotrophin levels. The variability can be as high as 4% among black Nigerians and as low as 0.2% among the Japanese.

Allophenic embryos are the product of two early-cleavage (usually 8- to 16-cell stage) embryos that have been aggregated together to form a composite chimaera. The earlier-dividing blastomere of the 2-cell embryo contributes more cells to the inner cell mass than the later-dividing cell. The disproportionate contribution to the inner cell mass by the more advanced blastomeres has been exploited to manufacture, for example, sheep/goat chimaeras in which the primary fetal component is derived from the goat but the sheep trophoblast constitution is compatible with development to term in a sheep foster mother.

There is evidence that allophenic regulation can also occur spontaneously in humans. These allophenic individuals have two genetically different cell types within the same body, each with its own set of genetically defined characteristics. The simplest explanation for the existence of such a phenomenon is that fraternal twins fused to create a single composite individual.

MATERNAL RECOGNITION OF PREGNANCY

Pregnancy is an immune challenge to the mother but the fetal allograft has to be tolerated by the potentially hostile maternal immune system. It is well established that implantation is not only maternally controlled but, in addition, embryo-derived signals are essential to ensure receptivity, decidualization of the endometrium, protection from immunological rejection and the continued maintenance of luteal function. In a few laboratory and domestic animals a great deal is known about the hormonal requirements for implantation and early embryonic development but much less is known about these requirements in primates. Successful implantation and normal development of transferred embryos has been achieved in women with ovarian failure treated with sequential administration of oestrogen and a combination of oestrogen and progesterone to stimulate the follicular and luteal phases of the ovarian cycle, respectively. In the normal hormonally primed uterus, bidirectional signalling commences soon after fertilization. Many types of signal have been identified including platelet activating factor, prostaglandins, histamine-related factors, steroids, proteins, metabolic products and immune-active factors. It seems, however, that a concerted action is necessary since no single factor can alert the mother to the presence of her embryo.

An essential requirement for the maintenance of pregnancy is that the normal luteolytic events leading to the demise of the corpus luteum are intercepted by embryonic signals so that the functional activity of the corpus luteum will be prolonged. In primates, a rising chorionic gonadotrophin (CG, a glycoprotein hormone produced in the syncytiotrophoblast) level is the signal that rescues the corpus luteum during the cycle in which pregnancy occurs. Human CG (hCG), first isolated and identified in 1927 from the urine of pregnant women, has a strong luteotrophic action due to its structural similarity with LH (sequence comparisons reveal that there is greater than 90% homology between the hCG and hLH genes). CG and LH are one of a group of four glycoprotein hormones that are structurally similar, with the other two being FSH and TSH. Each consists of two non-identical subunits, α and β. The β-subunits differ more extensively than the α-subunits and are responsible for conferring different biological activities on the hormones. The predominant function of hCG is the same as that of LH; namely, the stimulation of steroidogenesis. hCG can be detected in the maternal circulation 1 week after the midcycle LH surge, which coincides with attachment and the commencement of trophoblast growth. After its initial appearance in the maternal circulation, the concentration of hCG increases very rapidly until about the 8th week of pregnancy, remains constant between weeks 8 and 12, before declining to a relatively constant level for the remainder of pregnancy. The early and rapid rise in hCG concentration in maternal blood makes it a useful marker for the diagnosis of early pregnancy. Antibodies are specifically directed against the β-subunit which confers immunological as well as biological specificity. CG, in addition to its established role in sustaining the corpus luteum of pregnancy, is involved in other functions related to fetal development and maintenance of pregnancy, including intrauterine immune privilege, placental metabolism and fetal gonadal development (Chapter 9).

In human pregnancy there is a general reduction in immune responsiveness, although pregnant women are basically healthy with little clinical evidence for systematic immune suppression. Immune suppression is locally effective at the maternal–placental interface, that is, recognition of antigens on the villous syncytiotrophoblast exposed to the maternal blood is made difficult. Antigen presentation by the conceptus is controlled by several mechanisms over the duration of pregnancy. For example, the early embryo prior to hatching is covered by the zona pellucida and later the syncytiotrophoblast epithelium is covered by a sialomucin coat which protects many of the surface antigens from immune detection and attack.

In addition, high local levels of progesterone lead to an anti-inflammatory response, further blunting immune effectiveness.

Maternal immune tolerance typically develops during pregnancy but several types of disorder of human pregnancy have a possible immune etiology. For instance, certain recurrent spontaneous abortions have been linked to maternal autoimmune conditions and preeclampsia, a pregnancy disorder characterized by poor placentation, may also be caused by immune factors. An immune attack against the trophoblast can cause pregnancy loss or poor placentation. Likewise, an immune attack can also be directed against specific tissues. Exposure to incompatible red cells, for example, provokes an immune response causing fetal haemolysis. The best known haematologic disorder affecting pregnancy is fetal anaemia due to rhesus or Rh red cell antigen in which maternal and fetal blood incompatibility exists. Rh-negative women become sensitized to the fetal Rh antigen acquired by a small feto-maternal haemorrhage. Once the immune response has been evoked, the maternal Rh antibodies produced will cross the placenta and destroy the Rh-positive red blood cells of the developing fetus. There are several different types of red cell antigen systems causing fetal anaemia from isoimmunizaton but the rhesus system was the first described (by Landsteiner and Weiner in 1940). It is also the first fetal disease for which *in utero* therapy (fetal blood transfusion) became available and the first for which prophylaxis (removal of maternally circulating antigens with Rh immunoglobulin given post- and antepartum) has been successful.

GENOMIC IMPRINTING

Mammalian embryos need both maternally and paternally inherited pronuclei for normal embryonic development to proceed. In every vertebrate class, except in mammals, the phenomenon of virgin birth or parthenogenesis occurs. The fact that parthenogenesis occurs in various species of fish, amphibians, reptiles or birds is conclusive evidence that development can occur in the absence of a paternal genomic contribution. With the development of efficient and safe methods for the transfer of mouse pronuclei, it was established that the embryo must have one pronucleus derived from each parent because mammalian pronuclei are not functionally equivalent during embryogenesis. Subsequent experiments, again in the mouse, determined that the maternal component performs a unique role in the development of the embryo and that the paternal component performs a unique role in the development of the extraembryonic tissues.

This interesting epigenetic effect is termed genomic or parental imprinting and implies that some form of selective marking or imprinting has taken place during the gametogenesis of sperm and oocytes. Expression of the imprinted gene is dependent on its parental origin so the embryonic and the extraembryonic cells differentiate from different lines of precursor cells. For example, identical genes carried on the paternal chromosomes fail to support normal development of the inner cell mass and, conversely, if carried in the maternal chromosomes they fail to support normal development of the trophoblast. Current research makes use of the transgenic mouse to determine the mechanism by which differences in DNA result in imprinting. In some transgenic mice the foreign DNA exhibits both structural and regulatory features specific for the parent from which it is inherited. For example, when chromosome 11 transgene is paternally derived the offspring are larger than normal siblings, while maternal inheritance produces smaller offspring. This suggests that 'transgenes' may be expressed at greater levels when inherited from the father than the mother. This higher expression is associated with lower degrees of methylation of cytosine residues. There are differences in DNA methylation during oogenesis and spermatogenesis, but whether methylation is a primary cause of imprinting is not known. It may be that parental environment dictates the chromatin structure including its methylation pattern which, in turn, influences gene expression.

If such a fundamental process as genomic imprinting occurs during mouse gametogenesis and is the inherent cause of failed parthenogenesis, then there should also be evidence for imprinting among other mammalian species. An interesting example is the development of hydatidiform moles in the human, the result of fertilization by two haploid or by a single diploid sperm producing a diandric conceptus with a triploid genome of 69,XXX, 69,XXY or 69,XYY. Such embryos become androgenic due to failure of the maternal genome to adequately participate in development and this results in a conceptus characterized by pronounced cyto- and syncytiotrophoblastic growth. The trophoblastic hyperplasia and embryonic dysgenesis of moles are strongly reminiscent of the development of diandric (two male pronuclei) mouse embryos. The hydatidiform mole is of special interest since it has a high capacity to transform into a choriocarcinoma, an invasive and potentially life-threatening tumour.

Some human hereditary diseases seem to support the universality of mammalian imprinting because certain mutations affect development differently depending on whether the mutation is maternally or paternally derived. In the human, for example, certain forms of diabetes are more likely to be expressed in children if they inherit the mutant gene from their fathers. Another interesting insight comes from two forms of human mental retardation, Angelman syndrome and Prader–Willi syndrome. Both syndromes are disorders of chromosome 15; however, patients with Angelman syndrome are hyperactive while those with Prader–Willi syndrome are slow moving and overweight. In Prader–Willi syndrome, the mutation is inherited from the maternal chromosome, whereas in Angelman syndrome the mutation comes from the father.

General references

Bourne, G. H. (ed.) (1990). Aspects of some vitamins, minerals and enzymes in health and disease. *World Review of Nutrition and Dietetics*, Vol. 62. Karger, Basel.

Chard, T. (1992). Pregnancy tests: a review. *Human Reproduction*, **7**, 701–710.

Colbern, G. T. & Main, E. K. (1991). Immunology of the maternal-placental interface in normal pregnancy. *Seminars in Perinatology*, **15**, 196–205.

Drobnis, E. Z. & Overstreet, J. W. (1992). Natural history of mammalian spermatozoa in the female reproductive tract. *Oxford Review of Reproductive Biology*, **14**, 1–45.

Gordon, J. W. & Bradbury, M. W. (1991). Genomic imprinting: a gene regulatory phenomenon with important implications for micromanipulation-assisted in vitro fertilization (IVF) – review. *Journal of In Vitro Fertilization & Embryo Transfer*, **8**, 5–14.

Hartshorne, G. M. & Edwards, R. G. (1991). Role of embryonic factors in implantation: recent developments. *Baillière's Clinical Obstetrics & Gynaecology*, **5**, 133–158.

Landsteiner, K. & Weiner, A. S. (1940). An agglutinable factor in human blood recognized by immune sera for Rhesus blood. *Proceedings of the Society of Experimental Biology & Medicine*, **43**, 223.

Miller, D. S. & Seifer, D. B. (1990). Endocrinologic aspects of gestational trophoblastic diseases. *International Journal of Fertility*, **35**, 131–153.

Mitchell, M. D. (ed.) (1990). *Eicosanoids in Reproduction.* CRC Press, Boca Raton, FL.

Muramatsu, T. (1990). *Cell Surface and Differentiation.* Chapman and Hall, Suffolk, UK.

Ober, C. (1992) The maternal-fetal relationship in human pregnancy: an immunogenetic perspective. *Experimental & Clinical Immunogenetics*, **9**, 1–14.

Owens, J. A. (1991). Endocrine and substrate control of fetal growth: placental and maternal influences and insulin-like growth factors. *Reproduction, Fertility & Development*, **3**, 501–517.

Schatten, H. & Schatten, G. (eds.) (1989). *The Molecular Biology of Fertilization.* Academic Press, San Diego, CA.

Surani, M. A. (1991). Genomic imprinting: developmental significance and molecular mechanism. *Current Opinion in Genetics & Development*, **1**, 241–246.

Varner, M. W. (1991). Autoimmune disorders and pregnancy. *Seminars in Perinatology*, **15**, 238–250.

Wassarman, P. M. (ed.) (1991). *Elements of Mammalian Fertilization*, Vol. 1. CRC Press, Boca Raton, FL.

9 Maternal physiology during gestation and fetal development

The presence of a growing fetus in the uterus is an extra-physiological load on the mother and to cope with this extra load the maternal homeostatic regulators of gestation are set at new levels. The pregnant woman experiences adaptive changes in body composition, cardiovascular function and metabolism in order to support the adjustments which enable successful development and delivery of a new individual. This requires that the mother provide an intrauterine environment compatible with satisfying fetal needs. Essential mechanisms relating to the transfer of nutrients and oxygen from the mother to the fetus and the removal of heat, carbon dioxide and nitrogenous wastes from the fetus need to be established. Initially, the maternal organism responds to ovarian hormones during the menstrual cycle in anticipation of implantation and, subsequently, it adjusts to hormonal and neuronal signals originating from the maternal, placental and fetal compartments to synchronize the progress of development. The placenta is an extra source of hormones to supplement the neuroendocrine and immune systems. At the end of gestation, homeostatic controls are critical in mediating maternal–fetal wellbeing because parturition involves the sudden disruption of the maternal–fetal exchange system. At birth the commitment to reproduction is continued and involves a different set of adjustments resulting in lactation and parental behaviour.

THE PLACENTA AS A MATERNAL–FETAL INTERFACE

The placenta is an organ designed for the transfer of heat and matter between the maternal and fetal blood flows and, as such, is a sophisticated example of biological devices known as exchangers. As described in Chapter 8, the original trophoblastic capsule becomes underlaid with mesoderm and is transformed into the chorion which becomes part of the placenta. The basic step in the formation of the human placenta is the appearance of chorionic villi which are the functional units. As seen in Fig. 9.1,

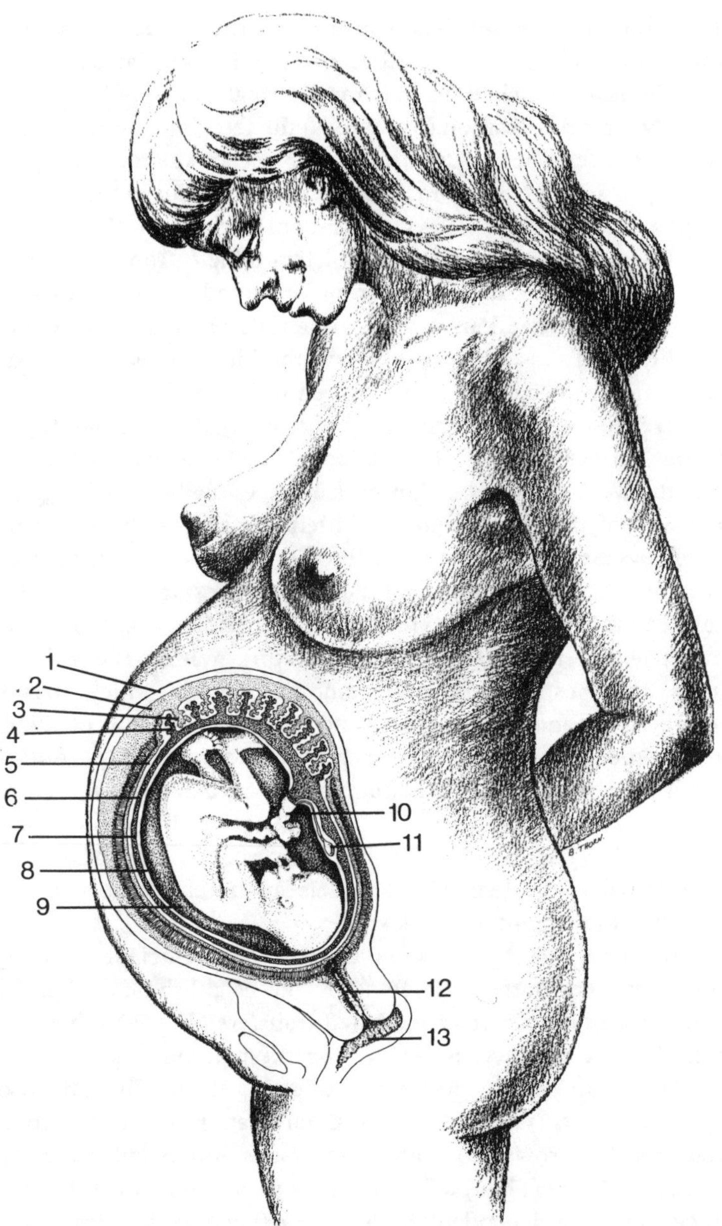

Fig. 9.1 The mother and her fetus. (1) Myometrium, (2) endometrium, (3) decidua basalis, (4) chorionic villi in maternal sinus – chorion frondosum, (5) uterine cavity, (6) chorion, (7) chorion laeve, (8) amnion, (9) amniotic fluid, (10) umbilical cord, (11) yolk sac, (12) cervix, (13) vagina.

villi of the chorion penetrate deeply into the uterine tissue whilst enlarging and branching extensively. They become highly vascular as fetal blood vessels become established in the mesodermal cores of the chorionic villi. Shortly after fertilization (3–4 weeks) the fetal blood circulates under the drive of the fetal heart (the heart and blood circulatory system is the first fetal organ system to become functional). The villi, in turn are bathed by maternal blood which flows continuously through the intervillous spaces. The uteroplacental arteries which originate from arterial trunks and supply blood to the haemochorial placenta and the uterine wall, open directly into the intervillous spaces. Materials are readily interchanged between the blood vessels of the villi and the intervillous spaces. At first, the whole surface of the chorion is covered with villi. However, those villi on the surface towards the uterine cavity gradually thin out and degenerate leaving the surface smooth. This smooth side is known as the chorion laeve. As the fetus grows, the chorion laeve eventually crowds against the opposite wall of the uterus almost obliterating the cavity of the uterus. Full development of the placental villi only occurs in a restricted area of the chorionic vesicle, the so-called chorion frondosum which constitutes the fetal part of the placenta and results in the discoidal shape of the human placenta (placenta means 'flat cake' in Latin). Within the villi are the capillaries, branches from the larger umbilical vessels, which carry blood between the fetus and the chorion frondosum. The portion of the uterus that houses the choronic villi and makes up the maternal portion of the placenta is referred to as the decidua basalis. Structurally, therefore, the placenta has two components, the fetal chorion frondosum and the maternal decidua basalis; functionally, the placental circulation involves two distinct parts, the fetal blood vessels in the chorionic villi and the maternal blood in the intervillous spaces.

The properties of the placenta promote the efficient exchange of molecules between mother and fetus by the establishment of metabolic diffusion gradients and carrier-mediated transfers. The effectiveness with which gradients can function depends on the placental blood flow rates, placental metabolic activity and factors affecting them. The fetus uses the uptake of nutrients via the umbilical circulation to fulfil two major requirements: for growth to build new tissues and as substrates to fuel energy metabolism. The pacing of growth is controlled by peptide growth factors and is dependent on the interaction of genetic and epigenetic factors. Human, and to a lesser extent all primate, growth is characterized by delayed development and extended periods of growth. Delayed differentiation and maturation in favour of prolonged linear growth is directly related to the large brain size prior to birth. The human undergoes maximal brain growth during the last trimester of gestation and the first 6 months after birth. Slower brain growth occurs until the infant is approximately 4 years old with little or no increase in brain size thereafter.

Transport of respiratory gases

Gases such as oxygen, carbon dioxide and carbon monoxide move across the placenta by simple diffusion. However, the transfer of oxygen is made more complex by its interaction with haemoglobin, by the total area available for diffusion and by the rates of blood flow in the uterine and umbilical circulations. Increase in uteroplacental blood flow during pregnancy results primarily from the formation and growth of the placental vascular bed. In addition to establishing an adequate uterine blood flow, the maternal circulation provides for fetal heat loss. During pregnancy, cardiac output increases by an average of 40% (from 5 to 7 l/minute) and this is achieved by increases in both heart rate (tachycardia) and stroke volume. Plasma and blood cell volumes are also expanded. Increased cardiac output and expanded blood volume provide an extra force to push into the tissues nutrients needed to meet the heightened demands of pregnancy. However, to prevent the blood pressure from rising dangerously, the vascular resistance is correspondingly reduced, resulting in the maintenance of a normal blood pressure. Additionally, the increased peripheral circulation to the hands, the upper arms and legs allows the extra metabolic heat produced by the fetus to be convected out. The expanded blood volume also acts as a safeguard against the blood loss associated with parturition which amounts, in average vaginal deliveries, to approximately 500 ml.

The resting breathing rate is not changed during pregnancy but near term the respiration efficiency is increased by an average of 30%. This is achieved by an increase in tidal volume which approximates to an extra 3 l/minute. Under normal circumstances, following maternal cardiorespiratory and metabolic adjustments (see next section), oxygen transfer from the mother to the fetus can be maintained at rapid and continuous rates. Yet, despite this, the fetus exists in an environment in which the diffusion gradient across the villous membrane is shallow compared to that across the adult lung. Additionally, the fetus has to share the available oxygen with the placenta which has a high metabolic rate.

Several homeostatic mechanisms have evolved which ensure stable oxygen transfer and consumption despite the relatively shallow oxygen gradient across the placenta. These can be summarized as follows.

Villus lining As well as placental growth and increased total villus surface area, the thin epithelial lining of the villus may thin even more by stretching, thus increasing the diffusion efficiency. Typically, the pressure of oxygen in the blood of the maternal sinuses is 30–40 Torr (1 Torr is 133.3 Pa) higher than that in the fetal capillaries. As pregnancy progresses, the permeability of oxygen and other nutrients transported through the placenta increases, reaching a maximum about 6 weeks before parturition. The surface area available for exchange between mother and fetus decreases in the last month of gestation because the ageing microvilli, overlying the villi, become less dense.

Haemoglobin dissociation curves Fetal haemoglobin has a higher affinity for oxygen than has adult haemoglobin; that is, the fetal oxyhaemoglobin dissociation curve is displaced to the left of the maternal oxyhaemoglobin dissociation curve at normal body temperatures and pH close to 7.4 (Fig. 9.2). An important consequence of the differences is that the fetal blood can be near saturation even if a substantial gradient exists between the oxygen pressure of the two bloods. Reading from Fig. 9.2, for example, at a pressure of 50 Torr the maternal blood is about 85% saturated with oxygen while the fetal blood will still be able to pick up oxygen to at least 85% saturation even if the oxygen tension on its side of the placenta has

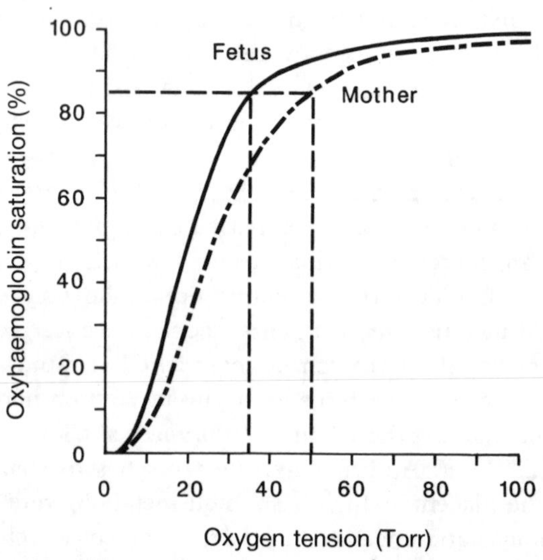

Fig. 9.2 Oxygen dissociation curves of fetal and maternal whole blood in the human. (1 Torr is 133.3 Pa.)

dropped to 35 Torr. At very low oxygen tensions, fetal haemoglobin carries about 30% more oxygen than adult haemoglobin. A decrease in pH in the fetal tissues displaces the oxygen dissociation curve to the right facilitating the unloading of oxygen.

Haemoglobin concentration The haemoglobin concentration of fetal blood is about 50% greater than that of the mother. This allows an increased amount of oxygen to be transported to the fetal tissues. At birth the haemoglobin level is approximately 17 g/100 ml blood. During the first few postnatal weeks as the older erythrocytes are broken down, the adult concentration of 11–12 g/100 ml is reached. The iron released from the worn-out fetal blood cells is stored in the liver and recycled because the milk is naturally very low in iron (1–2 mg/l) and cannot supply the requirements of the fully breast-fed infant.

Oxygen debt Fetal tissues can build up a temporary oxygen debt at times of oxygen depletion as this alters the pattern of energy metabolism from oxidative phosphorylation to glycolysis. This oxygen debt must, however, be repaid quickly as the ensuing accumulation of lactic acid produces metabolic acidosis. Simultaneously, accumulation of carbon dioxide causes respiratory acidosis with the pH of the blood and tissues falling even further. The 3-month fetus, for example, cannot withstand a deficiency of oxygen for more than 20 minutes before its store of liver glycogen is utilized and acidosis has reached toxic levels.

Removal of carbon dioxide For each mole of oxygen metabolized by the fetus 0.94 mole of carbon dioxide is produced as a metabolic end-product to be eliminated via the maternal circulation. Dissolved carbon dioxide (CO_2) within both fetal and maternal erythrocytes is in equilibrium with bicarbonate ions (HCO_3^-), a reaction catalysed by carbonic anhydrase. Normally only a low pressure gradient (4–8 Torr) for carbon dioxide exists across the placental membrane, but this is sufficient to allow adequate diffusion of carbon dioxide from the fetal blood into the maternal blood. Carbon dioxide is highly soluble in water which facilitates its diffusion through the aqueous phase of the placental membrane. Other fetal wastes which are transported across the placenta are carbon monoxide, urea, bilirubin and bile salts, most of which diffuse passively into the maternal circulation and amniotic fluid.

Hypoxia

Despite the evolution of adaptive mechanisms for the effective transport of oxygen, the fetus is still at risk when the oxygen level in the uterine environment falls. It has been suggested that oxygen deprivation may be responsible for more than 30% of the deaths of all stillborn infants and a major cause of intrauterine growth retardation (IUGR). Developments in the technology of ultrasound scanning since its introduction in 1975 have made non-invasive measurement of fetal breathing, body movements, heart rate and flow velocity in the blood vessels possible. These advances in the assessment of fetal health have been particularly useful in evaluating the effects of many pharmacological agents and drugs on the utero-placental-fetal circulation. An exciting development in diagnostic ultrasound is the use of colour Doppler in conjunction with transvaginal sonography. The transvaginal colour Doppler is a single probe which can provide clear images of deep lying fetal tissues. However, cautious rather than over enthusiastic use of this technique may be prudent since its operation requires high intensity ultrasound and vibration energy. Using Doppler technology, characteristic changes in the pattern of development, brought about by impaired utero-placental circulation with consequent malnutrition, can be observed in experimental fetuses.

The fetus has powerful adaptive responses to oxygen deprivation stress. Using baro- and chemoreceptor reflexes as well as adrenally derived circulating catecholamines to monitor falling oxygen tension, it can engage in a survival defence strategy. Fetal movements, for example, consume significant amounts of oxygen and a reduction of these when oxygen supply is limited aids in maximizing delivery of the available oxygen to vital organs. In the oxygen-deprived fetus, cardiac output is depressed but blood flow to the brain, heart and adrenal gland is maintained as much as possible at the expense of blood flow to other non-essential organs. Such a redistribution of blood in favour of vital organs makes the fetus better suited to survive intact *in utero*; however, if normal circulation is not soon re-established, compensatory disturbances may result in permanent functional defects not only in the individual but, if the developing germinal cell line is involved, in the subsequent generation. A significant association between increased cardiothoracic ratio and IUGR in neonates with birthweights of 2000 g or less has been established. Growth-retarded infants also have a higher than normal risk of suffering from a tendency to hypertension and ischaemic heart disease in adult life. Studies of English and Welsh populations revealed an especially strong relationship between low birthweight and stroke where the combination of poor prenatal and postnatal growth led to high death rates 60 or so years later. Men who weighed 2500 g or less at birth and 8 kg or less at 1 year had a relative mortality rate of 220. (Death rates were expressed relative to a national average of 100 standardized for age and sex.)

Many stressful stimuli, both physical (malnutrition and drug consumption, for example) and psychological (maternal pain and anxiety, for example) cause oxygen deprivation and IUGR. Drugs and xenobiotic agents can cross the placenta by diffusion. In general, water-soluble drugs with molecular weights under 800 readily diffuse across the placenta as do hydrophobic drugs. Lipophilic, low molecular weight anaesthetic gases, such as nitrous oxide, halothane and methoxyflurane, diffuse rapidly across biological membranes including the placenta and are found in the fetal circulation within seconds of maternal exposure. There is no mechanism by which the placenta can discriminate between a toxin and a nutrient, but it is able to metabolize certain foreign compounds, providing a mechanism for detoxification as long as the products themselves are non-toxic (Chapter 19). Xenobiotics such as cigarette smoke alter normal nutrient transport, particularly oxygen, predisposing the fetus to starvation. For decades now there has been no doubt that maternal smoking is related to retarded and impaired fetal development, yet, due to nicotine's strong addictive qualities, many women do not give up smoking despite health warnings on their children's behalf. Estimates of the incidence of smoking among women vary considerably but according to conservative US figures 20–25% of smokers continue to smoke during pregnancy despite active anti-smoking campaigns. The trend in western societies is of a steady increase in smoking among women of child-bearing age, especially the younger age groups.

Maternal smoking effects the fetus and survival chances of the newborn by the artificial induction of intrauterine hypoxia which, in turn, is responsible for fetal morbidity and IUGR. Cigarette smoke contains about 2000 different compounds with approximately 10% being particulates containing nicotine and tar. Tar is a general term for polycyclic aromatic hydrocarbon products such as phenols, benzopyrenes and benzenes which are carcinogenic. The remaining 90% contains significant quantities of carbon monoxide, carbon dioxide, cyanides, various hydrocarbons, aldehydes and organic acids. Carbon monoxide has a higher affinity for haemoglobin than has oxygen, and since carboxyhaemoglobin has an increased affinity for oxygen, the unloading of the already scarce oxygen supply is impaired. Nicotine stimulates maternal release of adrenaline causing uterine vasoconstriction which further decreases placental perfusion and oxygen availability in the intervillous spaces.

Chapter 11 examines in detail the significance of IUGR and associated risk factors.

Transport of nutrients

Although a favourable glucose concentration gradient exists between the maternal and fetal arteries, glucose is conveyed to the fetal circulation by facilitated diffusion, that is, transport is enhanced by a specialized carrier mechanism. Under basal conditions, approximately 28% of glucose that enters the near-term placenta is actively transferred, with the balance used to drive placental metabolism. Therefore, the amount of glucose supplied to the fetus via the placenta is substantially less than that needed to sustain fetal energy metabolism, even under optimal physiological conditions. Oxidative metabolism is additionally fuelled by amino acids and lactate. Lactate oxidation represents approximately 70% of lactate utilization, with the rest entering the carbon pool for the production of carbohydrates. The human fetus has a high glucose requirement for fetal lipogenesis since at birth the body composition has a high concentration of fat. Another specific human requirement for glucose relates to the exceptionally high brain:body weight ratio with the large brain being a major site of glucose consumption.

It is generally accepted that intelligence is inherited, and the ancestors or fate are blamed: if it is deficient 'the child takes after dreary old aunt Flo who obviously does not have all her marbles'. Although each individual's intellectual potential is limited by heredity, many fail to reach their genetic potential due to adverse epigenetic factors which affect intellectual development as much as they do physical development. For instance, an association is established between gestational ketonaemia in the diabetic mother and a lower IQ in her offspring. Anecdotal reports suggest that during the early 1900s some obstetricians tried to reduce maternal mortality at delivery (associated with large babies) by routinely restricting the mother's caloric intake during pregnancy. Although a smaller head circumference does decrease the chance of maternal complications at parturition the reciprocal data examining the effects of this practice on infant intelligence and mortality were not apparently collected.

The rate of glucose utilization is a function of both glucose and insulin concentrations with the glucose effect predominating. In species such as ours where the feeding cycle is interspersed by fasting, short-term changes in plasma insulin concentrations are of minor importance in fetal glucose utilization. Long-term changes in insulin concentrations can, however, alter this utilization. For example, the chronically high maternal–fetal transplacental glucose concentration gradient in the gestational diabetic mother significantly increases glucose transfer to the fetus resulting in accelerated fetal growth and fat storage.

Many amino acids are actively transported across the placenta because the total amino-nitrogen concentration is greater in fetal plasma than in maternal plasma, as are the concentrations of many, but not all, individual amino acids. The transfer rates of protein and small peptide hormones across the placenta into the fetus is very low with the exception of certain antibodies and some pathogenic organisms. Immunoglobulin G (IgG) passes from the mother to the fetus by receptor-mediated endocytosis and cellular transport systems. By this mechanism, immunity to diseases which are common in the maternal environment and to which the mother has developed an immunity are transferred to the fetus during pregnancy and reinforced during breast-feeding. Some bacteria and viruses can gain access to the fetal circulation and it is possible that an infection that is not dangerous or produces no ill effect in the mother can have severe consequences for the fetus. The rubella (German measles) virus is such an example. Rubella causes deafness, cataracts and heart defects if the fetus is exposed to the virus at critical periods of development. Fortunately these tragedies can be eliminated by maternal vaccination. A major unsolved problem is the perinatal transmission of the human immunodeficiency virus (HIV) to the fetus during pregnancy and delivery. An infected child is very likely to die of AIDS by 2 years of age (Chapter 17).

The fetus requires relatively high levels of fatty acids to enable the development of adipose tissue. Free fatty acids can be synthesized *de novo* or can be obtained by placental transfer as the human placenta is relatively permeable to free fatty acids, glycerol and ketoacids. Cholesterol, carried in lipoprotein particles, is taken up by a receptor-mediated process.

Transport of water, electrolytes and vitamins

The fetus requires more moles of water per day than any other solute and this is transferred freely across the placenta with large fluxes in both directions. There is also a bidirectional extraplacental exchange between the chorian laeve and the amnion that accounts for approximately 10% of the total. Additionally the fetus exchanges water with amniotic fluid through swallowing and production of urine.

Many monovalent cations, anions, trace metals and vitamins are transported by a mixture of simple diffusion (sodium and chloride) and energy-dependent carrier-mediated mechanisms (calcium, phosphorus, iron and water-soluble vitamins such as ascorbic acid).

THE PLACENTA AS AN ENDOCRINE ORGAN

A well-functioning placenta maintains overall equilibrium between the fetus and the mother by providing an extra source of hormones for the gestational unit and pacing, in a complementary manner, the initiation, maintenance and termination of pregnancy. The placenta also plays a critical role in the regulation of fetal growth and in the direction of appropriate signals for the timing of parturition. Hormonal communication within the placenta involves endocrine, paracrine and autocrine systems; that is, once released, a particular hormone can act on the placenta itself or on targets within the maternal and fetal bodies. The basic biological mechanisms operative in intraplacental control have already been outlined as they related to the control of ovarian and testicular function (Chapters 4 and 5).

The 1980s has been a decade full of major advances in the elucidation of intraplacental control mechanisms with the discovery and characterization of numerous regulatory peptides, growth factors, hormones, cytokines and hypothalamic-like neurohormones similar in variety and complexity to the hypothalamic–pituitary–target organ axis. Because it is not possible to discuss in depth the whole field in this overview chapter, the following sections outline some of the better investigated endocrine regulatory systems in human pregnancy.

Gonadotrophin-releasing hormone (GnRH) and chorionic gonadotrophin (hCG)

GnRH progressively increases during the first 24 weeks of gestation and remains relatively constant thereafter. The most thoroughly examined function of GnRH is its ability to stimulate the secretion of hCG. Other activities of GnRH include the regulation of placental steroid and prostaglandin production. After 6–7 weeks of gestation, the corpus luteum of pregnancy is no longer necessary as the placenta can supply all the steroid requirements needed until term.

Apart from its early involvement in the rescue of the corpus luteum to maintain progesterone production (Chapter 8), hCG is also involved in fetal regulation of testicular and possibly adrenal steroidogenesis. During sexual differentiation, hCG regulates Leydig cell proliferation and testosterone biosynthesis before the fetal hypothalamic–pituitary–gonadal axis becomes functional. The role for hCG in the stimulation of fetal adrenal steroidogenesis is more controversial; however, the regulation of dehydroepiandrosterone (DHEA) production, by the fetal adrenal zone before ACTH becomes active at 16 weeks of gestation, seems well documented. DHEA and DHEA sulphate (secreted from the specialized fetal zone of the adrenal) are precursors for the placental synthesis of oestradiol and oestriol. Placental progesterone serves, in turn, as the precursor for the synthesis of fetal adrenal ketosteroids, most importantly cortisol. Maternally derived adrenal androgens, for example dehydroandrosterone or androstenedione, can likewise be aromatized to oestrogens on reaching the placenta, although 80–90% of the androgen substrate is of fetal origin. There are also several other biological functions not related to the stimulation of steroidogenesis which involves hCGs. These include stimulation of thyroid activity and relaxin secretion.

Inhibin and activin are present in the placenta at relatively high levels. These peptides affect gonadal physiology by modulating FSH release (Chapter 5). There is evidence that placental inhibin, by inhibiting GnRH, has an indirect control over the local release of hCG from the placenta. Conversely, activin plays a role in hCG secretion by stimulating GnRH secretion. However, circulating hCG levels during pregnancy may not only depend on changed equilibria in GnRH, inhibin and activin secretion patterns because other peptides and hormones (such as opioid inhibition) can also modulate GnRH secretion.

Corticotrophin-releasing hormone (CRH), adrenocorticotrophic hormone (ACTH) and β-endorphin

CRH is present in the human placenta from 7 weeks of gestation but it increases sharply during the last 5 weeks of pregnancy. CRH, induced by stress, stimulates the release of chorionic ACTH, glucocorticosteroids, β-endorphin and prostaglandins, which are all involved in normal labour and parturition. Abnormally high CRH levels are found during preterm labour and in stress-induced IUGR. It is not clear how much reciprocity exists between placental and fetal CRH; however, the fetal hypothalamic–pituitary–adrenal axis is one of the earliest integrative systems to develop as it is required to co-ordinate many neuroendocrine and enzymatic activities critical in growth and maturation. The fetus feels and responds to stress *in utero*. Chapter 13 describes the adapative response to stress, its physiological control and prenatal environmental influences which can modify postnatal physical and emotional states.

Placental opioid peptides (β-endorphin in particular) have other important biological functions in gestation relating to mood, appetite modulation and stimulation of placental lactogen (hPL) release (see next section). Increases in plasma β-endorphin levels during labour and parturition reduce the level of pain due to its strong analgesic effect and prevent the risk that psychological distress may cause a decrease in the labour-promoting functions of prostaglandin, ACTH and cortisol.

Extrahypothalamic neurotransmitter-mediated pathways in the CNS influence CRH release in both the placenta and the pituitary. However, during pregnancy, in contrast with the non-pregnant state, CRH is insensitive to negative feedback inhibition by glucocorticosteroids, thereby maintaining elevated plasma cortisol levels throughout pregnancy. Additionally, prostaglandin and glucocorticosteroids stimulate placental CRH which may be a specialized device in the regulation of labour (Chapter 10).

Placental lactogen (PL)

PL is a hormone that possesses lactogenic activity similar to that of prolactin (PRL) and, since it also shows GH-like activity and is produced early in pregnancy, it has also been termed chorionic somatomammotrophin. hPL is present in the syncytiotrophoblast as early as the 2nd week after fertilization and subsequently its concentration increases steadily until about week 34 of gestation after which it remains relatively constant. The GH-like activities of hPL involve the regulation of intermediary metabolism thus promoting fetal growth; the PRL-like activities regulate mammary gland growth, differentiation and lactogenesis. A major regulative function of hPL involves the adaptive shift of maternal energy utilization in favour of fatty acid mobilization and consequent sparing of glucose. The hPL-mediated increase in the basal rate of lipolysis is also accompanied by an increased sensitivity of adipocytes to other lipolytic stimuli coinciding with increases in fetal glucose uptake and utilization. The physiological significance of this shift implies that in the fasted state increased rates of lipolysis provide free fatty acids that can be utilized as a source of energy by the mother, thereby sparing glucose for use by the fetus. In the fed state, when maternal blood glucose concentration is high, the increase in glucose uptake and utilization by adipose tissue ensures that energy stores, in the form of triglyceride, will be available during subsequent periods of fasting. These characteristics establish hPL as an important regulator mediating fetal growth at times of food restriction. Under moderate nutritional deprivation, fetal growth is protected at the cost of maternal nutritional status, thus reducing the impact of fetal energetic stress. The adaptations of intermediary metabolism that occur during pregnancy are brought about by the actions of many hormones in addition to hPL, including oestrogens, progesterone, GH, PRL and glucocorticoids. hPL serves a critical role in periods when very large amounts of GH- and PRL-like activity are required, such as during starvation when an increase in total GH-like activity would improve the availability of nutrients for the fetus.

Placental lactogen is stimulated by CRH, β-endorphin thyrotrophin-releasing hormone (TRH) and GH-releasing-like substances. All these hormones are synthesized in the human placenta and support pituitary function. A weight reduction programme, popular in the 1970s, used prolactin-like compounds administered intramuscularly to bring about fast and effortless weight loss due to catabolism of adipose tissue reserves whilst on a low calorie intake. The nutritional stress caused by the series of 30–40 injections can, however, effectively deplete the body's reserves of all nutrients, jeopardizing health and lowering resistance to opportunistic infections.

NUTRIENT UTILIZATION DURING GESTATION

During gestation the maternal organism undergoes finely tuned metabolic adjustments aimed at preserving maternal homeostasis while, at the same time, providing for the growth and development of the fetus and placenta. In the human, a healthy pregnancy with a good chance of delivering a normal birthweight infant requires a total weight gain of about 10–12.5 kg. For a 12.5 kg gain the approximate distribution of the extra weight can be apportioned to 3.5 kg increase in maternal fat, 2.5 kg increase of intra- and extravascular fluid, 1.5 kg increase in uterus and mammary glands, 0.5 kg in amniotic fluid, 1 kg in placenta and a 3.5 kg term fetus. The increment in energy intake to cover these extra requirements is approximately 300 kcal/day (1260 kJ/day) throughout gestation. The efficiency with which ingested food is converted to usable energy varies according to the eating pattern and the food digestibility, caloric density and relative proportions of carbohydrate, fat and protein in the diet (the average heats of oxidation for carbohydrate, protein and fat are 4.2, 5.6 and 9.5 kcal/g (17.6, 23.5 and 39.9 kJ/g) respectively).

The additional energy requirements during pregnancy are met by a mixture of the following maternal adjustments.

(a) Increased food (gross energy) intake
(b) Increased utilization of maternal energy stores
(c) Decreased resting metabolic rate (RMR)
(d) Increased energy absorption
(e) Decreased external work.

Maternal metabolism

The human organism has evolved over the millennia in an environment where the food supply has fluctuated widely so it is not surprising that the total food intake changes very little or not at all in pregnant women when an adequate supply of food is available. Efficient utilization of the nutrients consumed effectively reduces the need for additional nutrients from the diet during gestation. At times of food shortage, adjustments in metabolism can further contribute to nutrient conservation so as to cover the RMR increase necessary near term when the fetal energy requirements are maximal. RMR changes, compared with prepregnant or postpartum states, follow a biphasic pattern, with the RMR initially remaining at the prepregnant level, or declining for a variable period of weeks, before rising significantly during the third trimester. The shape of this RMR curve varies widely among different populations and social groups with the prepregnancy RMR being critically important as a reference point in the body's estimate of the affordable cost of the future pregnancy. For example, a decrease in RMR early in pregnancy is sufficient to balance the third trimester increase. Studies of middle-class women possessing social privileges (Sweden and USA) revealed that the RMRs were about 35% higher in the third trimester compared to their non-pregnant levels, while women living in the depressed area of Glasgow (Scotland) had a much smaller (7%) rise after a reduction in RMR during the first two trimesters. Studies in 1983 on Gambian multigravidae established a direct relationship between the nutritional status of the mother and the degree of metabolic adjustment experienced during pregnancy. Two populations of women were studied; unsupplemented, consuming approximately 1500 kcal/day (6300 kJ/day), and supplemented, consuming 1950 kcal/day (8190 kJ/day). The unsupplemented women's RMR dropped below preconception levels between 10 and 25 weeks of gestation and rose thereafter to reach an 8% increase at 37 weeks. In the supplemented women, by contrast, no drop in RMR was seen in early pregnancy and the average increment during the last 8 weeks reached 17%.

Different levels of metabolic adaptation are important, initially for augmenting maternal energy stores when the demands of the fetus are small and subsequently for redirecting substrates and energy from mother to fetus. The first half to two thirds of pregnancy is the anabolic phase, characterized by weight gain due to the expansion of blood volume, deposition of fat and other reserves. The second half or one third of pregnancy is the catabolic phase, during which maternal blood glucose concentration decreases and maternal metabolism utilizes other substrates such as lactate and fat, thus sparing glucose for the fetus. The nutritional quality of the diet and the associated maternal metabolic response is reflected in placental structure and size. In one US study, placentae of women from a socioeconomically depressed and undernourished population weighed 15% less than those of middle-class women. There is a correlation between placental mass and the circulating levels of metabolically active placental hormones such as hPL, hCG, oestrogen and progesterone which promote a redistribution of available substrates. Low socioeconomic classification is significantly correlated with IUGR which, in turn, is perpetuated by the poverty cycle.

Severe malnutrition is rare in affluent communities, although self-imposed mal- and undernutrition does exist. Eating disorders range from quite severe as in anorexia nervosa or anorexia athletica (compulsive exercising) to moderate or mild such as in over-zealous adherence to low-cholesterol, high-fibre diets and fear of obesity (Chapter 18).

Energy absorption

During pregnancy the absorptive efficiency improves so that additional nutrients and energy can be extracted from a given quantity of ingested food. Normally the efficiency of energy absorption from the intestine is high (92–98%, depending upon the substrate); however, the cumulative effect of a small change in absorptive efficiency can contribute significantly to meeting the caloric demand of pregnancy. For example, if an intake of 2526 kcal/day (10 609 kJ/day) at an assumed 95% efficiency of absorption was increased only 3% to 98% efficiency, this would yield an estimated increment in metabolizable energy over the course of gestation in excess of 21 200 kcal (89 040 kJ), or an increment of 76 kcal/day (319 kJ/ day).

A possible mechanism mediating the increased energy efficiency during pregnancy is an increase in mucosal surface area, achieved by increasing the height and number of intestinal villi. Other mechanisms for increasing absorptive capacity involve changes in specific transport systems, increases in pancreaticobiliary secretions, intestinal blood flow and reductions in smooth muscle peristaltic activity, thus lengthening the transit time available for digestion and absorption. Hypertrophy of the small intestine has been observed in the third trimester of pregnancy and during lactation in humans and other mammals. Elevated levels of gut hormones, thyroxine, glucagon, insulin and glucocorticosteroids are associated with intestinal enlargement and villus hyperplasia.

An increase in energy uptake can also be achieved by increasing the residence time of ingesta in the gut. The mouth to caecum transit time is significantly prolonged in the second and third trimesters, thus increasing the opportunity for absorption. This effect is attributable to an inhibitory effect of progesterone and opioid peptide hormones on intestinal smooth muscle and may account for the increased incidence of constipation during pregnancy.

Nausea and vomiting occur in the majority of pregnant women with varying severity and duration. A US statistic report states that 89% of pregnant women suffer from 'morning sickness' and in approximately 15% of these nausea continued until delivery. It is likely that metabolic, endocrine and gastrointestinal functional changes are involved in its etiology, although the specific cause(s) of the illness is not known. However, a higher incidence of spontaneous abortion occurs in women without symptoms of nausea and vomiting, suggesting a link between steroid hormone increase and digestive problems. Food aversions and preferences during pregnancy may likewise denote steroid-mediated physiological requirements. Primates are good at food selection; studies of baboons, for example, demonstrate that they can always balance their nutrient budgets no matter what is seasonally available. They naturally go for the diet that maximizes energy intake, includes a minimum of all other nutrients, but which is below their maximum tolerable level for plant toxins. At times, baboons also seek out fruit and legumes rich in phytooestrogens (eating habits may affect their reproduction!). Food selection is also apparent from analyses of the nutritional content of the natural foods of Australian and African hunter–gatherers where energy and all other nutrient requirements are usually well balanced.

Activity

It is often assumed that pregnant women compensate for the increased cost associated with having a heavier body by reducing their level of activity. Surprisingly, in industrialized and developing countries, studies of activity patterns during pregnancy reveal that modification of activity is not universally used as a means of preserving energy balance. Occupation did, however, affect the activity pattern; homemakers with children and working women expended more energy than unemployed women without children, that is, women perform the same tasks when pregnant as when non-pregnant. If any change in activity occurred, especially during the third trimester, it was in the duration and intensity of work, not in the type of work, and in the reduction of leisure activities such as walking. The price paid for the continuation of heavy physical work into late pregnancy is reduced infant birthweight. Studies from Taiwan and Gambia show that birthweights of those born during the labour-intensive summer harvest months are lower by about 150–550 g when compared to those born during the less busy season. There has also been an increase in the number of women in industrialized countries who maintain an exercise programme throughout pregnancy. One survey showed that the energy intake of pregnant women who jogged regularly throughout gestation did not differ from that reported for sedentary women. Given an adequate diet, however, voluntary physical activity improves the efficiency of energy utilization and general health. The Gambian women on supplementary feeding were more active when compared to the non-supplemented group. The supplemented women benefited by improving their general health and ability to fulfil the social activities required of them.

General references

Ahmed, M. S., Cemerikic, B. & Agbas, A. (1992). Properties of functions of the human placental opioid system: a review. *Life Sciences*, **50**, 83–97.

Barker, D. J. P., Osmond, C., Golding, J., Kuh, D. & Wadsworth, M. E. J. (1989). Growth *in utero*, blood pressure in childhood and adult life, and mortality from cardiovascular disease. *British Medical Journal*, **298**, 564–567.

Bocking, A. D. & Gagnon, R. (1991). Behavioural assessment of fetal health. *Journal of Developmental Physiology*, **15**, 113–120.

Bourne, G. H. (ed.) (1987). Energy and nutrition of women. *World Review of Nutrition and Dietetics*, Vol. 52. Karger, Basel.

Bozynski, M. E., Hanafy, F. H. & Hernandez, R. J. (1991). Association of increased cardiothoracic ratio and intrauterine growth retardation. *American Journal of Perinatology*, **8**, 28–30.

Conley, A. J. & Mason, J. I. (1990). Placental steroid hormones. *Baillière's Clinical Endocrinology & Metabolism*, **4**, 249–272.

Cosmi, E. V., Luzi, G., Gori, F. & Chiodi, A. (1990). Response of utero-placental fetal blood flow to stress situation and drugs. *European Journal of Obstetrics & Gynecology & Reproductive Biology*, **36**, 239–247.

Herrera, E. & Knopp, R. H. (eds.) (1992). *Perinatal Biochemistry*. CRC Press, Boca Raton, FL.

Owens, J. A. (1991). Endocrine and substrate control of fetal growth: placental and maternal influences and insulin-like growth factors. *Reproduction, Fertility & Development*, **3**, 501–517.

Petraglia, F., Volpe, A., Genazzani, A. R., Rivier, J., Sawchenko, P. E. & Vale, W. (1990). Neuroendocrinology of the human placenta. *Frontiers in Neuroendocrinology*, **11**, 6–37.

Prentice, A. M., Whitehead, R. G., Watkinson, M., Lamb, W. H. & Cole, T. J. (1983). Prenatal dietary supplementation of African women and birth weight. *Lancet*, **1** (March), 489–492.

Thornburg, K. L. (1991). Fetal response to intrauterine stress. *Ciba Foundation Symposium*, **156**, 17–37.

Winick, M. (1989). *Nutrition, Pregnancy and Early Infancy*. Williams & Wilkins, Baltimore, MD.

10 Parturition and lactation: hormonal control

A PARTURITION

Parturition is the end point of a succession of endocrine events involving maternal, fetal and placental interactions. The major hormones involved in the onset and maintenance of human parturition are oestrogens, progesterone, relaxin, oxytocin, prostaglandins, catecholamines, cortisol and β-endorphin. Oestrogens, relaxin and prostaglandins promote cervical ripening; prostaglandins, progesterone, oestrogens and oxytocin regulate myometrial activity. Catecholamines and cortisol help regulate the energetics of uterine contraction, and β-endorphin acts as a pain modulator. The release of β-endorphin (which is substantially reduced by epidural anaesthesia or by analgesics) is a response to the stress of labour and mirrors plasma cortisol levels; that is, plasma β-endorphin levels rise during labour, reach a peak at delivery, then fall to non-pregnant levels within 24–48 hours thereafter.

Pioneering work carried out in the 1970s by Liggins and his collaborators (National Women's Hospital, Auckland, New Zealand) provided evidence that in sheep the fetus plays a major role in initiating its own delivery. Subsequent research demonstrated that fetal adrenal cortisol triggers the cascade of maternal endocrine changes that, in turn, promote myometrial responsiveness to prostaglandins and oxytocin. However, the role of the fetal pituitary–adrenal axis in the onset of parturition varies from pivotal, as in the sheep, to uncertain in other species such as primates. In humans (and monkeys) exogenous glucocorticoids fail to induce parturition, suggesting that the mechanism determining gestation length and the onset of parturition is more complicated than in sheep. The length of pregnancy in a given species is determined by the fetal genotype and can be confined to a strict timing schedule, as in many seasonal breeders, or be more variable. In the human the gestation period varies between 37 and 42 weeks. In addition to developmental variability, premature or postmature births are not unusual. Preterm labour, defined as birth before 37 complete weeks of pregnancy, occurs in 5–9% of all deliveries in developed countries and can be a major problem in clinical obstetrics. In developing countries and amongst lower socioeconomic groups, the incidence of preterm labour is higher. Immaturity of the newborn infant, as a result of IUGR and/or preterm delivery, is associated with neonatal mortality and morbidity. Factors such as low socioeconomic status (for example, nutritional deficiencies, social stress), advanced maternal age, inappropriate lifestyle (for example, smoking, multiple pregnancies), chronic low-grade infection and stressful life events during the current pregnancy all predispose to premature births. The risk of preterm birth in the second consecutive live birth is 15% in women who have had a preterm first birth and rises to 32% for a third consecutive live birth if the previous two births were preterm.

Since human gestation involves a prolonged period of rapid fetal growth, it may be that the option of preterm labour evolved as a necessary protective strategy against placental insufficiency and the possibility of fetal and maternal demise. Fetal adrenal participation in human parturition is likely, however, through the supply of precursor androgens for the placental synthesis of oestrogens. Fetal dehydroepiandrosterone (DHEA) and DHEA sulphate provide 80 to 90% of the androgen precursor for the placental synthesis of oestradiol and oestriol (Chapter 9). It may be that fetal production of DHEA and the resultant increased placental oestrogen synthesis inhibits the conversion of fetal pregnenolone to progesterone, causing a crucial shift in the localized oestrogen/progesterone ratio which, in turn, triggers the final preparations leading to labour. DHEA synthesis is controlled by fetal CRH and ACTH. ACTH is the main steroidogenic stimulus of the fetal adrenal cortex, which produces androgens in the first two trimesters but in the third trimester responds to ACTH by producing cortisol.

UTERINE CONTRACTIONS AND CERVICAL RIPENING

The safe delivery of the fetus is dependent on the development of rhythmical, sustained and co-ordinated contractions of the uterine smooth musculature and accompanying dilatation of the cervix at a time when the fetus has sufficiently matured for independent life. Throughout pregnancy the myometrium remains relatively quiescent and exhibits gentile contractile activity identified as non-propagating localized contractions ('A' waves) and regular 10–15 minute spaced propagated Braxton Hicks ('B' waves) contractions. The Braxton Hicks contractions are painless and increase muscle tone and development. During pregnancy the uterus is prepared for the muscular activity required during labour by a significant increase in total muscle content. The upper part of the uterus contains the largest proportion of muscle, which is arranged in an outer longitudinally and obliquely orientated layer of fibre bundles and an inner layer of circularly orientated fibre bundles. By contrast, the lower part of the uterus contains mainly circularly orientated fibres. When labour begins, the upper segment contracts and becomes thicker while the lower segment thins out, thus facilitating the entry of the presenting fetal part into the cavity of the pelvis. The individual muscle fibres are joined by a collagen-based ground substance and gap junctions. The ground substance provides the means by which contractile tension is transmitted along the length of the muscle bundle; the gap junctions provide low-resistance pathways for the transmission of electrical and molecular information. Development of well co-ordinated synchronous contractions during labour requires cell-to-cell coupling accomplished by these gap junctions. Throughout pregnancy the number of gap junctions is low or absent but this number increases dramatically at term. During parturition the function of the different parts of the reproductive tract must be integrated to ensure the expulsion of the fetus without compromising the perfusion of the placenta and the circulation of the fetus.

During pregnancy, the cervix remains firm and closed, providing a seal against external contamination and holding in the contents of the uterus. The cervix is mainly fibrous connective tissue composed of smooth muscle, collagen and connective tissue matrix. As the onset of labour approaches, biochemical changes promote a 'ripening' process of the cervix which results in softening so that when contractions begin it can stretch and dilate allowing the fetus to be propelled through. Although most of the proteolytic catabolism of cervical tissue has taken place before the onset of labour, during labour there is additional elastase activity that contributes to the destruction of connective tissue and accelerates the ripening process. A close co-operation between the factors responsible for generating enhanced myometrial contractility and developing cervical compliance is necessary to effect normal, uncomplicated delivery. Collagen is also a constituent part of the amniotic membrane and women presenting with premature rupture of membranes have a significant decrease in the membrane collagen content when compared with women delivering without premature membrane rupture.

HORMONAL EFFECTS ON THE MYOMETRIUM AND THE CERVIX

The uterus and cervix have two differing roles. The fetus must be held safetly in the mother through all the activities that she undertakes during pregnancy and it must be propelled safely out of its mother at the end of gestation during the birth process.

Progesterone, oestrogens and relaxin

The relative quiescence of the uterus during most of gestation is due, in large part, to progesterone. This hormone has a dual mechanism by which it suppresses uterine activity: directly via its own receptors and indirectly via oestrogen receptor inhibition. Direct mechanisms of action may be achieved through the uncoupling of excitation–contraction due to a diminished muscle cell permeability to calcium, the suppression of prostaglandin synthesis and inhibition of the development of gap junctions vital to the effective propagation of contractions. Another function of progesterone is the inhibition of tissue breakdown, ensuring that the cervix does not become pliable before term.

During late pregnancy, oestrogens promote uterine contractility through the stimulation of uterine contractile protein synthesis, prostaglandin synthesis and the stimulation of gap junction development between adjacent myometrial cells. Other vital functions of oestrogens are the stimulation of the ripening of the cervix by enhancing the binding of relaxin to the cervix and changes in the coagulation properties of blood. The extensive and deeply penetrating vascular placental exchange system, established during pregnancy, is abruptly destroyed during labour and delivery so mechanisms must be in place to protect the mother against undue blood loss at the placental site. One such mechanism is the sustained uterine contractions which occlude the major blood vessels. In addition, protective changes in the coagulation index of blood, ensuring the fast blockage of injured vessels, develop during pregnancy and the oestrogens increase blood clotting rates by stimulating the production of coagulation factors. The role of oestrogens in parturition is best described as 'priming', that is, they help to create the conditions favourable for labour but do not themselves play an active role in parturition.

The presence of a hormone causing separation of the pubic symphysis by relaxing the pelvic ligaments before parturition was first detected in rodents in 1926. However, in the human the existence of this hormone, relaxin, was in doubt until its isolation in the 1970s. Human relaxin was finally sequenced in 1983. Relaxin is a polypeptide hormone with structural and functional similarities to insulin and nerve growth factor and is at its highest concentration during the first trimester of pregnancy, although it is detectable in the circulation throughout gestation. It is produced by the corpus luteum and the placenta and is transported via the circulation to its target tissues such as the cervix, myometrium, decidua and breast connective tissue. Relaxin's main action is the facilitation and remodelling of target connective tissues to allow for structural changes in these organs during pregnancy and parturition. The best known changes are those which allow relaxation of the pelvic girdle (pubic symphysis) to facilitate the passage of the fetus during labour and the enhancement of cervical ripening and extensibility. Relaxin works in sequence with oestradiol and prostaglandins to bring about ripening; failure of ripening results in non-progressive labour and the need for Caesarean delivery. A secondary mode of action of relaxin is the inhibition of premature labour by inhibiting myometrial contractility until near term. This is achieved both by direct action and by inhibition of oxytocin release from the posterior pituitary.

Prostaglandins, oxytocin and catecholamines

Prostaglandins are the universal regulators of myometrial contractility and are pivotal to the onset and progression of labour; in this way they are the final common mediators of parturition. Prostaglandins not only induce uterine activity (administration of prostaglandin inhibitors, such as aspirin or indomethacin, effectively suppresses uterine contractions) but also ripen the cervix. Prostaglandins and their relatives, collectively known as eicosanoids, are synthesized from essential polyunsaturated fatty acids obtained from dietary sources, or produced from fatty acid precursors such as linoleic acid. Prostaglandin E_2 (PGE_2) and prostaglandin $F_{2\alpha}$ ($PGF_{2\alpha}$) are the major uterotonic compounds. Since their effects are predominantly local, all intrauterine tissues are capable of producing prostaglandins with the main sites of production being the decidua and the myometrium. Both these tissues produce PGE_2 and $PGF_{2\alpha}$ and, at term, the fetal membranes (amnion/chorion) produce mostly PGE_2. As mentioned above, prostaglandin synthesis and release is stimulated by oestrogen. However, fetal membranes and decidua synthesize other factors that may influence prostaglandin and steroid hormone release; for example, locally produced peptide hormones, neuropeptides and cytokines may modulate local mechanisms regulating uterine contractility.

Traditionally, prostaglandins have been used as clinical agents in the induction of labour and termination of early pregnancies. However, their use presents problems (abnormal uterine contractions and adverse effects on the fetus) because they are physiologically very powerful. Most of the major problems have now been overcome with the development of prostaglandin analogues usually delivered intravaginally or by injection. These modern prostaglandin analogues are very effective in the induction of labour and termination of early pregnancies (Chapter 15). Prostaglandins are also valuable as ripening agents for the 'unfavourable cervix' but other agents which ripen the cervix without inducing uterine contractions are preferable.

Oxytocin is the most potent known endogenous uterotonic agent as it is capable of eliciting uterine contractions at term in concentrations as low as 5–20 pmol/l but, since it is a neuroendocrine hormone, it is normally released in short pulses. The hormone's importance in breastfeeding is discussed later in this chapter (p. 179) so this discussion is restricted to its role in the onset of labour. The maternal levels of circulating oxytocin change very little during the onset and progression of labour. The major factor influencing the action of oxytocin is the 100–200 fold increase in myometrial oxytocin receptor concentration (and consequent increase in uterine sensitivity to oxytocin). This dramatic increase occurs shortly before and at the onset of labour. It has been suggested that endogenous oxytocin is responsible for Braxton Hicks contractions as ethanol, an inhibitor of oxytocin release, abolishes the contractions and stops early labour. The rise in receptor concentrations, induced by the changing oestrogen/-progesterone ratio, lowers the threshold for stimulation of uterine contractions and consequently triggers early labour. Oxytocin receptors are sensitive to steroid hormone regulation with oestrogen promoting and progesterone diminishing oxytocin release and receptor synthesis. Oxytocin, in turn, stimulates the production of PGE_2 and $PGF_{2\alpha}$ from the decidua. The prevailing theory states that in early labour oxytocin is driving the contractions whilst in active labour prostaglandins become the major force in myometrial activation. Fetal secretion of oxytocin may also be important in regulating the process of labour since oxytocin can be found in the fetal pituitary gland from 14 weeks gestation and the amount progressively increases up to term. Oxytocin secreted by the fetus can gain access to the decidua directly or through the fetal membranes to stimulate maternal prostaglandin production and signal the initiation of labour.

Clinically, acceleration of slow labour can be achieved by an intravenous infusion of oxytocin. The 'active management of labour', popular in the 1960–70s, recommended the routine stimulation of uterine activity with oxytocin during the active phase of labour in all primigravidae women whose cervixes failed to dilate at 1 cm/hour or faster. The objective was for all women to deliver within 24 hours. This practice is controversial because of risks to the fetus of neonatal jaundice due to erythrocyte destruction and bilirubin production, and of dehydration of the mother due to oxytocin's antidiuretic properties.

The output of fetal and maternal catecholamines increases during labour but this increase appears to be a consequence, rather than a cause, of labour. A moderate amount of stress is beneficial to the progress of labour and to the fetus by promoting adaptive responses to labour and preparation for birth. Normally, when the uterus is relaxed, blood flow through the placenta depends on the pressure difference between uterine arterial and venous blood. However, when the uterus contracts and the pressure within the uterine cavity increases, the placental vascular bed is restricted creating a hindrance to the flow of blood. The decreased placental blood flow accompanying uterine contractions must be countered by an adequate protective action by the fetus if oxygen deprivation and consequent fetal distress is to be avoided. In the adult, catecholamines prepare the body for action and energy expenditure (Chapter 13); in the fetus, however, the catecholamine response also serves another purpose more appropriate to the circumstances of life in the uterus. Because the sympathetic nervous system does not function fully in adult mode until around birth, the fetal adrenal medulla produces more noradrenaline than adrenaline in response to stress (exactly the opposite to the adult situation). Noradrenaline in the hypoxic fetus triggers a vagus response causing a drop in heart rate and oxygen expenditure and prevents fetal hypoglycaemia. As the fetus matures, increasing circulating cortisol directs adrenal medullary catecholamine production more in favour of adrenaline. As far as the hypothalamic–pituitary–adrenal cortex axis is concerned, the fetus has a mature reactivity towards stress and the presence of abnormally high levels of β-endorphin can be taken as a reaction to fetal suffering and distress. In addition, heightened stress leads to excessive maternal adrenaline production, which is associated with longer labour, lower uterine contractile activity and a shunting of blood away from the uterus and placenta. Dysfunctional labour eventually leads to maternal exhaustion and fetal distress. Overproduction of catecholamines, exceeding physiological bounds, causes poor postnatal adaptation and/or illness in the neonate.

Much of the stress of labour, particularly that imposed by outside social/medical intervention, is preventable. However, inherent maternal illness can disrupt the normal maternal/fetal adaptive responses to labour. Possible long-term epigenetic effects as a result, for example, of excessive consumption of physiologically potent drugs may affect normal labour in unknown ways. In a human study, caffeine consumption equivalent to six or more cups of coffee per day throughout pregnancy caused a significant increase in breech presentation. Laboratory research highlights similarities between this study and findings in the rat. It was observed that rats exposed *in utero* to caffeine, in the relative dose range of human intakes, experienced significant delays and difficulties in the progression of labour and parturition. Ineffectual labour, in turn, compromised the mothers and their offspring, many of which were stillborn. The severity of the outcome was dose dependent and is a matter of grave concern, especially since a subsequent generation, bred from individuals exposed to the drug *in utero*, was also affected; therefore, the drug-consuming habits of the rat grandmother had serious consequences for her children and grandchildren. Caffeine is the most widely used psychoactive drug in the world and, mediated by the CNS, affects every physiological system in the body. It is readily obtained from coffee, tea, chocolate, cocoa drinks, some cola soft drinks, analgesics and other over-the-counter drugs (Chapter 19).

The role of catecholamines in the initiation of parturition, if any, is not clear. It is of interest to note, however, that adrenaline can function as a cofactor in the stimulation of prostaglandin synthesis. Circulating levels of catecholamines rise with advancing gestation and as the fetal adrenal gland matures the content of adrenaline, noradrenaline and dopamine in the amniotic fluid also rises.

OVERVIEW

Human parturition is the result of the sequential maturation of an endocrine communication system between the mother and the fetus. A simplified summary, illustrated in Fig. 10.1, is as follows. The initiation of parturition has to be preceded by the facilitatory actions of progesterone and oestrogens. During the final 5 to 6 weeks of pregnancy, oestradiol production increases, changing the intrauterine progesterone: oestrogen ratio which in turn promotes prostaglandin production, the formation of gap junctions, cervical ripening and the synthesis of receptors for oxytocin, and prostaglandins. Whether fetal adrenal steroids play a facilitatory or definite role in the initiation of labour in humans is not yet established. The fetus does, however, have a co-ordinating role through

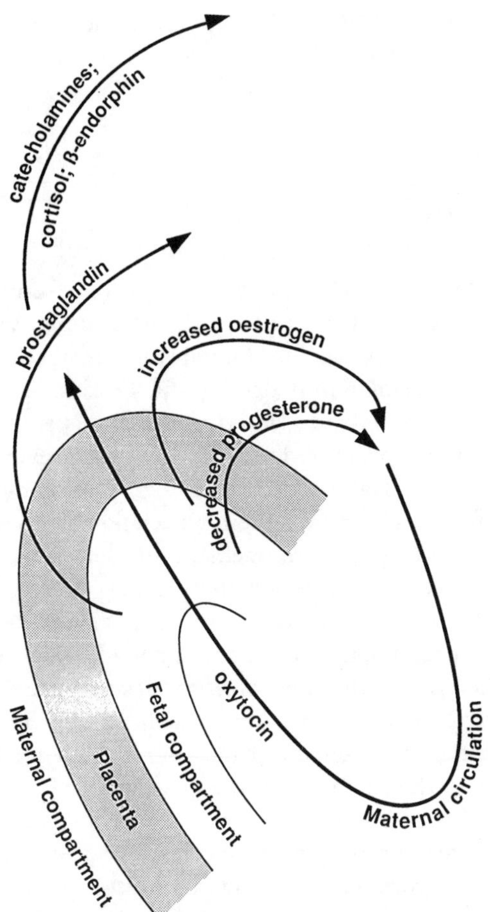

Fig. 10.1 Schematic representation of the hormonal chain of events leading to parturition. Increased oestradiol production shifts the intrauterine progesterone:oestrogen ratio which promotes prostaglandin and oxytocin production. Catecholamines, cortisol and β-endorphin release promotes adaptation and pain control emphasizing the oneness of physiological and psychological phenomena.

influences such as placental oestrogen production, myometrial activity (by mechanical distension of uterus and cervical excitation) and stimulation of oxytocin and prostaglandin synthesis. The stimulation of the myometrium during labour results from an interaction of oxytocin and prostaglandin, with oxytocin dominating the initial phase of labour (and perhaps triggering contractions) and prostaglandins essential for the progression of labour. Controlled catecholamine and cortisol release promotes maternal and fetal adaptation with β-endorphin regulating pain and mood thresholds.

B LACTATION

The mammary glands are thought to have evolved from the specialized skin area in the abdomen used for incubating the young, which developed in the early endothermic therapsids. Therapsids were mammal-like reptiles living from the mid-Permian to the end of the Triassic periods, 240 to 225 million years ago. The sweat glands of the therapsid-incubation area enlarged and provided specialized secretions which prevented desiccation of the eggs and the newly hatched offspring. The mammary glands in present-day mammals still retain their association with the skin, as the characteristic tubulo-alveolar arrangement of the endocrine tissue has structural similarities with the secretory units of sweat glands. The anatomic diversity of present-day mammary glands is the result of evolutionary pressures relating to the number of offspring, the optimal placement for the young, the minimization of exposure to damage from the physical environment and the impairment of efficient maternal locomotion. The number of mammary glands ranges from a minimum of two, as in humans, up to the 25 found in the possum.

Despite the diversity in gross anatomy, the histological features of the mammary glands are similar in all species. As seen in Fig. 10.2, the secretory tissue is composed of numerous alveoli, each consisting of a hollow ball of secretory cells surrounded by a network of contractile myoepithelial cells and capillaries. The alveoli are responsible for the synthesis and secretion of milk while the myoepithelial cells have a contractile function and are concerned with the movement of milk from the alveoli into the ducts prior to milk ejection. The alveoli are embedded in a stroma of loose connective tissue, forming lobules separated by dense interlobular connective tissue and connected to larger non-secretory ducts and sinuses.

STORAGE OF MILK WITHIN THE MAMMARY GLAND

Although lactation is triggered by the termination of pregnancy, it is anticipated both anatomically and physiologically from early life. During

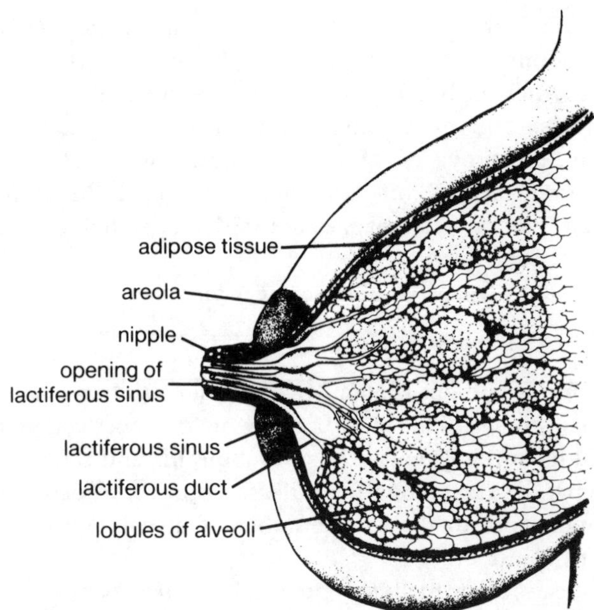

Fig. 10.2 The internal structure of the lactating mammary gland illustrating full development of the lobules of the alveoli and the branched ductular system.

adolescence and for several years thereafter the breasts mature influenced by ovarian steroids. This continues under the stimulus of the hormones of pregnancy and is characterized by the growth and maturation of the lobulo-alveolar epithelial tissue and proliferation of the duct system. By the third month of pregnancy the alveoli begin to secrete a thick liquid resembling colostrum and this is maintained under the influence of placental lactogen. During the third trimester, the breasts become organs capable of synthesizing and delivering milk, a process termed lactogenesis. Colostrum secreted during the first week postpartum is especially high in concentration of antibodies that convey passive immunity to the offspring. Subsequently, the secretion from the mammary gland rapidly becomes a fluid that provides all the nutrients required for growth and survival of the newborn. The co-ordinated secretion of a number of hormones controls mammary growth, lactogenesis and maintenance of lactation. After parturition, the volume of milk secreted increases to a maximum and then gradually declines, paralleling the offspring's graduation to a solid diet.

The number of myoepithelial cells increases during pregnancy under the influence of oestrogen, with oxytocin receptors appearing on the myoepithelial cells shortly before the onset of parturition. This correlates with the development of uterine sensitivity to oxytocin (see p. 179). Milk secreted in the alveoli becomes distributed between the alveolar lumina and the duct system of the mammary gland. In species that suckle intermittently, the provision of storage within the gland is vital for the adequate maintenance of milk secretion because, if the alveoli become overdistended, biochemical changes are initiated, signalling involution of the secretory tissue.

From the point of view of milk storage, mammals can be divided into two groups.

(a) Those species without obvious modifications (other than the alveoli) to increase storage. This group represents the most common situation and includes species such as the rat, rabbit, pig and cat. In the absence of any special adaptations for milk storage, the capacity is more or less proportional to the total number of alveoli present within the mammary gland.
(b) Those species in which the duct system has become modified to provide storage space. The most striking example of this group is provided by the ruminants. For example, in the cow there is a special cistern, immediately above the teat, which can hold between 100 and 400 ml of milk. In the primates (including women), the ducts leading into the nipple are dilated to form sinuses which provide a small amount of extra milk storage.

The nipple provides a grasping structure for the attachment of the offspring, restricts leakage from the gland between nursing (there is no sphincter for retention of the milk as milk will leak from an unsuckled nipple during milk ejection) and relays the sensory sucking stimulus to activate the milk-ejection reflex. Within the nipple are 10–12 galactophores or lactiferous ducts. Surrounding the base of the human nipple is the pigmented areolar region. Smooth muscle fibres are arranged radially and circularly within the areola around the base of the nipple; when contracted, these serve to elevate the nipple. Nipple-seeking, although helped by encouragement from the mother, is an innate behavioural pattern of the neonate (Chapter 11). Successful location of the nipple depends on tactile sensation, but olfactory cues are also important. The human baby forms its teat by drawing both the nipple and areola region (including the underlying milk sinuses) into the buccal cavity and suckles in bursts of 1 to 2 minutes interspersed with pauses of 20 to 30 seconds. Milk flow from the breast correlates with sucking activity and as feeding progresses milk intake declines as sucking episodes become shorter. Most of the milk is obtained within the first few minutes of the nursing episode.

LACTOGENESIS

The ability of the mammary gland to synthesize and secrete copious quantities of milk begins in the immediate periparturient period, increases for a variable period of time postpartum, and then gradually decreases. The secretory activity parallels the demand of the offspring for milk, with the suckling intensity being an important determinant of lactational performance. The process of suckling stimulates lactation in two ways.

(a) By the reduction of the inhibitory effects of increased intramammary pressure due to milk accumulation.
(b) By generating the powerful stimulus that causes marked increases in secretion of the hormones that are required for the continued synthesis of milk.

There is no single hormone that initiates lactation; rather, during the third trimester of gestation, a cascade of endocrine events occurs which prepares the mammary gland for the secretion of milk. Lactogenesis is a two-stage process. During the first stage, which occurs during the last trimester of pregnancy, the alveolar cells differentiate cytologically and enzymatically. The second stage consists of secretion of milk, beginning 1–4 days before parturition to 1–3 days after parturition and continuing until maximal milk secretion occurs. The successful initiation of lactation requires increased secretion of prolactin in conjunction with acute reductions in circulating progesterone around the perinatal period (high levels of progesterone explain, in part, why lactogenesis is suppressed during most of gestation).

Prolactin is the key hormone controlling milk production. The entire process of lactogenesis, however, requires multiple hormonal interactions. For example, growth of the mammary duct system is dependent on oestrogen synergized by the presence of GH, prolactin (circulating prolactin gradually increases throughout gestation coincident with increased secretion of oestrogens), placental lactogen, thyroid hormones and cortisol. Prolactin's role is in the induction and synthesis of the primary milk protein casein and the regulation of the mobilization of fat. The protein composition of human milk is quite different from that of cow's milk and contains 70% of its total protein in the form of whey (a mixture of proteins including α-lactalbumin, lactoferrin, lysozyme, albumin and immunoglobulin) and a smaller amount (30% compared to the cow's 80%) in the form of casein. Lysozyme helps destroy bacteria and is present in human milk at 3000 times its concentration in cow's milk. Fat, mostly in the form of triglycerides, provides 40–50% of the energy in human milk. The quality (the ratio of saturated to unsaturated fatty acids) of the lipid in breast milk changes with the mother's diet; for example, the percentage of linoleic acid can vary from 1% to 45% depending on how much linoleic acid the mother is consuming. Prolactin functions in conjunction with GH, cortisol and insulin which are the major components of the integrated endocrine control of metabolic homeostasis. These hormones affect lactation by partitioning nutrients between competing demands. For instance, GH favours protein degradation and lipolysis, transferring the products toward milk production, whilst insulin replenishes maternal fat stores from carbohydrates. Prolactin also mediates parental behaviour, the subject of the next chapter. The term homeorhesis is used to describe the endocrine regulation of this long-term nutrient-partitioning phenomenon.

Administration of natural or synthetic oestrogens or of oestrogen–progestogen or oestrogen–androgen combinations inhibit secretion off milk in a variety of species, including the human. Oestrogens have been used clinically to inhibit lactation in women not wishing to breast-feed their babies. However the progestogen-alone oral contraceptive (mini-pill) does not seem to significantly reduce lactational performance once lactation has been established. All ingested steroidal substances are secreted in breast milk, but at the dosage of progestogen used the amount transferred seems small and without short-term effects. The possible long-term effects, however, are difficult to access. On the other hand, there is no doubt that breast-feeding in women, through raised prolactin levels, delays the resumption of ovarian cyclicity and the menses. The duration of the delayed fertility is very variable, both between and within different populations. Lactational amenorrhoea may last from 2 to 3 months to 3 to 4 years, with the duration depending on the pattern of suckling, the more frequent or intense the suckling, the longer the delay. Both the basal level and the amount of prolactin released in response to suckling decreases with time. The decline is more noticeable, however, in women where there is a relatively low frequency (3–5) of daily breast-feeding compared with a high frequency (more than 10) of breast-feeding episodes maintained throughout lactation. Because the start of the preovulatory LH surge during the normal menstrual cycle occurs at night, in the majority of women it may be that nighttime suckling is more likely to suppress the return of fertility. By contrast, in non-breast-feeding women, ovulation takes place within 4 to 8 weeks postpartum. A proportion of women ovulate during breast-feeding and may become pregnant without any intervening menses.

The many advantages accruing to the breast-feeding mother may be summarized as follows. There is acceleration of the postpartum uterine involution, a contraceptive effect, emotional well-being due to the knowledge that the baby is supplied with all nutritional and immunological needs and the opportunity for mother–child bonding. The milk's cheapness and sterility, together with a provision of some protection against breast cancer, are additional pluses.

THE MILK-EJECTION REFLEX

The process of milk ejection (the active expulsion of milk from the mammary glands also sometimes referred to as 'let-down' or 'draught') is a prime example of a process involving the integrated activity of the nervous and endocrine systems. The neuroendocrine control of milk ejection is regulated by a simple reflex arc in which the afferent limb is formed by ascending pathways transmitting the suckling stimulus from the nipples up to the hypothalamus and the efferent limb is formed by oxytocin neurons projecting to the posterior pituitary. Activation of the oxytocin neurons by the suckling stimulus causes release of oxytocin. This, on reaching the mammary gland, stimulates the contraction of the alveolar myoepithelium which forcefully expels the stored milk. Most of the milk (90%) can be removed within 4 minutes of the onset of suckling and many women show an adaptive conditioned reflex, responding to cues such as the sight, sound or smell of their babies by oxytocin release. Oxytocin release in women most frequently occurs before the baby is actually attached to the nipple. The evolutionary advantage of developing a mechanism for conditioning of the milk-ejection reflex is that it may serve to shorten the nursing episode and thereby reduce danger from predators. Suckling does not influence oxytocin release alone but also exerts widespread adaptive effects on somatosensory sensation, sleep–wake cycle, food intake and the secretion of other anterior pituitary hormones in addition to prolactin. Oxytocin is released by vaginal distension and coitus. The effectiveness of stimulating the reproductive tract to entrain milk ejection has been known and exploited for centuries in dairying practice amongst tribal communities.

The response of the mammary gland to oxytocin is also regulated by the autonomic nervous system and is perhaps related to the evolution of ancient survival strategies. Sympathetic nervous and adrenal activation (stress) causes vasoconstriction of the vessels supplying the mammary glands and prevents oxytocin from reaching the myoepithelium. This mechanism forms the basis for the blockade of the milk-ejection reflex during anxiety and other severe forms of stress.

Impairment of the milk-ejection reflex in women

The incidence of breast-feeding within western cultures markedly declined from the 1920s to the 1960s, when only about 20% of mothers leaving the hospital were breast-feeding their babies. However, since the 1970s this decline has reversed, with 50–90% of all infants being initially breast-fed, and its popularity is still increasing. The impetus for this change was partly due to 'back to nature' movements within western society and partly from the general recognition that breast-feeding is superior to bottle-feeding on nutritional, biochemical, hygienic, psychological and contraceptive grounds. The change has been brought about by avoiding practices that discourage breast-feeding, such as supplementary feeding, strict feeding schedules and long periods of separation of the mother from her child. Even in this positive atmosphere not all mothers who wish to breast-feed are able to establish lactation and others are compelled to give up prematurely because of problems. The major reasons for failure to establish successful breast-feeding include painful breast engorgement, inadequate milk production, sore nipples and diminished motivation when supplementary feeds are introduced. There is a strong link between nervous temperament and the failure of breast-feeding which relates, at least in part, to the disrupting influence of anxiety on the normal functioning of the milk-ejection reflex. Folklore and child-care books abound with good advice as to how a mother should maintain a calm temperament and avoid emotional disturbance during lactation; not always an easy order to follow!

As mentioned above, severe stress leads to adrenal medullary activation and adrenaline-induced cutaneous vasoconstriction. This in turn causes a peripheral block to milk ejection by the prevention of oxytocin action on the myoepithelium. Milder emotional disturbances may also affect breast-feeding by disrupting the central release of oxytocin through inhibitory pathways that prevent excitation of the oxytocin neurons or oxytocin release from the neurohypophysis. A clinical trial has demonstrated that intranasal oxytocin spray, when used before nursing, increased the average milk yield by more than 20% and, more importantly, reduced from 20% to 5% the numbers of mothers producing less than 20 g of milk per feed. It is, of course, preferable to eliminate anxiety-related problems at the source rather than artificially inducing milk-ejection. Additionally, sedatives, analgesics and other drugs given to the mother during labour may have a negative influence on the initiation of breast-feeding by suppressing oxytocin release during suckling and/or inhibiting lusty feeding by the infant. A study of suckling activity, using artificial teats, showed that in infants whose mothers had been given sedatives there was significant depression in the rate of sucking (from 50 down to 20 sucks/minute). In addition the negative pressures obtained were lower and these effects persisted over a 4-day period.

Poor nipple protractility (flat or inverted) and sore nipples may also cause failures of breast-feeding. Sore nipples are sometimes the secondary result of other problems in milk transfer leading to excessive non-nutritive sucking by the baby. Painful suckling, in turn, will produce an emotional inhibition of the milk-ejection reflex and/or stress-induced release of cortisol. Milk transfer can also be disrupted by disturbances in the suckling behaviour of the baby (such as cleft palate, which will prevent the baby from applying suction) and prematurity. The premature baby cannot suckle adequately and has to be fed by means of a stomach tube, preferably with the mother's own milk obtained by a breast pump after giving exogenous oxytocin. Perhaps the most important source of disturbances in suckling behaviour is that babies quickly learn to reject the breast if milk is more easily obtained by suckling the artificial teat of a bottle. Hence, if supplementary bottle-feeding is introduced this will make the baby restless on the breast, leading to anxiety in the mother which further jeopardizes the chances of a successful nursing period.

Similar factors causing significant disturbances of milk transfer in wild animals have determining power in the selection of species' survival. For example, the emotional inhibition of the milk-ejection reflex caused by fear could be a significant problem, especially in species that have their habitat restricted so much that they are unable to find a quiet spot for the rearing of their young. The same problems can arise in captive animals brought in from the wild.

LACTATION AND NUTRITION

Since human milk is a unique substance that supplies complete nutrition and at the same time provides protection against infections, breast-feeding, even if only for a short time, is to be encouraged. Lactation is, however, an energy-intensive process. At the height of milk production 600–800 ml may be withdrawn, and with milk containing approximately 0.8 kcal/ml

(3.2 kJ/ml), an extra 500–600 kcal (2000–2400 kJ) are necessary, daily, just to produce the milk that the infant is consuming. In addition, milk is extremely rich in certain nutrients which must be supplied from the diet or from the maternal stores. For instance, lactation can be a considerable drain on maternal protein supplies especially in vulnerable groups such as strict vegetarians consuming marginal amounts of protein and the malnourished in developing countries. Repeated pregnancies and short birth intervals accentuate a poor nuritional status. Unless during pregnancy the mother has deposited 3–4 kg fat as an energy store for the extra demands of lactation, she will suffer depletion of her nutritional reserves. Milk volume and composition (except in severe cases) are preserved at the expense of maternal tissue. For instance, the mother's calcium requirement is higher during lactation than even during the last months of pregnancy (1.5–2.0 g of calcium each day are necessary to achieve positive calcium balance). During lactation an equal amount of calcium should be consumed from dairy food (600 ml of milk daily is sufficient to cover her and her infant's demands), kale, cabbage, turnip, greens, certain nuts such as almonds and red meat, or be supplemented. Calcium deficiency during lactation will, however, have little effect on the calcium content of the milk as the deficit is simply withdrawn from the maternal bone and the resulting effects may become obvious only years later with the development of osteoporosis. If lactation is promoted in women who are severely malnourished, the quality of the milk will deteriorate and become watery. Inadequate nutrition has also an adverse effect on the psychological health of the mother with the possibility of depression leading to infant neglect.

All of the nutrients needed by the mother can easily be acquired by consuming a balanced diet. No vitamin and mineral supplement is necessary provided the food is adequate; however, one vitamin, B_{12}, will be deficient if the mother is a strict vegetarian since this vitamin is found only in foods derived from animal sources. It must either be consumed as a supplement or be obtained in certain foods fortified with it, for example cereal products and yeasts. Additional fluid is another important need of the lactating mother since she must replace an extra 600–800 ml of fluid per day (human milk is approximately 88% water). Skim or partially skim milk is an ideal replacement since it is rich in protein and calcium. Water and juices are also good fluids but certain fluids should not be consumed in large amounts, particularly those high in caffeine which readily passes into the milk. With improved ability to detect very small concentrations of drugs it has been shown that almost all maternally ingested drugs will pass into breast milk. In cases of chronic maternal drug consumption, the infant may accumulate the drug because of a reduced elimination capacity. The human neonate, for example, lacks the adult ability to readily demethylate caffeine and its immediate dimethylxanthine metabolites, which are themselves metabolically active.

Ethanol consumed in large doses decreases oxytocin release in the mother and is transferred to the infant in amounts that may be as high as 20% of the maternal dose, such babies are characteristically lethargic and heavier. Occasional alcohol intake, especially of low-alcohol beverages, is not, however, associated with adverse effects on the infant and may have a calming effect on the mother. Nicotine is present in the breast milk of smoking mothers and Δ^9-tetrahydrocannibol from marijuana is concentrated in breast milk. It is ironical that the numerous advantages of breast-feeding must be balanced against mounting concern for the increased exposure of many infants to drugs and environmental pollutants transferred via breast milk.

General references

Barr, H. M. & Streissguth, P. (1991). Caffeine use during pregnancy and child outcome: a 7-year prospective study. *Neurotoxicology & Teratology*, **13**, 441–448.

Ben-Jonathan, N., Laudon, M. & Garris, P. A. (1991). Novel aspects of posterior pituitary function: regulation of prolactin secretion. *Frontiers in Neuroendocrinology*, **12**, 231–277.

Casey, M. L., Cox, S. M., Word, R. A. & MacDonald, P. C. (1990). Cytokines and infection-induced preterm labour. *Reproduction, Fertility & Development*, **2**, 499–509.

Jeffrey, J. J. (1991). Collagen and collagenase: pregnancy and parturition. *Seminars in Perinatology*, **15**, 118–126.

Koren, G. (ed.) (1990). *Maternal-Fetal Toxicology: A Clinician's Guide*. Dekker, New York.

Mitchell, M. D. (ed.) (1990). *Eicosanoids in Reproduction*. CRC Press, Boca Raton, FL.

Pollard, I. & Claassens, R. (1992). Caffeine-mediated effects on reproductive health over two generations in rats. *Reproductive Toxicology*, **6**, 541–545.

Steer, P. J. (1990). The endocrinology of parturition in the human. *Baillière's Clinical Endocrinology & Metabolism*, **4**, 333–349.

Taverne, M. A. M. (1992). Physiology of parturition. *Animal Reproduction Science*, **28**, 433–440.

Uldbjerg, N. & Ulmsten, U. (1990). The physiology of cervical ripening and cervical dilatation and the effect of abortifacient drugs. *Baillière's Clinical Obstetrics & Gynecology*, **4**, 263–282.

Van Landingham, M., Trussell, J. & Grummer-Strawn, L. (1991). Contraceptive and health benefits of breastfeeding: a review of the recent evidence. *International Family Planning Perspectives*, **17**, 131–136.

11 Parental behaviour and the physiology of the neonate

THE PERIOD OF GESTATION

The technical term, derived from the Greek and Latin terms, for the newborn is neonate, and most researchers agree that we can refer to a baby as a neonate until the end of the first month of independent life. The neonate is a curious mixture of competence and helplessness. As we have seen from previous chapters, the growing, developing embryo/fetus is far from a quiescent entity passively accepting the events taking place in its protective uterine environment. Rather, the fetus is an active, dynamic individual who is increasingly engaged in the regulation of its own development. Between fertilization, when their genetic make-up is set, and birth, babies accumulate a load of experiences that will contribute in the shaping of who they are and what they can bring to their new life. The importance of the continuum from the single cell to the newborn child, from the youngster to the adult, cannot be overemphasized because throughout the course of these different phases, involving a wide variety of interactions with the environment, the individual becomes a social being with a sense of self. As already noted, primate fetuses remember their early physiological environment *in utero* and prepare, under the prevailing conditions, for the future. The degree of maturity at birth varies from offspring to offspring, therefore the perinatal period spans the continuum from late fetal to early postnatal development.

The significance of gestation as an experience for the mother must also not be neglected. Normally both the mother and her newborn adapt easily because the relationship established during pregnancy continues to develop postnatally. Mother and child are already familiar with each other and their mutual responses to the environment. Sounds penetrate from the outside world to the fetus within, and movement of the fetus is felt by the mother. If a mother says that her fetus is particularly active, her expectation is of an active baby, and usually she gets one. Pregnant women have recurring periods of activity and periods of rest to which the fetus accommodates; for example, fetal activity patterns closely follow the maternal day/night activities and endocrine rhythms. Maternal environmental factors, such as the circadian rhythms of CRH, cortisol, melatonin, uterine contractions and body temperature, are significant in the entrainment of the fetal biological clock. Thus mother and child share rhythms of sleep and wakefulness, eating and socializing (see p. 202). The baby's capacity for social interaction begins to develop in the 3 months before birth. By 28 weeks, the fetus' hearing functions are much as they will be after birth, and the fetus hears and responds to speech, learning the rhythm before the words. The newborn actively selects and filters visual and auditory stimuli and can recognize its mother's voice and the voices of those close to her. Experiments demonstrate that a neonate 3 or 4 days old prefers a recording of its mother's voice reading a story to that of another woman reading the same story. Instinctive babycare is based on near-universal experiences, singing to, rocking, swinging, swaddling and holding babies close; these show up in all cultures in various forms. Babies are soothed by singing – the rhythm of singing is probably the way fetuses hear their mothers' voices *in utero* – and by being held firmly, reflecting the uterus as a tight swaddle with the contractions of pregnancy providing frequent 'hugs'. Babies are soothed by being held close to the heart, and the adult heart beat rate of 72 beats per minute has been exploited commercially by the manufacture of teddy bears that play a recorded human heart beat at this rate. The rocking chairs, cradles and hammocks used all over the world duplicate the soothing rock of the pelvis when the mother walked during pregnancy. Babies seem to need to be rocked or walked, which is understandable as they recently left an environment in which they had been rocked for many hours each day. Early *in utero* experiences leave permanent legacies; in grief, children and adults alike may revert to a perceived time of security by rocking.

It is with mothers that fetuses have the most direct relationships, and it is on mothers that fetuses have the most direct effects. However, it is not only to mother that fetuses relate. An emotional relationship also develops with the ones close to her and the mother will often use this closeness to create a bond between the fetus and the rest of the family. For instance, it is natural for the father and siblings to reach out and feel the fetal movements and to make contact. Thus the process of socialization, which goes on for a lifetime, begins before birth. Social and physical contact is vital to the neonate. Many subtle long-term influences of physical deprivation during the neonatal period have been documented. For example, in all mammalian species studied to date, physical contact (handling) assists physiological mechanisms to function optimally, and a resulting acceleration of all parameters of growth and development is noticeable. Accelerated maturation following physical stimulation has also been reported for premature babies in intensive care (Chapter 13).

These psychological aspects of pregnancy may help to explain why the relationship that a surrogate mother has established with her fetus by the time she births cannot, in many cases, be ended abruptly without trauma. A gestational relationship may outweigh opposing claims of genetics and signed surrogacy contracts. The old-fashioned term 'with child' declares this distinction better than the contemporary term 'expecting'.

THE PHYSICAL ADAPTATION OF THE NEWBORN TO EXTRAUTERINE LIFE

From complete dependence on the uterine environment, the newborn undergoes a period of remarkable development. Some rapid adaptations to the new environment outside the uterus are necessary as the intrauterine environment relieves fetal organs of certain functions; for example, the placenta substitutes for both lungs and kidneys. The relative weightlessness of the fetus in the amniotic fluid precludes the development of muscular strength and the isothermal environment means that calories are not expended in maintaining body temperature. In addition, the gastrointestinal tract does not function in digestion. However, before birth all of these systems must reach a degree of minimum development sufficient to sustain life immediately after birth.

The respiratory system

During the last few weeks of pregnancy, the number of alveoli in the lungs increases and surfactant is produced which reduces the surface tension of pulmonary fluids, ensuring expansion of the alveoli and a sufficient exchange of gases at birth. Surfactant is a phospholipid protein mixture and the enzymes required for its synthesis are promoted by fetal corticosteroids. Surfactant is found in the human fetal lung from weeks 18 to 20 onwards but increases sharply in concentration during the last 2 months of gestation. To perfect lung function, from 26 weeks onwards the fetus spends an increased percentage of time (near term approximately 20 minutes out of every hour) making rapid respiratory movements. These breathing movements occur more often in sleep cycles and are purely diaphragmatic; they do, however, move small amounts of amniotic fluid in and out of the lungs. This activity acts to fine-tune the reflex neuromuscular co-ordination and provide the distension needed to promote lung growth in order to initiate breathing and to fill the lungs with air at birth.

There are many stimuli to encourage normal continuous breathing at birth. As labour approaches, the incidence of fetal breathing movements decreases, lung liquid production gradually decreases and, during labour, absorption of lung liquid begins. During the passage through the vagina, as the O_2 supply is decreased due to placental and cord compression, blood CO_2 level is increased, stimulating (at a critical level of P_{CO_2}) the respiratory centre in the medulla oblongata of the brain. The respiratory response to P_{CO_2} is increased by removal of the placenta, probably via loss of a placentally produced respiratory inhibitor. Approximately 85% of lung fluid is normally expelled from the mouth and nose during vaginal delivery, provided there is sufficient compression of the chest wall. The remaining fluid is absorbed by the pulmonary lymphatics and capillaries, or expelled within 48 hours. A baby born by Caesarian section may be slow to breathe because of low blood CO_2 levels and congestion of the lungs with fluid. Lung fluid will take a few days to clear in these babies and can cause respiratory problems. At birth, a strong baby usually gasps and then cries, which induces regular breathing within a minute. A good lusty cry is essential to produce the pressure needed to inflate all the alveoli. The longer it takes before all the alveoli are inflated, the more difficulties will present. This is always the most critical time in premature babies whose lungs and respiratory centres may not have adequately matured. Failure to produce sufficient surfactant has serious consequences for lung expansion, as is seen in neonatal respiratory distress syndrome (RDS), the symptoms of which can be linked to surfactant deficiency and are seen in approximately 1.5% of all newborn babies. The collapsed lungs of RDS infants cannot be adequately ventilated or effectively retain air and the physiological dead space is enormously increased. Lung maturation can be accelerated by administering glucocorticoids (usually the cortisol analogue betamethasone) over 4–5 days to the mother when the fetus is to be delivered between 28 and 32 weeks and weighs less than 1500 g. The administered steroid crosses the placenta and increases fetal surfactant production, but it can have adverse side effects (may exacerbate maternal hypertension for instance) and so its use may not be beneficial in all cases. Advances in the understanding of the structure and function of surfactant have led to the successful development and use of artificial surfactants that can be administered at birth. Surfactant replacement for the treatment of RDS is now widespread in Japan, Europe and the USA.

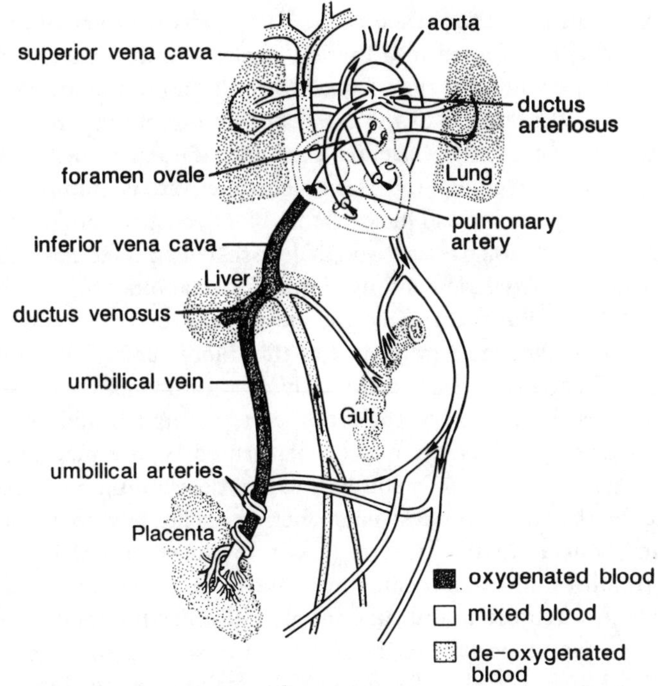

Fig. 11.1 The fetal circulatory system depicting the shunting channels: foramen ovale, ductus arteriosus and ductus venosus (see text for a description of the blood flow).

The circulatory system

The above respiratory changes initiate the changes in the neonate's circulation required for the conversion from the intrauterine situation to the postnatal and adult state. In mechanical terms, these involve the closure and obliteration of the shunting channels of the foramen ovale, ductus arteriosus and ductus venosus. As shown in Fig. 11.1, the umbilical vein that conveys oxygenated blood from the placenta divides near the fetal liver. The main stream enters the ductus venosus (which forms a by-pass around the liver) and the other stream enters the liver along with various branches of the portal vein. These streams recombine and mix in the inferior vena cava (IVC), which, due to the shunt, contains fairly well-oxygenated blood. On reaching the heart, the bulk of the fetal blood by-passes the non-aerated lungs by two routes. First, most of the blood from the right atrium directly enters the left atrium via an opening called the foramen ovale, thus by-passing both the right ventricle and the lungs.

The blood in the left atrium then mixes with a small volume of pulmonary venous blood and is forced into the left ventricle and thence to the main aorta. This carries blood to the coronary vessels and to the proportionately large head and upper parts of the body and represents the main route for oxygenated blood from the placenta. Secondly, the remainder of the blood in the right atrium enters the right ventricle and is pumped into the pulmonary artery. However, most of this blood, instead of going through the lungs and left heart, passes through a vessel called the ductus arteriosus which is linked to the descending aorta. Only about 10% of the blood leaving the heart flows through the lungs. Therefore, by utilizing the ductus venosus, the foramen ovale and the ductus arteriosus, blood flow through the liver and lungs occurs only to a minor extent in the fetus. This conserves the energy of the fetal heart, because functions normally required of the fetal liver are instead performed by the placenta and the mother's liver. Most of the blood within the descending aorta returns to the placenta (because the resistance here is relatively low) and is re-oxygenated, but a smaller portion passes to the abdominal viscera and lower extremities and then drains into the IVC. A portion of the blood that passes through the umbilical cord also supplies the fetal membranes.

After birth the blood flow to the lungs is increased as these inflate. This causes all the blood to flow from the right atrium to the right ventricle, and none passes through the foramen ovale or the ductus arteriosus. The foramen ovale is covered by a thin valve on the left atrial side and as long as the pressure in the right atrium is greater than that in the left atrium (which is the case in the fetus) blood flows from right to left, by-passing the right ventricle and lungs. However, when the lungs expand, the right heart can then pump blood easily through the lungs so that the right atrial pressure falls below that of the left and this reversed pressure differential closes the valve over the foramen ovale. The blood is then pumped out through the aorta and passes the other end of the ductus arteriosus. The ductus venosus and the ductus arteriosus become gradually occluded during the first few days or weeks of life. The remaining parts of the fetal circulation also become fibrosed because blood flow through them has ceased. If the neonate makes poor respiratory effort, the foramen ovale and the ductus arteriosus will remain open for some time and cause problems, as is often the case in premature babies.

When the above respiratory and circulatory changes have occurred, the newborn will be pink, although the hands and feet may remain blue for a day or two. The changes can be scored by the popular Apgar system (Table 11.1). Initial scores are for colour and respiratory effort, and if the baby's respiratory and circulatory changes have been completed satis-

Table 11.1. *The Apgar score sheet*[a]

	Score character			
Sign	0	1	2	Score rating
Heart rate	Absent	Slow (below 100)	Over 100	
Respiratory effort	Absent	Slow, irregular	Good crying	
Muscle tone	Limp	Some flexion of extremities	Active motion	
Reflex irritability (response to slapping soles of feet)	No response	Some motion	Cry	
Colour	Blue, pale	Body pink, extremities blue	Completely pink	
Scored at . . .		min after birth . . .	TOTAL: . . .	

[a] After the inventor, Virginia Apgar.

factorily, the muscle tone and reflex responses can be assessed. Most babies will score eight or more after 1 minute. A score of five or less denotes the baby needs immediate treatment and oxygen must be given as a delay could result in death or irreversible brain damage. The causes of a low Apgar score are mostly respiratory and may be due to hypoxia *in utero* due to compression, poor function or premature separation of the placenta. Drugs taken by the mother during labour can also depress respiration.

The repertoire of the newborn's reflexes, such as the grasp, suck, swallow, walking and moro or startle reflex, are checked and should all be normal. Jerky movement or lack of movement may denote damage to the nervous system. Other physical and biochemical details are also examined such as testing for congenital dislocation of the hip and phenylketonuria (PKU), a recessively inherited disorder characterized by a deficiency of an enzyme involved in phenylalanine metabolism and which results in mental retardation.

The gastrointestinal system

During life *in utero* the fetus has not used its alimentary tract for a digestive process, but near term it swallows approximately half a litre of amniotic fluid daily. This tests the swallowing reflex and helps to prepare the intestines for their digestive functions after birth. Probably the most important function of swallowing, however, is concerned with the control of amniotic fluid volume. Excessive amniotic fluid usually co-exists with impaired fetal swallowing. Swallowed water and electrolytes are easily reabsorbed, but before birth the lower part of the gut is filled with the residue of large molecules such as cell debris, bile salts, fatty acid and lanugo (fetal hair). As defaecation *in utero* does not normally occur, this material, called meconium should be passed within 24 hours of birth. If the passage of meconium is delayed, the bile pigments may be reabsorbed, adding to the workload of the liver and increasing the chance of jaundice. Passage of meconium also denotes that the alimentary tract is complete and working properly.

The renal system

At parturition, renal function must undergo a radical change, since the constant supply of water, sodium and other electrolytes through the placenta is lost. Although the placenta is the main organ of excretion, fetal kidneys are functional *in utero* and produce substantial quantities of hypotonic urine. Fetal urine is an important source of amniotic fluid protein and electrolytes, which together provide the osmotic pressure necessary to retain water within the amniotic sac. The total mass and osmotic pressure of the amniotic fluid is regulated by balancing fetal swallowing and fetal micturition.

Other metabolic adaptations and temperature regulation

On birth, the most immediate problem is the energy requirement to cope with ventilation of the lungs, digestion, excretion and temperature regulation. The newborn infant has a larger surface area to body weight ratio than the adult, and consequently there is a greater physical problem in maintaining the body temperature (the fall in body temperature of the newborn is initially extremely rapid). An accelerated build-up of energy reserves (glycogen and fat deposition) in the fetus occurs just before parturition. In the neonate, hepatic glycogen is used to maintain blood glucose until gluconeogenesis can be initiated; skeletal glycogen is used *in situ* to provide energy for muscular activity; cardiac glycogen provides energy to the heart, thereby protecting from anoxia in the immediate postnatal period; and lung glycogen plays some part in the differentiation of the alveolar epithelium. The amount of fat laid down in the fetus varies considerably depending on the species and is high in humans. Free fatty acids are the most important lipids for energy homeostasis of the fetus and are stored, as triglycerides, in various tissues including the liver and the two types of adipose tissue, brown and white. Adult humans have little brown adipose tissue but it is prominent in the newborn. Brown adipose tissue acts as a means of producing heat; ordinary (white) adipose tissue functions mainly in energy homeostasis of the whole body. Catecholamines, released on exposure to cold, stimulate triglyceride breakdown in brown adipose tissue. In white adipose tissue, lipolysis depends on a balance between insulin and catecholamine levels.

THE SIGNIFICANCE OF LOW BIRTHWEIGHT AND ASSOCIATED RISK FACTORS

Low birthweight (2500 g or less) is a major cause of infant mortality, with most deaths occurring in the neonatal period and most being a consequence of inadequate fetal growth. Inadequate fetal growth may result from prematurity (duration of pregnancy less than 37 weeks from the last menstrual period), or poor fetal weight gain for a given duration of pregnancy (IUGR) or both (Chapter 9).

The neonatal mortality rate in affluent societies has dropped substantially since the late 1970s, but it has not been due to a decrease in the incidence of low-weight births. Instead, this mortality decline has been accomplished primarily by improving the chances of survival of low birthweight infants through neonatal intensive care technology. For example, figures for the USA indicate that about 7% of all neonates weigh 2500 g or less, with about 1% below 1500 g. Early this century, two thirds of all infant deaths occurred in the post-neonatal period between 1 and 11 months of age and reflected environmental factors, commonly infections that were often accompanied by fever, diarrhoea and poor nutrition. This figure has approximately halved since then, but mortality due to the sudden infant death syndrome (SIDS) has increased dramatically in western countries, with SIDS now being a major component of mortality between

1 and 11 months (see p. 195). Thus, the contribution of low birthweight to total infant mortality and morbidity is relatively greater now than it was in the past. In developing countries, where infant mortality rates are still unacceptably high, low birthweight neonatal deaths account for 20% to 40% of the total. In addition to increasing the risk of mortality, low birthweight also increases susceptibility to a wide range of illnesses, learning disorders, behavioural problems and iatrogenic complications due to neonatal intensive care interventions. An association between low birthweight and congenital anomalies has also been established.

Because of the recognition that early diagnosis of risk factors decreases perinatal complications, considerable effort has been directed towards the categorization of the multiple maternal, placental and neonatal risk factors. Therefore, a large body of information exists about factors associated with low birthweight, and the population most likely to be 'at risk'. Many of the determinants relate to the helplessness engendered by vicious poverty cycles and include elements of physical and psychological stress. The following gives a simplified overview of the most prominent risk factors which can be simply categorized.

Demographic characteristics

These include low socioeconomic status, low level of education, disadvantaged ethnic minorities, childbearing at extremes of reproductive age and being unmarried. Teenage mothers and those 35 years of age or older have higher rates of low birthweight births than mothers in their twenties or early thirties. The risks among teenage mothers probably reflect other factors such as low socioeconomic status and high levels of stress, rather than young age itself. Unmarried mothers have consistently higher rates of low birthweight, also due to stress and generally low socioeconomic status.

Teenage pregnancies have become a particular worry in the USA. By 18 years of age, 25% of black and 10% of white American women are mothers; of those teenagers, 90% (black) and 40% (white) are unmarried.

Medical risks

These include a poor obstetric history, indicator diseases like diabetes and chronic hypertension, poor nutritional status, poor weight gain during pregnancy due to toxaemia/preeclampsia or iatrogenic prematurity. Other health care risks in this category are an interpregnancy interval of less than 6 months, multiple pregnancies (more than four) and inadequate prenatal care. The risk of low birthweight is substantially reduced among women who initiate prenatal care during the first 3 months of pregnancy.

Behavioural and environmental risks

These include smoking, alcohol and other substance abuse (see below) as well as unintentional exposure to various hazardous chemicals through occupation or environment. Women between the ages of 21 and 36 use drugs more than at any other time in their lives, and this period also coincides with their childbearing years. Intrauterine growth retardation and elevated frequencies of birth defects, associated with nicotine and ethanol consumption in particular, is a serious social problem (Chapter 19). The relationship between industrial pollution and premature birth is more difficult to define and not easy to escape; pollutants do not recognise behavioural boundaries.

Other risks

Other less-well categorized risks include psychosocial and physical stress, endocrine imbalances (particularly progesterone deficiency, Chapter 18), uterine irritability and spontaneous preterm labour.

Many individuals are exposed to several of the above risks concurrently and stresses with hypothalamic–pituitary–adrenal involvement are additive. Consequently, no single approach will solve the low birthweight problem but instead several types of programme need be undertaken simultaneously.

SUDDEN INFANT DEATH SYNDROME, MATERNAL DRUG ABUSE AND IUGR

While the etiology of SIDS remains a matter of speculation, a number of related epidemiological risk factors have been identified. Two identifiable risk groups are infants with intrauterine growth retardation and those from drug-abusing mothers. It is well established that infants born to drug-abusing mothers risk IUGR, immaturity and suffer a 5–10 times increased incidence of SIDS. When fetal growth is inadequate, the immaturity of the infant's respiratory system may result in recurrent episodes of apnoea (cessation of breathing) and chronic hypoxia. Brainstem immaturity, the presence of non-active immature lung surfactants, increased airway resistance and narrowing of the peripheral airways underlies much of the high mortality risk of IUGR infants. Brainstem dysfunction is often reported in apnoea of prematurity where the appropriate neuroregulatory responses to hypercapnia (high circulating carbon dioxide levels) and hypoxia (low circulating oxygen levels), are suboptimal. Sleep apnoea, which occurs during periods of deep sleep and lasts for 10 to 20 seconds, poses a recurrent risk of death or hypoxaemia threatening further neurological damage.

Since fetal breathing movements are essential for the normal development of the lungs and respiratory musculature, any drug that suppresses fetal breathing movements for prolonged periods will adversely affect lung growth and maturation, especially if the fetus is exposed repeatedly to the drug. Sadly, many drugs including nicotine and ethanol have been shown to cause a reduction in the incidence of fetal breathing movements, with the extent being significantly related to the rise in maternal plasma drug level. A Boston study found that neonates exposed to nicotine *in utero* had underdeveloped lungs, with lung function reduced by up to 50% compared to drug-free controls. Other studies have reported apnoea in infants who retained high plasma caffeine concentrations acquired prenatally. The significance of these disparate observations becomes apparent when the association between abnormal ventilatory patterns and future problems, such as SIDS, is considered.

It has been proposed that the severe stress of drug withdrawal in infants born to drug-addicted mothers aggravates episodes of apnoea which may result in SIDS. The major characteristics of the neonatal narcotic withdrawal syndrome are CNS excitation, altered gastrointestinal function and respiratory abnormalities including more frequent apnoeic episodes, longer duration of apnoeic events and more periodic breathing and seizures. Only some symptomatic infants can be effectively treated without medical intervention; the remainder require pharmacological therapy, adding to their already heavy chemical burden. It is ironic that both caffeine and theophylline have been used in the treatment of infants with apnoea. Although these xanthine drugs act as respiratory stimulants, decreasing the incidence of apnoea, their use has obvious drawbacks, not least their highly addictive nature. Unless the chaotic cycle of drug abuse during pregnancy is broken by education and management of the social problems of the at-risk population, an increasing number of infants damaged *in utero* will, if they survive, suffer long-term physiological and behavioural effects.

THE PRIVILEGE OF REPRODUCTION: INDIVIDUAL AND COLLECTIVE RESPONSIBILITY

Public health care and intense education are the most effective preventative measures because, in a fundamental sense, healthy pregnancies begin well before conception. Poor nutrition, smoking, moderate to heavy alcohol use and other stresses apply to people in general and can be dealt with before conception. The majority of common risk factors apply to both sexes but male-mediated deleterious effects on the offspring are less well documented and publicized. Paternal use of noxious agents may exert deleterious effects directly by genetically altering sperm or indirectly by exerting modifying effects on the environment surrounding the maturing spermatids. In the rat, consumption of ethanol, caffeine, morphine and methadone among others, has been shown to cause retardation of fetal growth and reduced neonatal survival. The evidence exists that humans are not exempt from similar effects on reproductive function associated with paternal exposure to noxious chemicals. Common non-adaptive lifestyles are often duplicated, with both mother and father contributing genetic and epigenetic burdens on the development of the fetus. Specific female risks prior to conception are inadequate weight for height, susceptibility to rubella, age, short periods between pregnancies and high parity. Medical and social risks during a pregnancy are also specifically female. Education about reproduction, contraception, pregnancy and the responsibilities of being a parent will alleviate some of the worse cases and eliminate a good proportion of marginal cases of preventable growth retardation. Not many parents knowingly want to handicap their offspring. Additional effort spent in preconception/prenatal education is socially desirable. As unmet needs are greatest among the poor and the young, these two groups are at high risk of having low birthweight infants. Infants weighing 2500 g or less at birth impose a large economic burden on the whole community, initially through the need to provide neonatal care and subsequently in health and disability care of surviving infants. The provision of adequate educational assistance and prenatal care can be covered from cost savings in the care of low birthweight children. Savings through prevention of low birthweight should, in the long-term, more than offset the additional cost of prevention programmes.

Approximately 4% of births in the USA involve some form of congenital abnormality, with a similar incidence in Australia. Infants born at very low weight (below 1500 g) also commonly face serious heart and/or lung impairments. Many of these infants are 'cured' by amazing techniques such as intricate respirators, sensitive monitoring devices and sophisticated surgery. The price, however, is high for the struggling infant, which may be left with iatrogenic conditions (blindness, brain damage or cerebral palsy), and for the community responsible for the expenses involved. Ethical considerations in regard to the growth-retarded fetus are numerous. As arguments and counter arguments to give or withhold life-preserving care rage, the number of potentially healthy but needlessly handicapped conceptuses does not decrease. In my view, if a humane community judges that failure to provide life-preserving treatment to a needy newborn constitutes child neglect, then failure to make adequate preventative provisions to minimize the need for intervention at birth likewise constitutes a failure in basic human rights. It is the children who are punished by social neglect and ignorance. Individuals must understand that the privilege of reproducing and responsibility towards the unborn child are linked. Likewise, a caring society requires adequate resource allocation for the prevention, enforcement and provision of services.

ADAPTATION TO MOTHERHOOD

The birth of a child results in an enormous change for the mother that affects her physical, mental and social state.

Physical recovery: the puerperium

The puerperium is that period in which the genital organs return to their non-pregnant condition. This takes from 6 to 8 weeks, coinciding with the neonatal period of the infant. The main changes may be summarized as follows.

The uterus and placental site

The uterus decreases rapidly in size after the expulsion of the placenta and membranes and continues to involute until, on average, by the 12th puerperal day it is below the level of the symphysis pubis. Breast-feeding helps the process of uterine involution and shape restoration due to the exercising effect of oxytocin. The placental site also contracts rapidly, superficial tissue is shed and a new epithelium soon covers the area. For the first few days after the birth, the mother discharges lochia which consists of blood and tissue debris from the placental site.

In affluent societies, infection rarely occurs (although this was not always so) because the new epithelium forms a barrier, but it may occcur if the tissues are poorly nourished, if labour has been traumatic or if pathogenic organisms present are particularly virulent. In developing countries no one knows how many women die in childbirth or as a result of puerperal complications: most are poor and live in remote areas so their death is considered of little political importance. A typical estimate of the maternal mortality rate, per 100 000 live births, in the developing world is 100–300, compared with 7–15 in developed countries. The shame is that many of these deaths would be preventible with appropriate access to family planning (eliminating unwanted pregnancies) and simple medical care. A large proportion of deaths occur because women are having more pregnancies than is biologically adaptive. Ironically, 40–60% of all married women in the Indian subcontinent, for instance, say that they do not want as many children as they are forced to bear but do not have access to contraceptives. In addition, for every woman that dies many more survive but suffer permanent damage to their health; for example, complications of infection such as pelvic inflammatory disease, fistulae, incontinence and chronic pain are common.

The cervix, vagina and vulva

After delivery the cervix is patulous, bruised and admits 2–3 fingers; this aperture reduces but never regains its pre-pregnant state. The swollen, bruised vagina, however, quickly regains its former elasticity. The perineum, which can be either overstretched or cut, has an excellent blood supply and by the 7th day of the puerperium should be normal.

The urinary tract

Initially a marked diuresis is normal due to cell metabolism following uterine healing and the dramatic fall in blood progesterone concentration. A transient irritation of the pelvic floor muscles and the bladder may also be experienced.

Blood clotting factors

During the first few days following delivery, blood clotting factors are raised, preventing excessive bleeding. However, concomitant with a drop in oestrogen levels, the volume and coagulability of blood are returned to normal by the end of the puerperium.

Mood disturbances: postnatal depression

Mild depression

Pregnancy is associated with an increased incidence of mental illness and mood disturbances in both the immediate antenatal period and postnatally. Social, environmental and biological components underlie this association. Some of the most significant changes in a woman's life occur after childbirth and adaption can be traumatic. Mild depression is an emotional state in which the mother may feel anxious and depressed for short periods but these quickly return to normal (the 'postpartum blues' syndrome). Mild mood swings affect up to 80% of women any time between 2 and 5 days after the birth of their babies. Sleep deprivation, extra demands of the offspring and the re-establishment of the non-pregnant endocrine balance can all induce this condition in otherwise healthy women. The marked changes which occur in circulating concentrations of many hormones during pregnancy and their dramatic fall with parturition has focussed research on hormones, in particular β-endorphin and cortisol. In one theory, mild depression is attributed to insufficient circulating β-endorphin and the consequent deprivation of its analgesic and euphoric properties. Women who show the most dramatic postnatal falls in β-endorphin levels are also the ones most likely to have mood disturbances. Since the placenta produces β-endorphin in increasing amounts throughout pregnancy, the effect of its sudden removal may be similar to that observed in narcotic withdrawal experienced by heroin addicts when their body's supply disappears. There may also be a relationship between a disturbed readjustment of the hypothalamic–pituitary–adrenal axis and postpartum blues.

From the hormonal theory of depression has evolved a reanalysis of the old belief that violent postnatal endocrine fluctuations may be moderated by eating the placenta (many other mammals routinely eat their placentae). Eating the placenta, however, is not a new, western invention since, for thousands of years, Chinese women traditionally were given placental broth to fortify them after labour.

Postnatal depression

Postnatal depression (also referred to as postnatal stress) strikes one in ten mothers and is of considerable severity. Unlike the 'postpartum blues', postnatal depression is not a temporary loss of emotional control and cannot be simply attributed to hormonal instability. In her first pregnancy and childbirth, a woman experiences, possibily, the most significant changes in her life, physically, mentally and sociologically. All these changes are potentially stressful. The first thing a mother loses is sleep. Breast-feeding also places heavy demands on the body and, unless a mother is eating well and drinking plenty of fluids, her body can become depleted of vital nutrients. Surveys indicate that the women most likely to suffer postnatal depression are women who have delayed having children until their careers are established and who have had little contact with other mothers or support from their families. Previously they had been competent, reliable, independent and emotionally stable (in fact the antithesis of the depressed stereotype) and now fall victim to the belief that everything is their responsibility. Social conditioning of 'female role' responsibilities may have an effect here because often there is an underlying sense of guilt felt by the mother of an irritable child which may, for example, be expressed in 'what am I doing wrong, what can I do differently?'. The father, on the other hand, externalizes the problem as 'what's wrong with the baby, what can we do to settle the baby down?'. Interestingly, a research team at Edinburgh University has found a link between premenstrual tension and postnatal depression, but little is understood of either condition. Nutrition, alcohol, smoking and other stresses may be contributing variables. Counsellors urge depressed mothers to ask for help from their partner, family and friends. Ideally, parenting is not a single responsibility but one to be shared among those close to the mother and her baby.

Postnatal psychosis

Psychiatric illness after childbirth has been observed and documented since its description by Hippocrates in 400 BC and is a severe mental illness of, mostly, unknown aetiology which can lead to violence such as suicide, child neglect and abuse. Fortunately, the condition is uncommon but still affects an estimated 1 in every 1000 deliveries. Typically, the mother suffers illness of such severity that hospitalization with full psychiatric control is necessary to assure her safety and/or the safety of her infant.

Postnatal or puerperal psychosis presents itself in two distinct syndromes (together with many cases in which one syndrome moves into or mixes with the other). The first syndrome is an agitated, confused state, often with delirious or schizophrenic symptoms, which begins between 3 to 10 days postpartum. The second syndrome is a depressive reaction with confusion and somatic complaints which usually begins insidiously after the 3rd week postpartum. If nothing is done, beyond protection from self-destruction, 70% of patients get well after a very difficult time. The remaining 30% of women stabilize at various levels of deterioration, or continue downhill to chronic schizophrenia and sometimes even dementia. If a woman who had postpartum psychosis recovers and has another pregnancy, the chance of a recurrence of the disease increases from 0.1% to 25%. The etiology of psychiatric symptoms may be idiopathic or the first indications of an intracranial lesion, infection, metabolic derangement or acquired immunodeficiency syndrome (AIDS).

Bonding and social relations

It is easy to see how social interaction within the family could be influenced by the baby's particular characteristics. Newborns differ in their motor activity, irritability and responsiveness, but they all engage in primitive social relations. Very soon the baby will recognise the facial and olfactory characteristics of persons mostly in contact with him or her, especially if the most constant attendant is the mother and she is breast-feeding. A steroid molecule (a possible progenitor of 4-α-androsterone) in human sweat concentrates in the mammary glands and acts as a pheromone aiding speedy mother–infant bonding. This compound has the property of altering mood much as sedatives do and may be a natural tranquillizer. Thus the process of socialization is aided by the effect of hormones, including pheromones, which gives the offspring a concept of self derived from the familial smell. Experiments demonstrate that non-human primate females reared in a socially-deprived environment make poor, aggressive and rejecting mothers. However, human bonding is very complex, utilizing many variables, but there may be possible parallels linking childhood distress and parental care. Simple needs for food, closeness and sucking comfort should be satisfied. Failure to satisfy these simple needs can lead to distress as the baby does not understand reasons, only needs. Learning appears to depend on the timing of reward and the repetition of the stimulus. Early learning is also linked to conditioning and may set subsequent behavioural patterns.

The most obvious method that the newborn uses to communicate with its social environment is crying. Quite young babies display different cries depending on whether the crying is stimulated by hunger, pain or anger. Communication also occurs when a parent responds to a crying baby; for instance, the mother's breasts respond by gushing milk when her baby has given a hunger cry. Mothers quickly learn to give their babies the stimulation that they seem to respond to. There are several effective ways to quiet a crying baby besides feeding. Most of these relate to the *in utero* experience already described and include rhythmic movement, auditory stimulation and swaddling.

In summary, behaviours relating to crying, sucking, smiling, clinging, eye contact, sound and movement can be seen as important mechanisms fostering attachment. They are also mechanisms signalling the baby's genuine need for parental care. Generally the parents are the most significant people in a baby's environment, with the mother usually being the primary attachment figure. The parents' interaction with each other and with the baby will form the basis of its social, emotional and personality development. The quality of these early interpersonal interchanges is therefore crucial.

General references

Behrman, R. E. (ed.) (1985). *Preventing Low Birthweight*. Institute of Medicine, US National Academy Press, Washington DC.

Bock, G. R. & Whelan, J. (eds.) (1991). *The Childhood Environment and Adult Disease*. John Wiley, West Sussex, UK.

Callan, N. A. & Witter, F. R. (1990). Intrauterine growth retardation: characteristics, risk factors and gestational age. *International Journal of Gynecology & Obstetrics*, **33**, 215–220.

Fleming, A. S., Steiner, M. & Anderson, V. (1987). Hormonal and attitudinal correlates of maternal behaviour during the early postpartum period in first-time mothers. *Journal of Reproductive & Infant Psychology*, **5**, 193–205.

Jones, C. T. (1991). Control of glucose metabolism in the perinatal period. *Journal of Developmental Physiology*, **15**, 81–89.

Kandall, S. & Gaines, J. (1991). Maternal substance use and subsequent sudden infant death syndrome (SIDS) in offspring: review. *Neurotoxicology & Teratology*, **13**, 235–240.

Maris, A. (1988). *An 'Epidemic' of Adolescent Pregnancy? Some Historical and Policy Considerations*. Oxford University Press, Oxford.

Mirmiran, M., Kok, J. H., Boer, K. & Wolf, H. (1992). Perinatal development of human circadian rhythms: role of the foetal biological clock. *Neuroscience & Biobehavioural Reviews*, **16**, 371–378.

Ogunkeye, O. O. & Otubu, J. A. (1990). Tests of foetal lung maturity, a review. *African Journal of Medicine & Medical Sciences*, **19**, 65–70.

Pillai, M. (1991). Behavioural studies of the human fetus. *Contemporary Reviews in Obstetrics & Gynecology*, **3**, 139–147.

Pollard, I. & Smallshaw, J. (1988). Male mediated caffeine effects over two generations of rats. *Journal of Developmental Physiology*, **100**, 271–281.

Porter, R. H. (1991). Human reproduction and the mother infant relationship: the role of odors. In *Smell and Taste in Health and Disease*, ed. T. V. Getchell, pp. 429–442. Raven Press, New York.

Smith, R. & Thomson, M. (1991). Neuroendocrinology of the hypothalamo–pituitary–adrenal axis in pregnancy and the puerperium. *Baillière's Clinical Endocrinology & Metabolism*, **5**, 167–186.

Sneddon, J. & Kerry, R. J. (1984). Puerperal psychosis: a suggested treatment model. *American Journal of Social Psychiatry*, **4**, 30–34.

Thurston, F. E. & Roberts, S. L. (1991). Environmental noise and fetal hearing. *Journal of the Tennessee Medical Association*, **84**, 9–12.

Tye, K., Pollard, I., Karlsson, L., Scheibner, V. & Tye, G. (1993). Caffeine exposure *in utero* increases the incidence of apnea in adult rats. *Reproductive Toxicology*, **7**, 449–452.

12 Decline in male reproduction and the menopause

There is a natural loss of fertility due to ageing in both sexes: the menopause in women and the more gradual reproductive decline in men. Ageing, as with all aspects of reproduction, has both its genetic and environmental components. This is particularly true for the menopause where a genetically influenced time frame is strongly modulated by the environment.

The ageing process in both males and females can be accelerated by the following influences.

- Cytotoxic agents and xenobiotics (that is, biologically foreign chemicals). For example, organophosphorus pesticides, certain anticancer drugs and components of cigarette smoke. Epidemiological evidence suggests that cigarette smoking has an anti-oestrogenic effect in women, perhaps as a result of increased adrenal stimulation. On the other hand, aromatic hydrocarbons in cigarette smoke may accelerate the ageing of oocytes.
- Occupational stress. For example noisy environments and the sterilizing effects of ionizing radiation.
- Nutritional stress. Nourishment is fundamental as it relates closely to the basic requirements of life and is closely correlated with fertility. Malnutrition increases the frequency, severity and duration of other stresses, such as infectious diseases.
- Disease. For example, gonorrhoea and the delayed effects of mumps on the reproductive system. It is often difficult to distinguish between the direct effects of disease and the consequences of drug treatment used to combat disease.
- Immunological disorders. For example, 3–8% of men produce antibodies to their own sperm. An autoimmune disorder resulting from antibody production against the gonadotrophin receptors has also been identified.
- Past endocrine experience. This clearly affects the status of the testis and ovary, as well as the hypothalamus, at any age. For example, in woman, oestrogen treatment, even within the physiologically normal

range, accelerates age-associated changes in reproductive function. This also holds true for steroids of adrenal origin where prolonged increased secretion of glucocorticosteroids accelerates age-related changes in men and women.

Exposure to steroids during sensitive periods in early development modulates the response of their target tissues to subsequent hormonal stimulation. Some of the variability within a population may relate to important events which include uterine position, exposure *in utero* to steroids and drugs possessing steroidal actions and ingestion of naturally occurring phytooestrogens during juvenile development (Chapter 16). Other diverse environmental phenomena inducing psychological stress or anxiety may also cause steroid hormone imbalances. Unfortunately, the origin of extraneous influences on steroid biochemistry in humans is harder to define and is, therefore, much more difficult to prevent.

Whilst many factors potentially shorten fertile life there is no known way of extending it significantly. In women, for instance, long-term inhibition of ovulation resulting from serial pregnancies or steroidal contraception has not been effective. By contrast, in rodents, natural ageing and follicle atrition is slowed by both hypophysectomy and underfeeding. The reason for the discrepancy between women and female rodents is not clear but may reflect differences in the reproductive strategy of short-lived species. These possess comparatively few follicles which disappear rapidly, in contrast with the parsimonious utilization of a larger store in longer-lived species.

NEUROENDOCRINE CONTROL OF SENESCENCE: ANIMAL EXPERIMENTS

The search for perpetual youth has occupied philosophers and scientists since time immemorial. Research directed at finding a long-lasting physiological approach to rejuvenation of individuals has been vigorously pursued without success. It has, however, contributed to our understanding of the ageing process. Transplantation experiments, mostly in rodents, have demonstrated that the development of deficiencies in the hypothalamus plays an important role in reproductive ageing and that the age-related decline in gonadal function is not the sole mediator of reproductive senescence. When the ovaries of senescent non-cycling rats are transplanted into young regularly cycling ovariectomized hosts, many of these young rats exhibited oestrous cycles. Similar results were obtained when the pituitaries of old non-cycling rats were grafted into young hypophysectomized rats. This demonstrated that neither the ovaries nor the pituitary are primarily responsible for the loss of cyclicity. Fetal hypothalamic tissue, grafted onto the third ventricle, can correct pituitary and gonadal deficiencies and so restore sexual function and fertility in impotent male and non-cycling female rats. These experiments supply further evidence that subnormal hypothalamic function is the primary cause of reproductive failure in aged rats.

The success of the hypothalamic grafting research (although limited) created much interest in geriatric medicine and subsequent research focussed on neurotransmitter function. It was found that in the hypothalamus of old rodents neuronal activity was impaired and nuclear binding sites for oestradiol and noradrenaline concentration were reduced. In view of the role of nuclear receptors in oestrogen receptor-mediated functions, this age-related change could underlie many aspects of the altered LH regulation. The preovulatory LH surge becomes progressively impaired with age, so that eventually LH surges are no longer readily induced by oestradiol. Strong evidence has accumulated for the hypothesis that impairment of the pathways afferent to the GnRH cells are responsible for fertility loss, since administration of drugs elevating brain catecholamines increases the average lifespan of rats and mice and promotes fertility. The decline in hypothalamic catecholamines with age appears to be due to loss or damage to neurons by metabolic, environmental and genetic influences. In truth, however, multiple neurotransmitter functions deteriorate with age and several areas within the hypothalamus, particularly the median eminence, may contain imbalances of neurotransmitter concentrations.

The relation of the hypothalamus to reproductive senescence in humans in unknown at present; however, hypothalamic function appears to be secondary to that of the ovaries in the maintenance of menstrual cycles. A few years prior to the menopause, menstrual cycles tend to become irregular, ovarian hormones are reduced and gonadotrophin secretion rises suggesting that in women ovarian failure is the main consequence of reproductive ageing. Fundamental differences in mechanism may again reflect species differences (or inadequate understanding of all the factors involved) and the rodent is probably not a suitable species to study as a model representative of humans. Of considerable interest are new studies demonstrating that with improved conditions of husbandry, primates live longer and may demonstrate a phenomenon of menopause much like that in women. Hot flushes were observed in postmenopausal rhesus monkeys and stumptail macaques as well as in young female monkeys after ovariectomy. Rhesus monkeys and chimpanzees manifest declining fertility during midlife before cessation of menstrual cycles. Menopause occurs in chimpanzees after the age of 40 years, that is, approaching the same age as in women. The similarity in reproductive physiology between humans and chimpanzees may make this primate a good model for studies of ageing and will certainly provide a vast improvement over the rodent.

Little is known of hypothalamic function in ageing men. Free testosterone secretion and spermatogenesis decrease with age, but sufficient testosterone and sperm are produced by the testes of healthy older men to maintain fertility well into the eighties. The immune system can also influence the neuroendocrine system. Immune competence and neuroendocrine function both decline with age but little is known of their interactions during ageing.

REPRODUCTIVE SENESCENCE IN MEN

Ageing in men is much more subtle than it is in women, as men often remain fertile into old age. They usually do not suffer abrupt changes in fertility or other sexual functions equivalent to the midlife climacteric experienced by women. There is no male parallel to the irreversible loss of fertility resulting from almost complete exhaustion of ovarian oocytes and the abrupt decrease in circulating gonadal steroids that occurs at menopause. However, during and after midlife, men can suffer from a range of reproductive impairments such that only a minority retain full sexual functions throughout their lifespan. Both men and women, provided their health is good, can remain sexually active into their eighties despite loss of fertility.

A major difficulty in analysing the decline in male reproductive function during midlife is that the changes are not clearly defined and may be a mixture of the following.

- Eugenic, that is, general features of ageing that are not related to any definable disease
- Pathogenic, that is, diseases of ageing or related to lifestyle (for example, heart disease, diabetes or the effects of drug abuse)
- Psychogenic, for example, depression, loss of spouse.

Sexual behaviour

Sexual performance and rates of sexual activity decline gradually (typically, less than 35% of men older than 80 years are sexually active). The aetiology of the decline in sexual behaviour is complex but, as mentioned above, it can often be linked to age-related disease. It is estimated that impotence occurs in 30 to 60% of adult diabetics and is a common side effect of alcoholism, as well as of drugs such as those that are used to treat hypertension. Depression may also have a significant effect.

There does not seem to be a clear link between the extent of sexual activity and levels of blood testosterone in old men or during the mid-life period when sexual activity begins to decline. In some men, testosterone decreases to castrate levels but there is a wide range of values and those of older and young men overlap. Testosterone is sensitive to environmental cues and is elevated as a consequence of sexual arousal and depressed as a consequence of stress. Deficiencies in testosterone are not usually relevant when assigning causes of impotence and alleged 'cures' such as extracts of animal testicles are completely ineffective. Many other variables such as experience of sex are important in the human male and sexual behaviour only rarely reflects circulating hormonal levels.

The testis

Leydig cell numbers commonly decrease with advancing age in some men but this is not universal and a large loss of Leydig cells can also occur during malignant disease and other protracted illnesses. In one study, the average number of Leydig cells was decreased by 44% in a group of 50–79-year-old men when compared with another group aged 20–48 years. Baseline and peak testosterone levels are generally lower in older men. Additionally, the tissue availability of testosterone begins to decrease because the sex hormone-binding globulin titre and the fraction of testosterone it binds increases with age, thus reducing the free hormone available to the target tissues. The testis produces new germ cells, to a variable extent, from puberty to the end of life and it is estimated that daily sperm production is more than 50% lower in older men. Paternity has been documented at least to the age of 95 years (about 40 years beyond the record in women), but there is an age-related reduction in the number of normal sperm, the daily production of sperm and in the conception rate. It is important to note in this context that paternal-mediated contributions to birth defects are higher (as gauged by the fetal death rate of the offspring, incidence of Down's syndrome and other dominant mutations) from older fathers who, therefore, have a personal responsibility toward their unborn children. Society too has a responsibility in teaching its citizens that creating new individuals is a privilege, not a right, and parental age should be considered when planning to reproduce.

The number of Sertoli cells in the human testis also exhibits an age-related decline analogous to that in Leydig cells. One postmortem study of men aged between 20 and 85 years who were apparently in good health prior to sudden death showed that young adults (20 to 28 years) had approximately 500 million Sertoli cells/testis, whilst older adults (50 to 85 years) had a mean of 300 million cells/testis – a significant decline. There is also a significant correlation between Sertoli cell number and daily sperm production. Sertoli cell alterations due to ageing or to injury are similar no matter what the cause and can be recognised morphologically. Sertoli cell damage due to disease or stress in younger men is reversible, unless chronic or very severe, whilst in older men such damage may not be reversible.

REPRODUCTIVE SENESCENCE IN WOMEN

Females lose the capacity to ovulate well before the maximum lifespan is reached and the traditional marker of reproductive senescence is the menopause, characterized by the loss of menstrual cycles. This occurs at a median age of 50–51 years but is earlier in stressful environments. Interestingly, the age at menopause has always been approximately the same. Aristotle, for example, notes that 'for the most part fifty marks the limit' of womens' reproductive capacity. The average maternal age at last birth is 40 years across the world, and few women anywhere are fertile after 50 years. The oldest fully documented birth was to Mrs Ruth Alice Kistler of the USA who (as recorded in the *Guinness Book of Records*) gave birth to a normal daughter in 1956 at the age of 57 years and 129 days. After the age of 30 years, fertility in women is known to decline from its peak in the early twenties and this fall precedes the age when menstrual cycles become variable. Normally, irregular menstrual cycles begin to reappear in the early forties and mark the approach of menopause. This 10-year period of change is known as the climacteric; taken from the Greek word *klimacter*, meaning critical record. The word menopause is recent and was introduced, in 1899, to define the event of menstrual cessation. Interestingly, the 25-year-long epoch of maximum fertility is flanked by periods of reduced fertility: the adolescent sterility and peri-menopausal sterility. The problem of the confounding effects of disease is less important in studies of reproductive decline in women than it is in men because the loss of fertility during midlife in women often occurs before any major age-related pathology becomes prominent.

Most studies report that 50% of first-trimester abortions are chromosomally abnormal in women of all ages and 15–20% of pregnancies are miscarried. Early pregnancy wastage, however, increases substantially in women in their thirties and is over 50% after age 40 (this figure is higher in countries with substandard hygiene, poor nutrition and a high rate of disease). The incidence of genetic abnormality in stillborn infants increases from 1.3% in mothers under 20 years to 8.8% in mothers aged 30 to 39 years and 35.8% in mothers over age 40 years. Trisomy of chromosome 21 (Down's syndrome) is the most common abnormal karyotype in liveborn infants.

Because the risk of a chromosomally-abnormal child increases with maternal age, many pregnant women of age 35 or older choose to have a prenatal genetic test. Testing maintains the option of therapeutic abortion in cases where the results reveal a genetic abnormality. Amniocentesis, the aspiration and culture of amniotic fluid cells during the second trimester, has been successfully used since the 1970s. The results of an amniocentesis are usually not available before the 20th week of gestation. The development, in the 1980s, of chorionic villus sampling (the transcervical aspiration and cytogenetic examination of chorionic villus cells) has provided an important alternative. Chorionic villus sampling provides genetic diagnosis during the first trimester, providing the option of a safer, less traumatic, first-trimester therapeutic abortion, but it may also be associated with an increased risk of spontaneous abortion. In experienced medical centres, however, both procedures are reported to be safe and accurate alternatives.

Chromosomal aberration is also a major cause of the high abortion rate seen after assisted reproduction in older women. The age-related decline in fertility can be corrected by ovum donation in conjunction with hormone therapy because the uterus in healthy women over 40 can adequately sustain pregnancies if endometrial maturation is artificially manipulated in the recipients. The ability to sustain a pregnancy to successful parturition, even in the absence of chromosomal abnormalities, also declines as maternal age increases due to a decline in host receptivity. Animal experiments, using embryos transplanted from young donors to young and old recipients, can differentiate the effects of ovum quality from uterine and other factors. It was found that older recipients consistently produced fewer live offspring than did younger recipients.

Sexual behaviour

The menstrual cycle is subject to modulation by social cues as is sexual arousal in men. For example, the frequency of sexual contact with men influences the variability of menstrual cycles in young as well as premenopausal women. Both long and short cycles are associated with reduced fertility, and social/sexual contact with men serves to increase the likelihood of cycles being ovulatory (Chapter 6). Human females have the potential for sexual arousal and orgasm at all ages, from childhood through the later postmenopausal years, because this is an essential aspect of continued bonding. Despite low oestrogen and androgen levels, sexual arousal in postmenopausal women may be only slightly below that of young women. The decrease of sex steroids at menopause often has consequences that can influence sexual behaviour; for example, intercourse may become uncomfortable or painful (dyspareunia) because of vaginal atrophy, back pain associated with osteoporosis and other pathophysiological changes. As in many other aspects of ageing, individuals differ strikingly in the severity of vaginal atrophy, which is not experienced by all women and may depend on individual differences in sexual activity, diet and genotype. Local application of oestrogen-containing creams can alleviate symptoms of vaginal atrophy.

The ovary

The reproductive decline is accompanied by a decrease in oocyte (follicle) number and an increase in gonadotrophin secretion with a preference of FSH over LH. Reduced oocyte number is the predominant cause of reproductive senescence but age at menopause is more sensitive to increases in the rate of atresia than to decreases in oocyte number. Loss of oocytes, due to removal of ovarian tissue, does not significantly alter the age of menopause as fewer than 0.001% of the oocytes originally in the human ovary reach maturity (the maximum number of expected ovulations is estimated to be less than 400). From approximately 7 million germ cells which are formed in the developing ovary, only approximately 2 million remain at birth and only 300 000 by menarchial age. This reduction occurs without ovulation and is attributable to follicle atresia. One view is that atresia is the dominant process of the ovary and the loss of follicles to atresia is the etiology of menopause. As the follicular pool decreases, the interstitial tissue continues to secrete steroids, now under increased gonadotrophic stimulation. The interstitial tissue secretes testosterone and androstenedione which may be aromatized in peripheral tissues to oestradiol and oestrone. Elevated androgen, or reduced oestrogen, levels impede the capacity for normal follicular development and may further accelerate follicular atresia. The CNS plays a significant role in the pattern of menstrual cycle irregularity during the menopausal transition because, with increasing age, the number of hypothalamic neurons decreases, resulting in changed GnRH pulse frequency. The pulsatile release of gonadotrophins tends to be slightly slower, which may impair follicular development, especially as a minimum (threshold) number of follicles is required to secrete sufficient oestrogen to trigger the preovulatory surge of gonadotrophins.

Despite individual differences, there is a general pattern of age-related variation in cycle length reflecting erratic follicle development. After age 40 years, abnormalities are seen in 30–50% of cycles, but many women still ovulate regularly. The decreasing follicle pool, with reduction in oestradiol and inhibin concentrations, brings about the selective increase in FSH. Higher tonic FSH levels may alter the velocity of follicular growth and number of ovulations since an age-related increase in dizygotic twinning has been documented. As FSH rises further, accelerated follicular growth may impair fertility by inducing anovulatory cycles with accompanying inadequate luteal phases. Typically, the length of cycles becomes increasingly variable as menopause is approached. During the transition to menopause (that is, the climacteric), cycles are shortened because of reduced follicular phases or are lengthened because of the failure of ovulation. Long, short and 'normal' (near to modal) cycles can be interspersed and ovulation may still occur immediately before the menopause.

Factors such as nutrition, race, parity and smoking influence age at menopause by, at most, 3 years either side of the normal median age. Some women, however, have a premature menopause, before the age of 40, and in a few below age 30. The incidence of premature ovarian failure before 40 years of age is estimated to occur in 1 to 3% of the general population. Ovarian failure is due to an acceleration of the naturally occurring process of atresia of oocytes which may be associated with different etiologies. Specifically, ovarian failure can occur as a result of genetic (for example, alteration in the number of X chromosomes), enzymatic (for example, defective maturation of follicles), endocrine (for example, abnormal forms of gonadotrophins) and physical (for example, environmental pollution) defects, or because of autoimmune disorders. A significant proportion (13–66%) of women with premature ovarian failure also have a variety of autoimmune diseases, suggesting that immunological defects may be associated with ovarian tissue dysfunction. Whether premature menopause was underdiagnosed prior to IVF programmes but is now being fully investigated or whether its prevalence represents a sinister effect of environmental pollution compromising the immune system remains obscure.

Possible reasons for menopause before the end of life

There are several hypotheses regarding the evolutionary advantage of the menopause.

(a) Menopause is advantageous in a long-lived species such as humans in that it ensures that defective ova are not fertilized. By contrast, sperm are produced throughout life and are not, apparently, subject to as great a risk of degeneration.

(b) Menopause ensures that children are born to mothers likely to live long enough to rear them. The human child is dependent on its mother for many years and, in the past, repeated at-risk childbearing, lack of hygiene, poor nutrition and little medical knowledge meant that women risked death with each confinement, placing themselves and the older siblings at risk.

(c) From a social perspective, menopause may have an adaptive advantage enabling some older women to pass on acquired wisdom and skills. By eliminating the special hazards of late childbirth, there was an improved chance of a few women living on and becoming wise, to the benefit of their tribe and race. For our hunter–gatherer ancestors, the only means of passing on knowledge was by example, demonstration or word of mouth, and elders of a tribe were given special status in return for knowledge and advice. This still holds true for many indigenous peoples, including the Australian aborigines. Margaret Mead, the American anthropologist, made this point in her book *Male and Female*. She observed that, in many societies, both prepubertal girls and postmenopausal women were treated as men and traditional female roles were dispensed with. Among the Tamils of Jaffna, for example, the climacteric is associated with positive gains in prestige, freedom, leisure and authority. The older woman, still healthy but freed from child-bearing, is in a good position to influence decisions of significance to her family and the broader community.

(d) According to a sociobiological argument, the menopause may have evolved under conditions which favoured the investment of the limited energy of older females in the nurture and protection of the offspring of kin, rather than in their own children. This would be a means of favouring their own genes carried by their grandchildren. This hypothesis is attractive since there is good evidence that life expectancy of even our earliest ancestors was sufficient for the co-existence of three successive generations. The maximum lifespan potential for the human being is estimated to be about 90 years and, at least for the last 100 000 years, some people survived to such an old age.

An alternative hypothesis is that during human evolution longevity has become extended and so the ovaries become inactive long before death. Whatever the original biological advantage of menopause, the present situation where a women contributes effectively to all aspects of society and is sexually receptive throughout her lifespan must be adaptive by strengthening the evolution of human social systems.

POSTMENOPAUSAL PHENOMENA

Menopause, like the menarche, is a definite developmental landmark associated with psychological, endocrine and biochemical adjustment. The climacteric is a time of spiritual as well as physical change, and provides a challenge by the opening up of new opportunities; it can be an exciting time of stock taking and making fresh decisions.

The most common postmenopausal physiologic phenomena can be divided into two general categories.

(a) Those that are specifically related to the castrate level of oestradiol and progesterone (ovariprival)
(b) Those that occur in the same age group of men but without a clear linkage to ovariant steroids, that is, age-correlated but not ovariprival.

Ovariprival menopausal changes

In general, the tissues that flourish at puberty, when ovarian steroid levels rise, will later regress after menopause. Menarche and menopause seem to be almost mirror images of each other.

The major ovariant-dependent changes can be summarized under the following categories.

(a) Hot flushes or flashes (see next section).
(b) Atrophy of genitourinary tract. After menopause, the ovariant production of oestradiol and progesterone virtually ceases and blood levels of these hormones approach that of the castrate. Ovaries may be completely atrophic or the interstitial (stromal) cells may secrete androgens. In addition, in postmenopausal women, there is at least a 50% increased conversion of adrenal dehydroepiandrosterone to oestrone and this, to some extent, compensates for the decline in gonadal hormone production with age.

(c) Atrophy of breast epithelial glands and ducts. This is usually not noticeable as the glandular loss is replaced by fat tissue.
(d) Autonomic nervous system instability. Because of the close relationship between the endocrine and autonomic nervous systems, many menopausal symptoms, such as headache, shoulder stiffness, joint pain, dizziness, ringing in the ears, crawling skin (pin and needles), short-term memory loss, lassitude and other unspecified aches and pains can be attributed to endocrine-induced autonomic instability and vasomotor disturbance.
(e) Regression of oestrogen-dependent benign tumours. The regression of uterine fibroids and fibrosis of the breast reflects oestrogen withdrawal: a positive aspect of the menopause.

There is no doubt that the climacteric can be a period of exceptional somatic stress reminiscent of the menarche; however, in most women the body's homeostatic mechanism does eventually readjust to its new endocrine set points. Many healthy women welcome the climacteric changes as powerful harbingers representing a new beginning. The climacteric syndrome can be culturally determined and its severity modulated by other pathological, environmental, socioeconomic and psychological factors. The eventual decline of gonadotrophin secretion in some postmenopausal women may be a consequence of a range of pathophysiological disorders rather than an intrinsic limitation in the capacity to produce GnRH. For example, cell or hormone receptor loss and fibrosis or a fundamental effect such as a missing metabolic signal may be involved. On the other hand, elevated secretion of LH and FSH may continue, undiminished, in some women even at 90 to 100 years of age.

Hot flushes

Hot flushes (flashes) and/or night sweats are experienced around menopause by an estimated 40% and 75% of western women. A hot flush involves a co-ordinated thermogenic episode that results in peripheral vasomotor dilatation and sweating in conjunction with a transient elevation of plasma GnRH, LH and catecholamines. Flushes can be extremely troublesome, sleep-disturbing events, often recurring at peak frequency every few minutes. In many women hot flushes cease spontaneously or gradually diminish, often within 5 years after menopause, although some suffer flushing episodes for decades, even to the end of their lives. Interestingly, surveys suggest that the hot flush is uncommon in most

Japanese women. In one survey only 20% experienced a hot flush at some time, and of those who reported positively, it presented few difficulties. Whether this difference reflects a cultural stoicism or a lifestyle amelioration is not clear, although most Japanese women regard menopausal symptoms as normal and transitional, heralding the latter part of life.

The diurnal hot flush frequency in some women has a morning peak (04.00–10.00 hours) and in others an evening peak (16.00–22.00 hours). A 12-hour periodicity also occurs. The physiological mechanism of the hot flush has yet to be elucidated, but it is likely that a common neural site regulates both the pulsatile release of GnRH and initiates hot flushes. Stress increases the frequency of hot flushes but does not precipitate it. The preoptic nucleus of the hypothalamus is involved in thermoregulation; this area also includes high concentrations of oestrogen-responsive neurons as well as GnRH-containing neurons. Since the flush is a vasomotor disturbance and very often coincides with stress, it may be mediated through catecholaminergic/opioidergic mechanisms which are important in the cyclic and pulsatile release of LH and the control of body temperature. The circadian periodicity of both plasma ACTH and glucocorticoids is well documented. Cortisol, for instance, is secreted in a pulsatile fashion with short secretory episodes normally occurring in a 24-hour period. The initial phase of sleep is usually associated with an absence of cortisol secretory episodes. Some hours after the onset of sleep, however, secretory activity begins with plasma cortisol levels reaching their highest concentrations about the time of waking. It is not known whether cortisol and/or proopiomelanocortin-derived hormones (Chapter 13) are involved in the hot flush. It may be noted that oscillations, pulses and rhythms are ubiquitous in biological systems, and during a period of massive physiological adjustment various normally separate hormonal oscillations may merge. The absence of hot flushes in prepubertal girls, despite very low plasma oestradiol levels (equivalent to those in postmenopausal women) suggests that the capacity for ovariprival hot flushes is acquired at puberty and is a maturational effect induced in the CNS by ovarian steroids.

Elevations of LH are not a cause of the vasomotor flush since flushes persist after blockade of LH release with GnRH analogues and since LH patterns are identical in postmenopausal women both with and without flushes. Temperature elevations during the flushes are correlated with much greater elevation of plasma GnRH than in asymptomatic women. Hot flushes are prominent a few days after ovariectomy in younger women and are also natural in some other primates. In the west, much effort is expended in alleviating the worst symptoms of the menopause by education and the recommendation of lifestyles which may alleviate some inevitable symptoms, as well as by the use of hormone replacement therapy.

Postmenopausal osteoporosis

There is substantial evidence linking oestrogen deficiency to postmenopausal osteoporosis. Oestrogen loss is followed by a significant reduction in intestinal absorption of calcium, and changed calcium metabolism may lead to increased bone catabolism, brittleness and weakness (the Dowager's hump in the elderly). Numerous studies demonstrate that, in the absence of oestrogen replacement, postmenopausal women suffering hip fractures have less endogenous free plasma oestradiol and testosterone than do non-fractured controls. Oestrogen replacement appears to reduce postmenopausal bone loss, decreasing the risk of fractures. However, increased porosity of bone will generally manifest itself sooner in individuals whose skeletons, at maturity, were light-weight, whose diet has been deficient in calcium and who have been sedentary. Excessive consumption of alcohol and caffeine and cigarette smoking accelerate calcium excretion and, for this reason, all have been regarded as lifestyle risk factors for osteoporosis. However, moderate alcohol intake ranging from three to six drinks weekly significantly increases (by stimulating aromatase activity) circulating oestrogen levels in postmenopausal women. Whether this increased oestradiol level is associated with beneficial changes in bone density is not known. Significant age-related bone loss is also observed in men whose much lower plasma oestradiol levels remain unchanged. Moreover, bone loss can be detected 5 to 10 years before menopause. Such observations indicate that many factors contribute to the aetiology of bone density loss in addition to postmenopausal oestrogen deficiency. Particularly relevant in this context is a sessile lifestyle coupled with inadequate nutrition. Physical activity is necessary to retain a strong musculoskeletal frame because weak muscles do not demand strong bones. A basic law of nature states that what is used remains or even hypertrophies, whilst what is not used becomes atrophied or even lost. It is a physiological nonsense to maintain that a well developed skeleton can be retained whilst the skeletal muscles atrophy. Evidence is now accumulating which demonstrates that moderate exercise not only prevents calcified bone loss but remineralizes decalcified bone. Quite apart from the relief from the symptoms of osteoporosis or the need for oestrogen replacement, exercise in elderly people (as in the young) works positively on the physical and emotional well being. Physically fit people are better able to cope with the full range of age-related deterioration. Additionally, an active lifestyle has to be supported by good nutrition, including calcium-rich foods, in order to supply the necessary raw materials for skeletal maintenance and efficient muscle function.

SEX STEROID HORMONE REPLACEMENT THERAPY

In a proportion of women, behavioural changes occur during the climacteric and postmenopause period which can include depression, tension, anxiety and mental confusion. These are probably linked to the need for psychological adjustment to a changing role/status or to worries about sexual attractiveness. Added to the psychological burden are the prominent physiological changes such as a decrease in or loss of libido, pain at coitus, hot flushes, fatigue, and osteoporosis in those with a genetic predisposition. All, or most, of these symptoms may be arrested by oestrogen replacement therapy (hence the basis for its present widespread use). Another case presented for postmenopausal replacement therapy relates to the belief that oestrogens appear, through their effects on lipid metabolism and inhibition of atherosclerotic plaque formation, to protect women from cardiovascular disease.

On the other hand, postmenopausal oestrogen replacement is associated with an increased risk of endometrial cancer and a possible increased risk of breast cancer and gall bladder disease. A positive dose–response relationship has been established between the risk of endometrial malignancy and both the duration of use and the strength of the medication used. That systemic oestrogens are associated with an increased risk of uterine cancer should not be surprising because of the effect of oestrogens in Müllerian tissues where they may either initiate or promote growth. Oestrogen, for example, promotes the growth of fibroid uterine cysts. Women on steroid replacement therapy may have many times the amount of oestrogen in their circulation than they had before the menopause. More modern transcutaneous methods of administration lower the level of oestrogen in the bloodstream, by-pass the liver and thus the need for biodegradation, and may be a successful method of replacing orally administered oestrogens. Since there is no uniform agreement about the safety and advisability of using sex steroid replacement therapy in older women, it seems that the decision is up to the physician and, ultimately, the individual, who must assess the validity of the counterindications. Meanwhile, steroid replacement therapy is increasingly prescribed as oestrogen alone, oestrogen/progesterone combinations, or with the addition of androgens. Testosterone alone has also been prescribed and reported as effective in normalizing decreased libido.

Like all things in life, the risks must be balanced against the benefits. As far as the menopause is concerned not all women have symptoms, nor do all benefit from oestrogen therapy. Those with symptoms (hot flushes, nervousness, joint and bone osteoporosis, atrophic vaginitis and dyspareunia) find that many symptoms are dramatically alleviated by oestrogen therapy. Steroid-induced resumption of menstruation, however, may counterbalance some of these positive effects. It may also be worth considering testosterone replacement, in physiological not pharmacological doses, to alleviate problems associated with libido or weak muscles. On balance, it may turn out that the solution, at least for large numbers of women, is through the active promotion of a healthy lifestyle in which stress is minimized, giving the body reserves to deal with the necessary physiological and psychological adjustments. In the final analysis, the climacteric is a normal part of growing older, it is not a deficiency disease. Exogenous steroid treatment may even interfere with the smooth establishment of the default system in which, with the help of the adrenal glands, a steroidal balance is re-established.

General references

Cohen, I. & Speroff, L. (1991). Premature ovarian failure: update review. *Obstetrical & Gynecological Survey*, **46**, 156–162.

Gavaler, J. S. & van Thiel, D. H. (1992). The association between moderate alcohol beverage consumption and serum estradiol and testosterone levels in normal postmenopausal women: relationship to the literature. *Alcoholism: Clinical & Experimental Research*, **16**, 87–92.

Heckerling, P. S. & Verp, M. S. (1991). Amniocentesis or chorionic villus sampling for prenatal genetic testing: a decision analysis. *Journal of Clinical Epidemiology*, **44**, 657–670.

Korenman, S. G. (ed.) (1990). *The Menopause: Biological and Clinical Consequences of Ovarian Failure, Evolution and Management*. Serono Symposia, Boston, MA.

Lock, M. (1991). Contested meanings of the menopause. *Lancet*, **337** (May), 1270–1272.

Paniagua, R., Nistal, M., Saez, F. J. & Fraile, B. (1991). Ultrastructure of the aging human testis. *Journal of Electron Microscopy Technique*, **19**, 241–260.

Stampfer, M. J., Colditz, G. A., Willett, W. C., Manson, J. E., Rosner, B., Speizer, F. E. & Hennekens, C. H. (1991). Postmenopausal oestrogen therapy and cardiovascular disease, ten-year follow-up from the Nurses' Health Study. *New England Journal of Medicine*, **325**, 756–762.

Swartzman, L. C. (1990). Impact of stress on objectively recorded menopausal hot flushes and on flush report bias. *Health Psychology*, **9**, 529–545.

Thatcher, S. S. & Naftolin, F. (1991). The aging and aged ovary. *Seminars in Reproductive Endocrinology*, **9**, 189–199.

Walling, M., Andersen, B. L. & Johnson, S. R. (1990). Hormonal replacement therapy for postmenopausal women: a review of sexual outcomes and related gynecologic effects. *Archives of Sexual Behavior*, **19**, 119–137.

Walsh, B. W. & Schiff, I. (1991). Menopause: advanced management strategies. *Current Opinion in Obstetrics & Gynecology*, **3**, 343–351.

Wise, P. M., Scarbrough, K., Larson, G. H., Lloyd, J. M., Weiland, N. G. & Chiu, S. (1991). Neuroendocrine influences on aging of the female reproductive system. *Frontiers in Neuroendocrinology*, **12**, 323–356.

Part two
Reproduction and social issues

13 Population dynamics, stress and the general theory of adaptation

A DEMOGRAPHY AND STRESS: ADAPTATION AND STRESS

Currently the two greatest problems facing humankind, problems which link environmental, health and social issues, are population growth and the urgent need for its control. As Paul Ehrlich (from Stanford University) has stated so succinctly: 'We are now in charge of the planet, but we are not behaving as if we know we are in charge of the planet'. Certaintly, we are not behaving in a manner that befits our special responsibilities, and, further, it seems that we are unwilling or unable to cope with our increasing numbers and the effect this increase has on our environment. The world's population is currently increasing at a rate of 90–100 million people a year, an increase of more than 10 000 people every hour. In 1990 there were over 5 billion people on this planet. As a species, it took 4 million years for the world's population to grow from just over zero to 2 billion in 1932. Then, in less than six decades, it reached the 1989 figure of 5.2 billion. The present projection tells us that there will be more than 6 billion humans by the year 2000 and 7 billion by 2010 (Fig. 13.1). The impact of a growth rate can be better appreciated by the time taken for the population to double its size. Table 13.1 shows the approximate doubling times at various growth rates. Growth rates of 2% (a doubling time of 35 years) is particularly relevant since large parts of the world have a growth rate close to 2%. Although there are distinct differences between the cultures of western Europe and north America on the one hand and Africa and Asia on the other, the biological problems facing all peoples are matched and require an equivalent urgent attention. The special biological problems for the industrialized countries concern air and water pollution, contaminated food and lack of open space. These all contribute to an increased incidence of stress-induced diseases, such as cardio-vascular disease and cancer. Stress-induced conditions are also

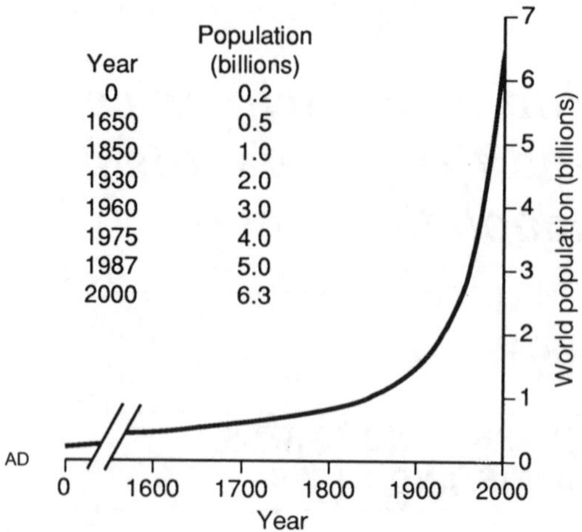

Fig. 13.1 Exponential growth of world population.

Table 13.1. *Population growth rates and doubling times*

Growth rate (%)	Doubling time (years)
4.0	17
3.0	23
2.0	35
1.0	69

called diseases of adaptation, that is, diseases of exogenous non-infectious etiology. The special biological problems for the developing, over-crowded countries concern lack of arable space, social instability, civil war, famine and disease; all contributing to an increased incidence of hopelessness and despair. According to the Food and Agriculture Organization of the United Nations, absolute poverty, however measured, is primarily the result of inadequate access to land to meet minimum needs. Of the 36 or so countries where poverty is the most severe, a significant proportion have already reached, or exceeded, the ratio of 'one man – one acre'. With lower average doubling times for their populations and the fixed nature of the land resource their future prospects are wretched. The need to effectively adjust to the changed human conditions and evaluate attitudes to birth control, rights of the unborn, and requirements of other species is obvious and crucial to the survival of all cultures. We may have already stolen from our children some of nature's future productivity due to irreversible environmental damage.

Before the 20th century, few people lived beyond the age of 75, while to-day more than 50% of the population in developed countries reaches age 75, and 25% live to age 85. It is estimated that in the USA, 35 million people will be older than 65 by the year 2000, posing major challenges in the social and medical care of the aged. Not even the seemingly under-populated places are safe from over-crowding. For example, since 1987 Australia's estimated annual growth rate has been in excess of 1.6%; this is the highest rate of any developed country and is approximately three times the growth rate of the United Kingdom. Even though the present population of 17 million is relatively low, this rate of growth (equivalent to a doubling time of 43 years) is far too high. If this growth rate is not curbed it will lead to a continuation of the environmental damage which began 200 years ago when Europeans first arrived on the semi-arid continent, so vulnerable to human interference. There are already many signs that Australia's future poverty will be in the loss of irreplaceable wilderness areas, the greatest resource of biodiversity (Chapter 14). Old-world farming methods have destroyed topsoil, salinated fresh water, demolished rainforests and coastal reefs, polluted seas and pushed to extinction unique indigenous flora and fauna at a rate worthy of entry in the *Guinness Book of Records*. It must be pointed out that Australia's growth is not due to births alone (natural population growth is around zero) but to an active and continuing immigration policy.

In addition to stark figures and growth predictions, we know that affluence can be as great a threat to the world resources as population size. To quote the Ehrlich formula: environmental impact = population × affluence × technological change.

It is the impact on the environment of a technology geared to supply each person's affluence that threatens the planet's future. Rich countries are a larger threat than less-developed economies because they consume more per capita. Additionally affluent societies have a greater access to the world's dwindling resources, reducing their availability to the less affluent. All western countries, possessing advanced technology, are disturbingly over-populated because their wealth is sustained by an economic strategy based on ever increasing consumer demand. 'The birth of a baby in Australia is a disaster for the world fifty times greater than the birth of a baby in Bangladesh' (Ehrlich, 1990 in the Milthorpe Memorial Lecture at Macquarie University, NSW). Population size and affluence created by existing technologies directly relates to our planet's physical deterioration.

The roots of our immediate problems lie in the size and rate of population growth. The result of this problem can, in part, be determined by an increased understanding of the physiology of adaptation, popularly known as 'stress' and its association with stress-induced social effects. The stress response is simply the homeostatic mechanism that achieves individual fitness and consequently increased survival of the species by the balanced integration of the nervous and endocrine systems.

THE LINK BETWEEN POPULATION DENSITY AND REPRODUCTION

The population density of a given animal species in a particular environment is rarely dependent on its breeding potential. Rather, densities fluctuate, sometimes widely, but tend to return to a relatively narrow range known as the carrying capacity. This tendency, in part, is governed by homeostatic mechanisms which increase in effect the greater the departure from the carrying capacity. These mechanisms, inbuilt into a specie's physiology to confer survival value, act through the neuroendocrine system via signals generated during social interactions. They have been collectively termed a 'density-stat', akin to the 'somatometer' discussed in Chapter 3 (p. 49). The importance of the density-stat is in the prevention of what is sometimes called 'overfishing'. If a population, through continued breeding, exceeds the limit that can be sustained by the resources of the habitat, then the subsequent generation will have its resources (especially food) seriously depleted. If, by chance, the maximum size of the population also coincides with particularly harsh environmental conditions then that species could seriously run the risk of extinction. To prevent this it has been suggested that social mechanisms such as territorial behaviour, dominance hierarchies and even such drastic measures as cannibalism have evolved to provide intrinsic control of population density.

The mode of action of the density-stat has been the focus of much debate, but its operation has been demonstrated both experimentally and in the wild and in all animal groups especially mammals. When individuals are placed in a confined space with shelter and unlimited food and allowed to breed freely, their breeding is controlled by that space and not by the availability of food or shelter. Calhoun in 1952 pioneered such experiments with rats. He found that his population of rats never exceeded 200 though the food and shelter provided would have supported in excess of 5000.

Subsequently, the connection between high population density and the activation of the general adaptation syndrome (GAS, described below) was made by Christian in laboratory studies of mice and field studies of wild rats. Christian demonstrated a direct link between population increase and adrenocortical activation, reduced fertility and immune deficiency. He systematically documented that crowded conditions interfered with practically every link in the chain of reproductive physiology and behaviour. For females, major consequences of crowding were disruption of the oestrous cycle, delay in the onset of pregnancy, reduced litter size, increased spontaneous abortion, increased neonatal mortality and failure in lactation. We now know that all were due to the direct or indirect effects of a reduction of GnRH secretion and increased secretion of glucocorticoids, the opioid β-endorphin and prolactin (Chapter 18). For males, the consequences were a fall in testosterone secretion, decreased sperm production and increased fighting. Fighting is the universal outlet for aggression which occurs when social systems deteriorate. One way of recognizing a normally functioning social order among animals is the general absence of destructive behaviour. For example a hungry rat in a tunnel with food at the end of it runs toward it, but if it is prevented from getting there it becomes angry and will attack innocent rats standing nearby. Christian's experiments also demonstrated a link between high population density and low resistance to disease in the weak and the young, who easily succumbed to infections. A stress-induced delay or suppression of puberty was also observed. Subsequent research widened Christian's findings and established a similar connection between high population density and stress for a large variety of mammalian species including primates. It is of special interest, however, that the adverse stimuli received in the course of social interaction which indicate density stress are not equally severe in all the individuals. In any population there are individuals who are particularly susceptible to stress. These are usually the young, the weak or those low in the dominance hierarchy. Of particular concern was the finding that the effects of crowding were perpetuated into a second generation. Christian's experiments demonstrated that the pups born to crowded mothers, but who were themselves allowed to mate in isolation produced growth-retarded young when compared to the offspring of control mothers not subjected to crowding stress (see p. 234). The link between crowding stress and adrenocortical activation has repeatedly been observed in many species other than rodents, including rhesus monkeys caged in groups and humans grouped together. Despite the fact that the same micro-evolutionary processes that are operational in controlling all animal populations are also significant for the human species, there is one major difference. By changing the world that surrounds us we have been able to transform the environment in which we live and the ways our population is organized.

It seems clear that the density-stat is driven by socially generated stress which, by activating the GAS, controls the size of a confined population through decreased fertility and increased mortality. The social interactions which activate the GAS involve competition between conspecifics in the establishment of social dominance and hierarchies, even before food and shelter shortages become apparent. Apart from the individual's status, the context of the social group has also to be taken into account when determining the outcome of social stress. This is particularly important as populations are not normally confined to a given space, as were the animals in Christian's experiments. Natural selection favours flexible survival strategies and behaviour patterns encouraging emigration and dispersal are winning survival strategies at times of social instability. Normal dispersal plays a powerful role in controlling population density in all living things. Field experiments conducted by Krebs and his colleagues on a free-ranging population of voles demonstrated that different genetic characteristics were favoured in different social situations. At times of severe population density stress, increased fitness may be achieved by giving up reproduction temporarily until better conditions for the rearing of the young are encountered. One option is to emigrate. In Kreb's studies, many of the émigrés were young females, that is, those whose fertility would, very likely, be restrained in Christian's closed system, but which would be exceptionally well fitted to reproduce rapidly in a freshly colonized habitat.

Historical perspective and current concept

No one can live without experiencing stress. A popular perception is that only serious disease or intensive physical or mental injury causes stress but this is not so. Many normal fluctuations in one's environment both physical (such as change of temperature) and psychological (such as change of ambience) may be enough to activate the body's stress mechanism to some extent. Stress is necessary for adaptation and, instead of being harmful, can also be the spice of life. Because of great individual differences, a stress which makes one person sick (distress) can be an invigorating experience for another (eustress). The essential thing is that the body must be prepared to adapt to changed circumstances by the initiation of an appropriate response.

Stress is essentially reflected in the total rate of all the wear and tear caused by life and, although it is impossible to avoid stress, a lot can be learnt about how to keep its damaging side effects to a minimum. It is argued that many common diseases can be avoided if the adaptive response to stress were more moderate and thus better controlled. For example, many nervous and emotional disturbances, high blood pressure, gastric and duodenal ulcers, certain types of sexual dysfunction, allergic, cardiovascular and renal derangements appear to be essentially diseases of inadequate adaptation. From an evolutionary perspective, the GAS is an ancient system which may have initially served to control population density but, in the course of time, acquired the ability to respond positively to many other stimuli as well as those which signal high population density. The primitive stress response was certainly further refined and incorporated into the evolutionary changes needed for the move from the sea to freshwater or onto land. It was the harsh conditions in those new environments which necessitated the development of a unique system for adaptation.

The physiological response to stress was first formally studied by Cannon in the 1920s. Cannon observed that many stimuli, both physical and psychological, caused the release of a humoral agent from the adrenal glands which raised blood pressure. The humoral agent, which he called sympathin, was later identified as a mixture of the catecholamines adrenaline and noradrenaline. Cannon's research eventually resulted in a description of the emergency (fight-or-flight) response to threatening environmental influences. This emergency reaction is brought about by the sympathetic nervous system acting in conjunction with the catecholamines from the adrenal medulla. Its function is to mobilize the body's resources for swift action (see p. 237).

The concept of 'stress' was then further developed and refined by Selye, whose studies culminated in 1956 in the description of GAS, which emphasized the role of the hypothalamic–pituitary–adrenocortical axis in the adaptation to stress. In 1936, Selye, then a young physician starting a research career at McGill University, Montreal, experienced major problems in the interpretation of his research findings. He had been injecting rats daily with a chemical extract to determine its effects and had identified consistent changes in the experimental animals: peptic ulcers, atrophy of the thymus and enlargement of the adrenal glands. To his surprise, however, rats in the control group which had been injected with saline solution alone showed identical changes. Selye focussed on what the two groups of animals had in common, the repeated injections, and wondered if the changes he had identified were actually a generalized

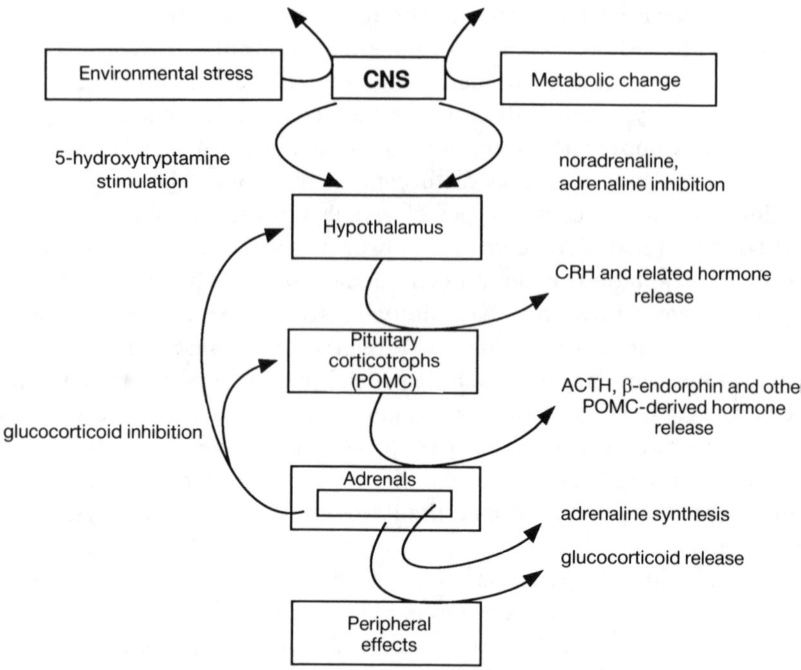

Fig. 13.2 Adaptation to environmental and metabolic changes. Following the initial alarm reaction, if the stress is prolonged it is the adrenocortical hormones which are chiefly involved. The release of the glucocorticoids (which include hydrocortisone, corticosterone and cortisol) from the adrenal cortex is under the control of ACTH, while the release of ACTH from the anterior pituitary is controlled by CRH and related hormones secreted by the hypothalamic neurons. CRH also stimulates the anterior pituitary to release β-endorphin and other POMC-derived hormones. The glucocorticoids, in turn, feedback to hypothalamic and pituitary centres to inhibit secretion of CRH, ACTH and their relatives, leading to a decline in glucocorticoids. ACTH can also stimulate the synthesis of adrenaline.

physiological response to unpleasantness *per se*. He later confirmed this to be the case. Selye borrowed the engineering term stress, signifying fatigue in metal under strain, to describe the body's response to any insult. Subsequently he defined stress as 'the nonspecific response of the body to any demand whether it is caused by, or results in, pleasant or unpleasant conditions'. Thus, the science concerned with stress physiology was born.

Figure 13.2 summarizes the contemporary understanding of the involvement of the hypothalamic–pituitary–adrenal axis in the homeostatic adaptation to environmental and metabolic changes. Corticotrophic cells of the anterior pituitary gland synthesize a common precursor glycoprotein molecule named proopiomelanocortin (POMC). POMC consists of three structural domains: ACTH and residing in the middle of the molecule, β-lipotrophin comprising the C-terminal sequence and an N-terminal fragment. These segments may be further cleaved to give rise to other biologically active compounds, for example the liberation of γ-lipotrophin and β-endorphin from β-lipotrophin. ACTH stimulates adrenocortical steroid biosynthesis, especially of glucocorticosteroids (mainly cortisol in humans). The glucocorticoids, in their turn, have a direct negative feedback effect on pituitary ACTH and hypothalamic CRH secretion leading to a decline in circulating glucocorticoids. CRH is the major hormone regulating ACTH secretion but a variety of other substances and neuropeptides (for example catecholamines and oxytocin) also stimulate the release of peptides derived from the adrenohypophysial POMC system. In addition CRH is also influenced by extrahypothalamic neurotransmitter-mediated pathways in the CNS. 5-Hydroxytryptamine is generally stimulatory and noradrenaline and adrenaline inhibitory, but other central neuromodulators may also participate in CRH control.

We now know that the endocrine response to a particular stress is a specific response mediated by multiple possible CNS pathways, not necessarily always activating the hypothalamic releasing hormones with pituitary involvement. Conversely, the synthesis of POMC and the secretion of its derived peptides is regulated by multiple factors at several levels of organization. Although Selye's basic concept of the GAS has been modified to account for stimulus-specific endocrine responses to stress, the CRH–ACTH–glucocorticoid sequence still remains the most important regulatory system governing the reaction of the body to environmental stress and metabolic change.

Since stressful stimuli (physical and psychological) can elicit a variety of endocrine responses differentially controlled by afferent neural pathways and peripheral input, these responses are presumably meant to be adaptive and aid the animal to best withstand stress. Many factors shape a stimulus-specific response. For example, not only an individual's social status but also the composition of the social group is important in determining the degree of stress experienced. Fighting in a previously stable group of rodents will occur if a stranger is introduced. Experiments involving monkeys likewise demonstrated that an important quality of a stressful experience was the exposure to a threatening or novel environment; environmental stimuli affected adrenal cortical activity only if novelty or distress was experienced. Therefore, stimulus-specificity of response according to individual experience is an important part of our understanding as to how identical environmental events may elicit different physiological responses. The individual nature of the response to life's events also applies to humans, as was graphically demonstrated in pioneering studies by Rose and his colleagues in the case of air traffic controllers. Controllers who had significant increases in cortisol with increases in workload had higher scores in peer ratings than controllers who did not respond hormonally to increases in the workload. The highly rated men were most admired by their co-workers for their ability to perform their jobs, and the high level of cortisol secretion was an index of the personal effort made to perform the job well. Therefore, the overall physiological response to environmental stimuli also includes the intensity of mental effort required to meet a challenging situation, even in the absence of a perceived physical threat.

It is one of the most characteristic features of the stress response that its various defensive mechanisms are generally based on combinations of three types of responses, namely attack (fight), retreat (flight) and passive tolerance. Survival depends largely upon a correct blending of attack, retreat and standing one's ground. To obtain the best results these three types of reactions must be perfectly co-ordinated in time and space so as to adjust our reactions to the changing demands of the situation at various times and in various parts of the body. When faced with an aggressor, it is by achieving this co-ordination with minimal distress that an individual can successfully defend itself.

COMMON MODIFIERS OF THE STRESS RESPONSE

The adaptive response to any stressor depends upon several variables. The following modulators are artificially separated since there is much overlap between each.

Sex Males are more likely to resolve a stress with physical force than are females. This difference may have resulted from differing roles in the division of labour in reproductive function. It has been suggested, for example, that the interaction of the male with his environment is heightened because it is he who has to cope with the burdens of social competition. Throughout the animal world it is the males who perform attention seeking behaviours, such as intimidating other males, attracting females and securing territory; or in human terms, it is they who struggle for rank in a social hierarchy, compete for favours from the opposite sex and fight for income and country. The female, who bears the main burden of reproduction, cannot squander her time and energy on social, often aggressive competition if her offspring are to be raised in a protected environment. So she typically takes on the social status of her mate. Chapter 7 deals in part with the evolution of sex roles developed to maximize the reproductive potential and the resulting sociobiological implications.

Genetic factors Genetic factors, other than sex, such as specific predispositions and vulnerabilities also determine which individuals succumb to stress and which do not. For example, a deficiency of adrenal 21-hydroxylase (which mediates the conversion of progesterone to deoxycorticosterone and aldosterone) may, depending on the degree of deficiency, interfere with a well-directed stress response.

Social factors and age As described previously, social factors such as group composition or position in the social hierarchy are important. Stress is more intense in individuals low down in the social hierarchy; position in the social hierarchy frequently depends on the age of the individual and varies with time, place and the composition of the social group. The importance of the group composition in humans is well demonstrated by the now famous observations on bomber crews whose adrenocortical activity was very similar in members of a particular group whereas the levels in different groups varied markedly. An undefined social factor amongst a particular mix of individuals was important in determining the level of adrenal gland activity. Many biological odours influence human social behaviour (Chapter 6), but whether the information is transmitted by a stress-reducing pheromone in this case is not clear. Steroid molecules have been identified which alter mood much as tranquillizers do and which may act as natural tranquillizers. A possible odorous steroid 5-α-androsterone, which contributes to the subliminal smell of clean, fresh skin and concentrates in areas such as the armpits and mammary glands, may be such a compound. Classical trials by staff from Guy's Hospital Medical School in London found that 5-α-androsterone sprayed on unreserved seats in a theatre were occupied in preference to the unsprayed ones.

Conditioning The physiological response of the body to some threatening external event also incorporates the memory of previous similar events and the perception of likely consequences. An individual's past history powerfully influences the physiological and behavioural response to stress. Conditioning results from physical and emotional experience.

Prenatal environment Prenatal influences on postnatal physical and emotional states comprise a special case of conditioning. For example, stress-induced maternal release of glucocorticoids and/or endorphins can modulate fetal development and play a critical role in many long-term disturbances of key functions controlling growth, maturation and steroid physiology. A well documented example of the way prenatal stress can affect certain reproductive parameters in the adult comes from experiments on rats carried out independently by Ward (Villanova University, PA) and myself. Taken together, the research demonstrated a similar dose-related effect on the sexual differentiation of male offspring whose mothers were exposed to a variety of stresses including crowding, malnutrition, anxiety, morphine and caffeine during gestation. Most marked was the reduction of the crucial rise in testosterone secretion by the fetal testes during the last week of gestation. If the stress was sufficiently severe, testosterone biosynthesis in the adult was permanently reduced (Pollard) and normal male sexual behaviour impaired (Ward) (see also p. 244).

Physical environment Environmental influences such as climatic extremes or the degree of pollution may modify the efficacy of a specific response to other stressors (see below).

Nutrition Stress places an extra load on the body which needs to be adequately compensated for nutritionally if an effective adaptive response is to be possible. Additional nutrients are mobilized in response to heightened energy utilization and the rapid synthesis of adrenal enzymes and hormones.

Cumulative effect Several stresses may simultaneously impinge on various parts of the body and, in proportion to their number, intensity or extent, can activate the GAS. Severe environmental pollution, for instance, adversely affects the immune system lowering resistance to disease.

In summary, the stress syndrome is the normal adaptive response allowing appropriate adjustment to the sum of all structural and functional changes which occur in an animal exposed to the variables of living. It is only when the response to stress is largely inactive and ineffectual, as is the case in prolonged states of anxiety or illness, that the body's systems may wear down in rapid ageing processes. In the short term during acute strain, the increased hormonal secretion facilitates a necessary adaptation to a demanding situation.

B DEMOGRAPHY AND STRESS: HORMONAL FACTORS

BIOCHEMICAL ASPECTS OF THE GENERAL ADAPTATION SYNDROME

The General Adaptation Syndrome (GAS) as defined by Selye brings the important element of time into the study of the physiological response to stress. A fully-developed GAS consists of three stages:

(a) The alarm reaction
(b) The phase of resistance
(c) The phase of exhaustion.

It is not necessary for all three stages to develop, as only the most severe stress leads rapidly to the stage of exhaustion and death. Most of the stressors which act upon us during a limited period produce changes corresponding only to the first and/or second stages. Transitory variations in stress levels have profound effects on our reproductive capacity. That fertility increases at special times of relaxation or heightened eustress is a common experience. Some knowledge of the major biochemical effects brought about by the stress hormones is essential in order to appreciate the importance of keeping our adaptive response within healthy limits. The following gives a simplified overview of the most prominent biochemical shifts which take place when the hypothalamic– pituitary–adrenal axis is activated.

The alarm reaction (or emergency response) produces profound alterations in energy metabolism which are mediated by the sympathetic nervous system to effect the release of adrenaline from the adrenal medulla and glucagon from the islets of Langerhans in the pancreas; it may also involve the release of ACTH. Glucagon, the most powerful hyperglycaemic agent known, accelerates glycogenolysis (the production of glucose by mobilization of liver glycogen) to provide the fuel for oxidative phosphorylation. Adrenaline stimulates triglyceride lipase activity in adipose tissue resulting in increased circulating free fatty acids (FFA) which may then be used for muscle metabolism. FFA utilization by the musculatory system saves the

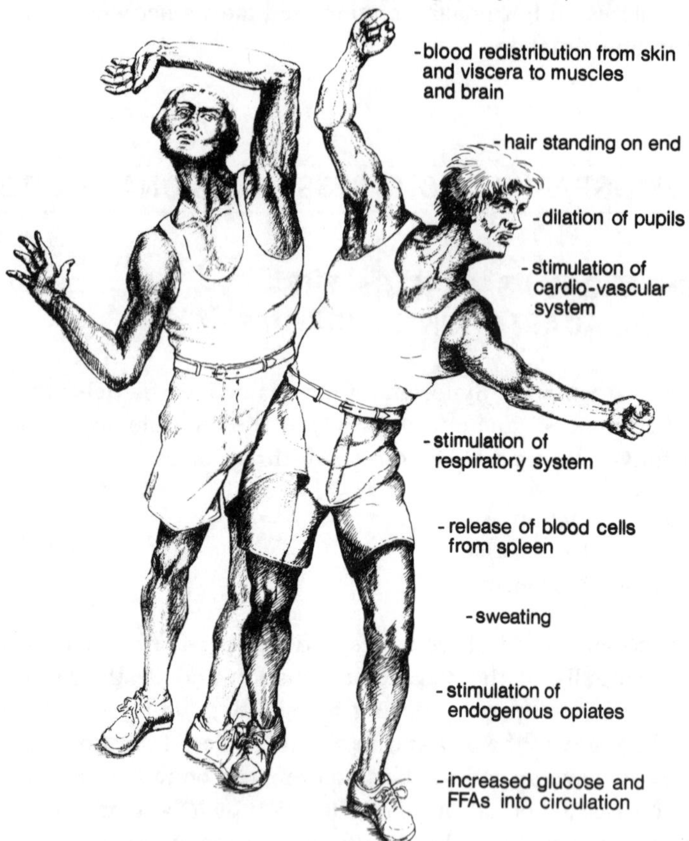

Fig. 13.3 The physiological effects of the alarm reaction mediated by the sympathetic nervous system to stimulate the release of adrenaline.

liver glycogen stores for use by other tissues with high glucose requirements, such as the brain. The main physiological effects of the alarm reaction are shown in Fig. 13.3. The listed effects all relate to the technique of combat and are aimed at increasing the fitness of the animal for violent and competitive muscular activity at times of threat.

If stress continues, its manifestations change as time goes on and the next stage of the GAS, the ACTH-dependent phase of resistance, occurs. Now the glucocorticoids are chiefly involved because they play a supportive role at times of special need and exert actions on almost every metabolic process. As with glucagon and adrenaline, the glucocorticoids are also concerned with energy metabolism. They promote gluconeogenesis (the transformation of non-sugars into sugars) and inhibit the utilization of amino acids for protein synthesis. In addition the glucocorticoids act synergistically with the two fat mobilizing POMC-derived lipotrophins (LPH) and provide additional FFAs. During the phase of resistance, many adaptive biochemical and behavioural changes take place; for example, an increased biosynthesis of enzymes in the adrenal glands adjusts to continuing and changing demands for adrenaline and glucocorticoids. Similar adaptive changes take place in the sympathetic nervous system and the brain. Although the major source of POMC-derived peptides is the anterior and intermediate (in species possessing it) lobes of the pituitary gland, POMC is also expressed by a number of extrapituitary tissues. Prominent among these are the hypothalamus and other CNS neurons, the testis, ovary, placenta, adrenal medulla and the immune system.

When considering the redirected energy metabolism under prolonged or chronic states of stress, it is not surprising that there is a complementary shutdown of anabolic functions directed towards growth, repair and reproduction in favour of mechanisms which promote readiness for immediate high-energy and high-priority action. It must be understood, however, that a well directed stress response increases reproductive fitness; only if stress is prolonged and non-productive does it affect fertility adversely. The terminal stage of the GAS, the phase of exhaustion, is entered only when the whole organism is exhausted through senility at the end of a normal lifespan or prematurely through accelerated ageing caused by the cumulative effects of differing stresses.

STRESS AND AGGRESSION: A VIEW FROM SOCIOBIOLOGY

Psychiatrists have known for a long time that stress originates from social feelings such as fear, hate, suppressed rage, anxiety, anguish or frustration. Animal behaviourists have observed that the solutions to stress are frequently similar in humans and other animals. Humans are highly socialized animals so many of the subtle ambiguous stressors in life are socially generated and may also be invented or exaggerated by the mind. Further the mind's perspective can alter the individual's physiology as profoundly as the external events themselves, and feelings such as hate, envy, fear and frustration all increase the likelihood of aggression. The postulate that war is an evolutionary asset was explored in Chapter 7. It was pointed out that by simply existing, one group can seriously threaten the resources of another and that the fitness of some individuals can be actively increased by warfare provided the cost of waging it is less than the benefits received. If severe crowding, poverty, inequality of resource distribution within and between boundaries is juxtaposed against the physiology of stress, then national instability and problems threatening global peace seem more understandable and wars predictable. Civil war can be a response to overcrowding in developing countries where gross social injustices exist, but it frequently degenerates into a power struggle among unstable hierarchies professing differing ideologies. Civil unrest also spreads to neighbouring countries and increases overall aggression in the vicinity. Similarly, within the more stable developed communities, social deprivation generates deep tensions which often lead to violence and a wish to destabilize the existing hierarchy. Some of the ethnic riots in the USA and Britain are, in the depth of emotion generated, civil wars kept in check by virtue of their small size and relative economic stability. As observed in Christian's experiments, in stable conditions it usually pays individuals to accept subordinate status rather than fight losing battles; however, it also pays them to enhance their position by taking advantage of existing instability. Two different survival strategies, each favouring the survival of the population's own genes, can be seen in times of increasing stress. The first is to fight to obtain high social rank and breed in spite of social upheaval, and the second is to flee to pastures new and breed there (the compromise for the non-reproductive population may be passive tolerance).

The universal consequences of war are the systematic denial of basic human and ecological rights with the senseless destruction of large areas of our planet, encouraged by the global weapons trade. As we move through the 1990s, the development of a personal, not institutional, responsibility based on a common human goal and a guilt for environmental destruction such as wars may give birth to a different community conscience displacing ruthless power. The most crucial issue facing us is misdirected power over the environment, and no-one wants to give up power. However, there is nothing that conventional military notions of threat and defence can do to give security for the environment. Vast territories are lost every year from environmental causes and if we do not quickly protect areas essential for maintaining life on earth, futile wars, in the context of environmental security, will be fought. In the last 20 years the planet has lost an estimated 500 million acres of forest, 300 million acres of productive land to desert, and 480 billion tons of topsoil; what is happening under the sea is hidden from view. Will we destroy what is left by fighting over it?

THE ENERGETICS OF FERTILITY: ETHICAL CONSIDERATIONS

As we have seen from animal experimentation, persecuted or deprived individuals have another option available to them apart from resistance. They can improve their fitness by dispersal. The option is increasingly being taken by people from overcrowded countries suffering from the aftermath of civil war or other devastation. Increases in the number of 'environmental' refugees should act as a global warning. In 1991 there were 17.5 million refugees, but if one also counted those displaced within their own country by civil war, a conservative estimate of the number was 35 million. However, emigration is only a viable option if there is somewhere to emigrate to. Mass movements of whole populations pose severe biological problems, not least the fertility of the émigrés. A recent example can be taken from the Vietnamese boat people stranded in holding camps in Hong Kong. The birth rate of these refugees surpassed the resettlement rate and, at one point, even the rate of influx of new immigrants. In July 1991, the 52 000 stranded represented four times the 1990 influx. The intolerable strain imposed on the host country led to the only possible option: compulsory repatriation back to Vietnam.

The fertility of the émigrés poses an apparent paradox if, as has been said, stress inhibits reproduction. However once the immediate threat to life is removed by the safe arrival at a camp, together with increased food consumption and decreased physical activity, a rebound in fertility invariably occurs. After all, Krebs' study did demonstrate that emigrating voles were exceptionally well fitted to reproduce rapidly in a freshly colonized habitat, and postwar baby boomers are a human example of rebound fertility after stress. The energy required for the shift from infertility to fertility is very small. Anovulatory women spend less energy during the luteal phase of their menstrual cycle, due to decreased progesterone secretion. It has been calculated that this postovulatory saving represents approximately 9% of the 24-hour energy expenditure. Therefore, a small increase in the total calorie intake may tip the balance and restore fertility.

Dietary choices have influenced the course of mammalian evolution. For example hunter–gatherers can balance their energy budgets in order to meet all their nutritional requirements. The Australian Aborigines, following a traditional way of life, are thinner than the minimum levels set by the World Health Organization but show no biological signs of ill health from this cause. Hunting and gathering is a regimen of frequent fasting to which our ancestors have become culturally and biochemically adapted. Studies in rats have demonstrated that high-intensity exercise produces decreased food consumption compared with that of a low-intensity exercise group, and both exercise groups ingested less food than no exercise controls. These exercising rats can, however, be kept in this physiological condition without signs of poor health. If the running regimen is initially such that the rats can barely maintain their weight, they in time gain weight unless exercise is further increased. Efficiency is achieved by a reduction in the resting metabolic rate. When the running requirement is suddenly relaxed there is a rapid 'catch up' growth and an onset of reproductive activity. Three factors combine to promote the rapid recovery of reproductive development.

(a) A great decrease in energy outlay
(b) An absolute increase in food intake
(c) The higher metabolic efficiency that was previously acquired.

Another biological factor may, in part, contribute to the observed rebound fertility. During periods of extreme stress, the energy costs of breeding reach a point where it is impossible for a female to breed without helpers. In this case it is adaptive to confine fertility to social group existence. Because the human is a social animal the species benefits by sharing the rearing of the young, which means that less time and energy needs to be spent by the parent. In refugee centres, there are those who care, that is, guard, feed and groom the children, thus creating a situation favourable for communal breeding. However, with such a strategy, there is not 'a world of fitness to win but a world to lose'. Herein lies the donor nations' moral dilemma, yet few aid agencies provide sex education and contraceptives. Donor assistance for family planning amounts to approximately US $560 million annually, or about 15% of family planning costs. But this accounts for only 1% of all developmental assistance and equates to about 0.4% of the donor's total budget. By comparison, military spending accounts for about 19%. Since the present state of the world affects everyone, we all have a right to know how our actions increase the threat to our environment and accelerate global misfortune. We also need to give serious thought to ethical considerations and responsibilities when communicating information or making policies. Up to now the bulk of the aid given is in response to a short-term emergency situation: the alarm reaction, to use the GAS analogy. The short-term policy of dealing, for instance, with famine will not change the future incidence of famine unless it is backed by a sustainable commitment – the stage of resistance. Short-term aid is wasted unless problems of civil unrest, corruption, illiteracy and over-population are not fully tackled by donors and recipients alike. Donors and recipients must then, through sustainable development, effect permanent change. Controlling population increase and providing for the needs of recipients and donors in order to establish peace are the areas of most urgent priority. Certainly, food donations backed by profitable arms sales will not achieve these objectives. The irony is that numerous surveys show that women, in developing countries especially, have no choice over their fertility and many are forced to have more children than they want. There must be something wrong with the rules of social morality if the unborn and the young do not have even the basic or minimum right of food and shelter? We desperately need a new community conscience or ethics to ensure that family planning programmes are not sabotaged by ethnic or religious objections.

The United Nations Convention regarding children's rights mentions the provision of educational services, the protection from abuse such as war and the implementation of the law. There are, however, no penalties for those countries breaking the law or, as in many African countries, those where the next meal is under threat for thousands of adults not to mention the children. The Convention is a good start but does not have the power to be effective where it is most needed. Many developing countries cannot afford the enormous finances required to attract the pharmaceutical companies and build the infrastructure necessary to set up the staff and clinics essential to control runaway fertility. No matter what food contribution is made by the richer nations, under the present demographic explosion nothing can be achieved. Global tension is rising because land is vanishing. Yet our careless attitude of mind does not put a stop to the growing military degradation of previously fertile land which is then simply abandoned. Linking aid funding for development with effective population control may prove helpful, at least in part, in reducing socioeconomic unrest and in directing resources towards constructive solutions that might otherwise be squandered on military or short-term measures. Specifically, donor agencies would fund technological improvements supporting health, responsible agriculture and family planning which are to be maintained under joint donor and recipient authority; recipients would work towards environmentally sustainable economic autonomy.

Developed nations, likewise, cannot continue to be bound by the mores of their time if they are to survive with dignity. Industrialized nations must deliberately and willingly limit wastage by curbing over-production and by not squandering precious resources. One of our greatest moral philosophers Aristotle distinguished in his *Treatise on Ethics* between the statement of a truth and the affectuation of an action; that is, it is not enough to *want* to do good, but to do good. Have we the will to tackle population growth, energy use and the divisions between the rich and the poor? Present global military spending, if redirected, can save our beautiful planet and give us environmental security.

WHERE DOES THE LIMIT LIE?

As already emphasized, the effects of stress are personal, and its potency depends on how a particular individual responds to it. The biological importance of the individual nature of perceived stress is clearly illustrated by studies on variable vulnerabilities to stress in a troupe of free-ranging baboons living on an African reserve. Sapolsky and his colleagues (of Stanford University) studied baboon dominance hierarchy and found that the function of key physiological systems differed between the dominant and the subordinate males. The secretion of testosterone when the animals were challenged is a good example. Despite the average resting testosterone levels being similar for all animals, under stress the levels diverged significantly between the dominant and subordinate males. As already described, testosterone levels fall in response to stress in all mammals due to a combination of an initial stress-mediated β-endorphin release (which by suppressing GnRH release suppresses LH and testosterone secretion) and a subsequent cortisol-induced decrease in the sensitivity of the testes to LH. Interestingly, however, it was observed that testosterone levels of the dominant males actually rose and remained significantly elevated for as long as an hour before declining, whereas those of subordinate males declined promptly. This brief initial testosterone rise gave the dominant males a survival and, therefore, social advantage since testosterone, as well as regulating aggression, increases the rate at which glucose reaches the muscles thus helping them to withstand a physical challenge. Since the hypothalamic–pituitary–gonadal axis was not implicated, a two-part mechanism was suggested to explain the divergent response. Firstly, the testes of the dominant males were less sensitive to the testosterone-inhibiting effects of cortisol and this slowed the decline of testosterone levels. Secondly, the rise itself resulted from the stress-induced release of catecholamines, which affect blood flow. For unknown reasons, the testicular vascular system of the dominant male was found to be particularly sensitive to the dilating effects of adrenaline and, as a consequence, the testes of dominant males received a greater blood supply during stress than did the testes of the subordinate males. The adaptive advantage, although

LH declines equally in both groups, is that any existing circulating LH can be delivered faster to the dominant male's testes causing a temporary increase in LH levels and increased testosterone release.

We have seen that cortisol is responsible for much of the double-edged quality of the stress response. In the short run it mobilizes energy, but its chronic over-production contributes to muscle wastage, fatigue, hypertension and impaired immunity and fertility. Ideally, then, cortisol should be secreted heavily in response to a truly threatening situation but should be kept in check at other times. This is precisely what occurred in the dominant baboons of Sapolsky's study. The resting cortisol levels of dominants is lower when compared to that of the subordinates yet will rise faster in response to a major stressor.

As emphasized throughout this chapter, a well directed stress response has survival value and increases reproductive fitness. It is the dominant animals in baboon society who have the breeding privileges. The reason for the differing response to stress in the baboons is not known but an important question is whether this difference can be solely attributed to genetic differences? Or might it be possible to gain a clue, or make a connection, between these adults and studies on early development. Does the permanently excessively high cortisol secretion during rest reflect long-term disturbances due to adverse prenatal influences? The nature/nurture argument has been with us for centuries but taking into account subtle influences that neonatal stress has on long-term mechanisms of steroid balance (see p. 234), new insights may be gained. For example, stimulation of very young rat pups (picking them up for a few minutes each day) assists physiological mechanisms to function optimally that not only increase bodyweight gain but accelerate development. Growth parameters which are hastened in the stimulated pups, compared to the untouched controls, include the earlier appearance of body hair, opening of the eyes, locomotion, various parameters of brain maturation and puberty. Accelerated maturation following physical stimulation has also been reported for infant monkeys and premature human babies. Several paediatricians/neonatologists now encourage the handling or stroking of low birthweight infants and it is reported that these infants respond positively whilst in intensive care. The benefits of early mental stimulation has long been expounded by educationalists who believe that a varied environment increases learning skills. Going back to the physiology of stress, Levine (of Stanford University) in the 1960s proposed a hypothesis that, during the critical period spanning adrenal maturation, the more varied the levels of circulating adrenocorticoids are, the greater the number of possible distinct levels of activity that can be displayed by the adult stress axis.

Indeed this hypothesis was strengthened in the 1980s by the finding that early handling of rat pups gives rise to a significant increase in the number of glucocorticoid receptors of the adult brain. A more flexible or finely tuned response to stress may be critical under certain circumstances and, as seen in the case of the dominant baboons, may bestow significant adaptive value on the individual. The extent to which infant stimulation is involved in improving the effectiveness of the stress response in humans is obscure, but it may depend on early critical environmental influences on the developing nervous system interacting with the genotype of the offspring prenatally. However, it is in the discovery of general principles, mainly from animal experimentation, that a greater understanding of our nature becomes possible.

HOW TO COPE WITH OUR INDIVIDUAL LIMIT

Aside from varied recurring stresses such as illness and pollution, many people, and particularly those from privileged backgrounds, are also victims of anxiety. Anxiety leads to insomnia, listlessness and fatigue and drives thousands of sufferers to take tranquillizers which, if taken over the long term, are damaging. However there is much self-medication, matching the potency of valium or librium without their adverse side effects, that can be done. If we are to utilize fully the potential we have, it helps to be in charge of ourselves and to learn to monitor our personal stress levels in order to make the necessary adjustments at times of increased demands. Since stress is personal – envigorating for some, devastating for others – the stress experienced depends not so much on what we do or what happens but on the way we take it. Characteristics such as self-confidence, reliance, trust, esteem and a flexible motivation springing from an inner strength all lower distress and heighten eustress. The highest achievers are typically self-actualized people who best cope with their share of stress. It helps to know your personal strengths and weaknesses and so adjust your life, as much as is possible, to function within these constraints. In addition, the routine incorporation of knowledgeable nutrition and relaxation (meditation, sport, music) and the ability to reduce other sources of stress when subjected to a certain particular stress conserves the body's energy for use in areas of maximum demand and effect.

However, the above statements may be useful for the privileged adults whose destiny is largely in their own hands. Since the importance of a particular group composition is understood and normally the most influential group in human society is the family unit, it follows that the dominant members of that group have extensive power in the setting of the general background stress level. This general background level, in turn, may condition the long-term reference set-point to stress of the subordinate members, particularly the children, living in the group. It is possible that the amount of anger and violence may be reduced if children were given the legal right of minimum standards of treatment. A new socially acceptable code of conduct will then depend on the acceptance of the privilege of reproduction rather than the right of reproduction. The investment of resources in the prevention of prenatal and postnatal child abuse will eventually improve our fitness, especially as the positive effects perpetuate into future generations and become apparent.

It would be ironic if we cannot use our exceptional brain, which has brought about the revolutions in technology and medicine, also to balance population growth and resource consumption with the new benefits we have gained. To live up to our responsibility of being in charge of the planet is the ultimate human challenge and, if successful, will be our highest achievement. Scientifically, we know how to control our numbers; new methods of contraception are increasingly being developed which give many different effective choices for population control. What remains is the will to use the technologies and reinstate a fair balance in the biosphere. 'Growth for growth's sake is the creed of the cancer cell';[1] we have nothing to gain and all to lose by imitating this behaviour. We need to co-operate and work together to make the environment habitable. Clearly, the fundamental issue is whether we will learn to use the natural resources in a sustainable manner and maintain an acceptable quality of the environment, or whether we will overwhelm the capacity of the natural systems with our human demands. As social creatures, we each have a very low fitness unless we co-operate with our fellows, and as animals we cannot survive if separated from the rest of the biosphere.

General references

Calhoun, J. B. (1952). The social aspects of population dynamics. *Journal of Mammalogy*, **33**, 139–159.

Cannon, W. B. (1929). *Bodily Changes in Pain, Hunger, Fear and Rage.* Branford, Boston, MA.

Christian, J. J. (1975). Hormonal control of population growth. In *Hormonal*

[1] Edward Abbey, Playboy Magazine, 1979.

Correlates of Behaviour. eds. B. E. Eleftheriou & R. L. Sprott, pp. 295–374. Plenum Press, New York.

Creel, S. R. & Creel, N. M. (1991). Energetics, reproductive suppression and obligate communal breeding in carnivores. *Behavioural Ecology & Sociobiology*, **28**, 263–270.

de Kloet, E. R. & Voorhuis, T. A. M. (1992). Neuropeptides, steroid hormones, stress and reproduction. *Journal of Controlled Release*, **21**, 105–116.

Ehrlich, P. (1991). *Healing the Planet*. Addison Wesley, Reading, MA.

Kalin, N. H. (1993). The neurobiology of fear. *Scientific American*, **268**, 54–60.

Krebs, C. J., Gaines, M. S., Keller, B. L., Myers, J. H. & Tamarin, R. H. (1973). Population cycles in small rodents. *Science*, **179**, 35–41.

Levine, S. & Mullins, R. F. Jr (1966). Hormonal influences on brian organization in infant rats. *Science*, **152**, 1585–1592.

Pollard, I. & Dyer, S. L. (1985). Effects of stress administered during pregnancy on the development of fetal testes and their subsequent function in the adult rat. *Journal of Endocrinology*, **107**, 241–245.

Pollard, I., Williamson, S. & Magre, S. (1990). Influence of caffeine administered during pregnancy on the early differentiation of fetal rat ovaries and testes. *Journal of Developmental Physiology*, **13**, 59–65.

Rivier, C. & Rivest, S. (1991). Effect of stress on the activity of the hypothalamic–pituitary–gonadal axis: peripheral and central mechanisms. Review. *Biology of Reproduction*, **45**, 523–532.

Rose, R. M. (1980). Endocrine responses to stressful psychological events. *Clinical Psychiatry of North America*, **3**, 251–276.

Sapolsky, R. M. (1990). Stress in the wild. *Scientific American*, **262**, 106–113.

Selye, H. (1976). *The Stress of Life*. McGraw-Hill, New York.

Ward, I. L. (1992). Sexual behavior: the product of perinatal hormonal and prepubertal social factors. In *Handbook of Behavioural Neurobiology*, Vol. II. *Sexual Differentiation*, eds. A. A. Gerall, H. Moltz & I. L. Ward, pp. 157–180. Plenum Press, New York.

14 Extinctions and the conservation of endangered species

A 'SURVIVAL OF THE FITTEST?'

> We have not inherited the Earth from our parents,
> we have borrowed it from our children.
>
> World Conservation Strategy

In the previous chapter, the link was made between population density-dependent stress and social issues of global concern. This chapter considers the retreat of the world's biota towards extinction as their habitats are destroyed or taken over by our own relentless human population expansion. The poem below is a child's plea for justice in response to the violation of other creatures' rights.

> I like the whale
> But he gets very pale
> When the whalers fail to harpoon him
> And he is not very glee
> When they go to sea
> He wishes he is free, don't forget him
>
> Morgan Pollard, aged 7, 1979

Wheneven small children feel the need to be directly involved with environmental problems on such a scale, so great a concern is demonstrated that history holds no precedent. As we know, the quality of a child's environment influences all aspects of its development, but the natural environment has a particularly powerful influence on emotional development and many children identify with their living surroundings. The wide natural world in its diversity, beauty and power gives humans that state of happiness that exists when the body and mind are together, alienation from it causes pain. Our earth must be respected and preserved for both itself and for future generations. We adult humans are knowledgeable enough to act as the planet's beneficient caretakers; if we fail, though, will the young forgive us our trespass against them?

ENDANGERED ANIMALS AND THE PHENOMENON OF SPECIES EXTINCTION

Endangered species are those threatened with immediate extinction. Vulnerable species are those likely to become endangered if subjected to uncontrolled predation or environmental change. Extinction is a normal dynamic part of evolution; indeed, during the Pleistocene, for example, the cold glaciations and the warm interglacial periods between them led to the extinction of many of our large marsupials such as *Zygomaturus trilobus*, an enormous browsing marsupial, and *Diprotodon australis*, the largest marsupial that ever lived. In addition to the effects of climatic forces, prehistoric extinctions in Eurasia coincided with the ascendancy of *Homo sapiens*. It may not be a coincidence that much of earth's megafauna, for example the aurochs (the progenitor of domestic cattle), musk ox, bison, woolly mammoth and the sabre-toothed cat, all disappeared about 20 000 years ago, at about the time that humans became skilful hunters. Every species comes into existence, evolves, may give rise to one or more other species and then becomes extinct. The vast majority of species that ever existed, therefore, are extinct, and all of those currently in existence have a limited 'life expectancy'. In the past, evolution has led to an increasing diversity of organisms and so, as genetic diversity flourished, mature, stable ecosystems also evolved. Today, however, we are concerned about reduction in diversity because extinction rates have increased rapidly in the last 200 years. Largely because their habitats are being destroyed, entire groups of organisms are forced to extinction at a rate greatly exceeding that of natural attrition and far beyond the rate at which natural processes can replace them. Ehrlich describes the systematic extermination of populations of non-human species as 'rivet-popping on Spaceship Earth'. Once a species' population is reduced to a critical point, the process of decline is accelerated as small populations have an impaired capacity to adapt to change due to loss of genetic variation. It is estimated that a minimum-sized population of approximately 500 is required to enable a species to keep pace with environmental change. However, if only a proportion of the population participates in breeding, as is the case in many mammals and birds, that number would need to be increased. Therefore, in the current artificial situation, instead of extinction of a species leading to its replacement by others which can more effectively utilize the available resources, we are faced with 'dead-end' extinctions and the loss of genetic potential which will take millions of years, if ever, to correct.

Since the 1990s we are losing species at a rate approximately 1000 times greater than the natural rate of extinction. We regularly receive news items concerning recent extinctions and consequently run the risk of forgetting the true impact through 'repetitive fatigue', even for the high profile species. The biggest concern is that we have not even begun to understand whether ecosystems are comprised mainly of assemblages of redundant subsystems, which can be randomly discarded at will, or whether we are destroying key systems without which the biosphere cannot function. Geneticists like David Suzuki warn us, as a 'matter of survival', that humans must halt or reduce dead-end extinctions and restore harmony with the natural environment by balancing increased human longevity with decreased fecundity. With accelerating extinction, evolutionary potentialities and the biosphere as a whole are degenerating. Common sense tells us that the limits of the genetic variability of populations, species and ecosystems are not boundless.

The right of other species to exist

This has nothing to do with balancing economic costs and benefits to humanity. Species, other than *Homo sapiens*, which may be the only other living creatures in the entire Universe also have a right to exist. To believe that humans beings are the only important form of life and to push all other organisms towards extinction is the ultimate form of foolishness and arrogance. War, destruction of the environment and greed are both morally and ecologically wrong. As sentient beings, we should expect some ethical behaviour from ourselves. Other species should not be pushed to extinction before their time because they, like us, struggled for and achieved evolutionary success.

Australia has the international record for the greatest number of extinctions in the shortest period of history and South Australia heads the national list of States. Early settlers in Australia saw the wilderness as a deadly enemy to be conquered and tamed, dominated and changed without pity or sympathy. It was Australia's isolation which gave rise to its unique flora and fauna but, because of this isolation, it is particularly vulnerable to the introduction of robust foreign species. Free from their traditional enemies, introduced species outcompete the indigenous fauna and flora. Another factor in Australia's habitat destruction is its semi-arid climate under which the land is unsuitable to foreign agricultural management

Fig. 14.1 Present restricted habitat of the numbat (*Myrmecobius fasciatus*) compared to its range at the time of European settlement.

techniques developed to suit European climatic and soil conditions. Abuses such as land clearance, monocultures and introduction of species which have become pests have threatened or pushed to extinction approximately 22% of arid-zone and 90% of desert-zone indigenous mammals. The introduction of foreign species has caused reductions in both the population and the range of many Australian fauna by predation, competition for habitats and introduced diseases. Habitat abuse by rabbits and sheep has led to the disappearance of the burrowing bettong and the eastern hare wallaby. Other threatened species are the numbat, bilby, honey-possum, koala, echidna and platypus, all of which score highly on biological uniqueness. The numbat, for instance, is now on the brink of extinction, restricted to a few small groups in south-west Western Australia (Fig. 14.1). Elsewhere, its numbers have been reduced by land clearing, inappropriate fire regimens and introduced predators. A lot of guilt has, recently, been felt about the plight of Australian indigenous species and co-ordinated efforts to keep species such as the numbat will probably save many from extinction. Efforts to establish the numbat in captivity began in 1987 with limited success. Also, the release of pollutants, in particular insecticides and other toxins (for example 1080 or monofluoro-acetic acid intended for rabbits but lethal to all mammals), has jeopardized the survival of many species. In total an estimated 4000 species of Australian plants and animals are now extinct. This tally may be a gross underestimate because many of the 'non-interesting' or small species may also have disappeared; all in a short 200 years.

The common goal of the farmers to monopolize all grasses for their herds or the entire photosynthetic activity for their crops has serious environmental consequences. Because of this exploitative farmer mentality, the displaced native vertebrates have nowhere else to go as any plant or animal not sanctioned by the farmer is promptly eradicated. Alien herds of mammals with cloven hooves break up the top-soil, predisposing the land to erosion and creating large expanses of treeless wasteland. It is not a lack of evolutionary imagination that results in all Australian species of mammals having the soft padded feet, the perfect solution for a semi-arid continent. Elsewhere, denuding the countryside of trees has permitted the brackish water table to rise, salting out vast areas of potentially fertile land. The irony of this attitude, despite the loss of fauna and flora, is that most farmers are at the same time extraordinarily inefficient producers; it is neither cost effective nor possible to achieve complete ecological monopolies.

River systems and the oceans need special attention because it is not always easy to see what is happening under the water. Species known to be severely threatened in the Australasian region are the great white shark, with only maybe 1500 to 2000 left worldwide; the southern right whale and the humpback whale; Queensland's dugong and the Murray River's trout cod. There is also the worrying trend that amphibians are rapidly disappearing worldwide. Amphibians are a life form most vulnerable to water pollution because of their dependence on water for breeding. The disappearance of our amphibians gives us a warning no-one can afford to ignore – that our freshwaters may no longer be able to sustain life. Might this be the harbinger of future changes?

THE VALUE OF GENETIC DIVERSITY

The essential ingredient of successful crop or animal breeding, whether to improve an existing stock or develop a new one, is genetic variability. To preserve genetic variability, populations within species must be saved from extinction. For example, plants are perpetually evolving new ways of fending off the animals and microorganisms that attack them. In domestic varieties this is done, at great cost, through artificial selection of privileged varieties grown as monocultures. Artificial selection is intimately related to the problems of extinction, as monocultures stand for genetically identical crops. Mixed herds that would take advantage of a broader spectrum of natural vegetation are a complex solution that might be ecologically successful. To herd native species, for example kangaroo farming, is probably a good idea for two reasons. Firstly, they are already adapted and so will not cause the destruction that introduced species usually wreak. Secondly, and perhaps most important, any animal gaining the prestige of being domesticated is safe from the threat of extinction. There are, however, an increasing number of enlightened farmers who work with the land to retain pockets of bush, gullies and river banks which provide habitats for many small natives. The encouragement of non-domestic animal species together with their associated native plants has rejuvenated land, restored balance and increasing farming efficiency.

I have singled out Australia because it is an extreme example of the damage that biological ignorance and arrogant disregard for the environment can do in a few human generations. Politicians focus on economic growth but disregard the shrinkage of the ecosystems in which our survival is embedded. In fairness, however, we have learned a little from the past and are now aiming to restore to some cultivated land a richer array of genetically distinct populations adapted to local conditions. Serious ecological issues have had political triumphs, for example the scientist MP Dr Bob Brown led the conservation movement campaigning on behalf of the unparalleled beautiful wilderness surrounding the Franklin River system in south-west Tasmania. This area now has World Heritage status and offers economic (through tourism) and inspirational rewards in perpetuity to the people of Tasmania, as well as safeguarding the habitats of many threatened species. The Labor ex-Prime Minister Bob Hawke, in one of his greener moods, initiated a programme of planting one billion trees in Australia by the year 2000. It is a country's people, however, who must support and lobby their leaders to actively direct resources towards peace and rehabilitation of the environment. We cannot exist without our environment, and our unchecked population increase at the expense of everything else will also cause our own extinction quite apart from any intrinsic right other species may have to existence.

CONSERVATION

The value of conservation to the world around and to the survival of the human species can be considered from different perspectives.

The economic argument

Humanity takes direct advantage of many other species at a level that is largely unrecognized by the vast majority of people, especially those in rich, high resource-consuming countries. The high opportunity cost of today's trend toward species extinction and environmental degradation cannot be calculated in monetary terms as its value has never been costed; however, common sense tells us that harvesting on a sustainable yield basis must be practised for a perpetual supply of anything. In contrast to sensible economic theory, we are living off our capital. The biosphere, that is, humans, domesticated animals and plants, and the creatures of the wild, is all part of a much larger natural ecosystem whose deterioration cannot be ignored on the balance sheet. We need to balance between the immediate short-term advantages gained by the sacrifice of species or populations and the long-term value of keeping them intact for their best use if preserved.

The economic argument for the need for conservation is widely recognized, but it is implemented, to a minor extent, only in some affluent communities, typically those in which population growth is close to zero or adjustable by planned immigration. The majority of humans are forced by poverty, mismanagement and population pressure into over-exploitation of their natural resources. A primary cause of both individual and ecological poverty is war, where the predatory state consumes resources to further its conflict and thus traps its inhabitants in the cycle of war debt. Economic conservation is concerned with the maintenance and improvement of human life-support systems by the rational use of renewable natural resources. For example, it is concerned with water supplies, soil fertility, control of pollution and harvesting of non-domesticated biological-based industries, such as forestry and fisheries, on the basis of sustainable yield. Conservation is also concerned with recycling, minimization of energy wastage and the management of finite resources of direct benefit to *Homo sapiens*. In 1990, the Australian Government initiated debate by the circulation of discussion papers on ecological sustainable development. The initiative consisted of draft proposals for the implementation of sustainable development in the various sectors such as agriculture, forestry, fisheries, mining, energy, manufacturing, transport and tourism.

Yet sustainable development, the balance of economic development and environmental protection, is in itself a major business opportunity not fully appreciated. Experts predict that costs incurred in cleaning up the environment and preventing further damage will, in future, exceed even the present staggering military spending. Companies in environmental industries can make huge profits from sales relating to environmental technologies. By proper care of our resources the economic argument incorporates into our lives the principles of equity and efficiency: equity, the just distribution of natural resources among present and future generations of users; efficiency, a natural resource should not be wastefully exploited. Sustainable development will also give rise to a more compassionate society.

The wildlife argument

Conservation of wildlife encourages the management of the earth's biosphere in such a way that any existing mix of species can remain until made extinct by natural processes. Altruistic concern for wildlife conservation, like economic conservation, is largely restricted to affluent societies. In over-populated societies, the overall consideration would be one of profit; for example, the silverback mountain gorilla in Rwanda or the African elephant in Kenya will only be preserved if they are worth more money alive than are their severed hands or tusks. This interpretation of conservation is still, however, based on human needs or responsibilities. It is increasingly asked, 'why save endangered species, and if so which ones?'. Animals are intrinsically selfish and we, with our power for manipulation of the environment, are the worst. Concern for the welfare of other species arises from enlightened self-interest or is an expression of genuine grief at the loss of our fellow species. Natural ecosystems are the product of many millions of years of evolution, and, like every other organism, human beings have played a part and are a product of that ongoing process. We are dependent on those systems for support.

An important aspect of conservation teaches that all human systems depend to some extent on non-domesticated animals and if something which is useful to us is destroyed then we have lost it for the future. For example, if we inadvertently kill off most insects, we have also lost many of our pollinators. Since a large portion of the world's forests, natural buffers against the increase in greenhouse gases, have already disappeared, marine algae, which are the major buffer against CO_2 increase, must be protected from destruction by pollution through special attention to the seas. It is also in the interests of farmers and horticulturalists to retain the maximum possible diversity of wild relatives of domestic cereals, fruits and vegetables. Such plants constitute a source of genetic variation which can, if necessary, be utilized in the future. Similarly, it is in the interests of pastoralists for the world to retain all wild species or subspecies of domestic mammals. The pharmaceutical industry supports scientific searches among wild plants and animals for chemical substances with useful drug properties. It is argued that the widest possible choice is needed to cover us for future changing conditions. For example, the tropical forests of the Amazon should be saved because of the immense value of the yet undiscovered foods, drugs and other resources that can be sustainably extracted from this ecosystem. Although this may be a morally selfish attitude, through the preservation of such an entire environment numerous other unique organisms can be preserved at the same time. As mentioned before, it is the older, more mature ecosystem which is most likely to provide future security, since its genetic diversity provides flexibility to respond to a changed environment. Rapid climatic changes, perhaps accelerated by human intervention, may lead to the loss of the species which cannot adequately respond to change, perhaps initiating conditions which are unstable for monocultures.

Another approach to wildlife conservation stems from feelings of biological affinity. The study of evolution leads to the realization that all organisms are related in varying degrees. Some humans feel that they have a responsibility toward fellow creatures proportionate to the degree of their human relatedness; for example, more responsibility for mammals than for birds than for reptiles than for fishes. The mountain gorillas of Rwanda's Mount Visoke, popularized by Diane Fossey and the movie *Gorillas in the Mist*, should survive if human beings really do care about their close relatives; sometimes nepotism can further a good cause! If something of this nature is saved for future generations to enjoy, it will be because millions of people have learned to care about wildlife and have insisted that the slaughter be stopped. Compassion and a sense of justice are strong elements here. Other products of evolution also have a right to existence, and humans needs are not the only basis for ethical decisions.

Reverting to the theme expressed at the beginning of this chapter, it is not only children who identify with nature. There is a deep feeling for other life forms that runs through all human societies. In many cultures that feeling has developed into a special relationship with other living things. Australian Aboriginal culture, for instance, is based on a striking connectedness with the natural surroundings. The 'Dreamtime' (Alcheringa) legends tell of the spirit world and teach that living people are the custodians of Earth Mother who has to be respected and protected at all times. When Aboriginal children are born they are given a child name; at puberty they are given an adult name designating living things for which they are responsible. With adulthood full environmental responsibility, as custodians of all living things, is expected. This description of the world in the present Australian context is Aboriginality and is akin to the view of nature, as seen in the past, by the Natural Theologists. People are more at peace in the natural world and for some contact with nature is an essential for survival. In this regard, Chapter 13 deals with the relationship of crowding, stress and aggression.

The biological argument

The biological argument for conservation is an indirect human-oriented economic argument. It arises from the healthy fear that if humans irreparably damage the environment and deliberately force many species to extinction, they are also attacking themselves and endangering their own survival. Geophysiological models exist which, if they approximate reality, equate diversity with global stability and resilience, making the previous two sections' justification for conservation small by comparison. In addition the benefit gained by a holistic ecological integration, in my opinion, is fundamental health through a shift in consciousness: living within the natural world not on it. Thus, the present environmental crisis may force us, to our advantage, to open up a new order of existence.

Gaia: a post-Darwinian evolutionary theory

Gaia: A New Look at Life on Earth by James Lovelock expounds the hypothesis that life – the biosphere – actively controls and adapts the atmospheric environment to its needs. That is, the physical and chemical condition of the surface of the earth, the atmosphere and the oceans are continuously made fit and comfortable for life by the presence of life itself. This is in contrast to the conventional wisdom which holds that life adapted to the existing planetary conditions as they evolved. Of Gaia's (named after the Greek goddess of the Earth) possible mechanisms or organization we have no knowledge; however, like children playing with adult equipment, we are bold enough to tamper with it. The theory may be supported to the extent that now it can be demonstrated, with the aid of numerical models and computers, that a diverse web of predators and prey facilitates a more stable and stronger ecosystem than a few more self-contained species or a shorter food chain of very limited mix. There are mechanisms controlling global homeostasis where independent homeostatic systems are merged and co-ordinated in the interests of the greater whole. That is, the world cannot be viewed as the sum of a series of complex systems because the co-operative network of all its systems has independent properties and powers much greater than the sum of its parts. Lovelock defines Gaia as 'a biological cybernetic system with homeostatic tendencies'. In this it is similar to the homeostatic mechanism underlying the body's adaptive response to stress which balances the energy needs of the whole organism by the master-control of the individual needs of its subsystems (Chapter 13). Thus, both Gaia and the GAS appear to be self-regulating, self-sustaining systems continually adjusting their physical, chemical and biological processes in order to maintain optimal conditions for life and, at the same time, their continued evolution. If the planet functions as a unitary system, its constituent parts, particularly the biosphere, co-operate in achieving a global homeostasis which can dynamically respond to internal and external challenges. Gaia is a single, natural, highly stable system which reflects the totality of spatio-temporal information passed on from generation to generation of living things.

The three principles of Gaia

If natural systems are committed to the elimination of randomness by virtue of the fact that they function, as has been suggested from cybernetics, it may be possible to identify mechanism(s) by which the basic stability or homeostasis is controlled.

Lovelock postulated that if Gaia existed she could be characterized by three major principles.

To keep conditions constant for all life Richard Dawkins has observed that both major and minor technological advances can be regarded as analogous to mutations. A body can cope with a certain number of mutations – the older we become the more mutations we carry but still function well. The cumulative effect of mutations is ageing, and ageing (the stress of life) weakens our ability to effectively homeostase and predisposes the body to deterioration. Most mutations, therefore, are deleterious to the organism; however, other mutations are the raw material driving evolution. Given more knowledge about the functional mechanism of Gaia, and our own human inventiveness, there may be room for optimism that increased understanding will give us, with the aid of beneficial 'technological' mutations, a better chance to facilitate, not harm, the essential homeostatic process. The ability to respond to environmental change is essential for survival on this changing planet. Now, as never before, with the greenhouse effect and other serious environmental problems having no ready solutions, the future could radically change.

Gaia has vital organs What we do to the planet may depend greatly on where we do it. It is suggested that the essential part resides in the tropics and is that which dwells on the floors of the continental shelves and in the soil below the surface. The destruction of reefs, such as the Great Barrier Reef, will have cascading effects not only on fish species throughout the tropical oceans but also on the shores and harbours now physically protected from erosion and wave action. If the Gaia hypothesis holds true, the reefs are also involved in the crucial task of regulating the salt content of the oceans by acting as the evaporation lagoons between the water and the tropical shores. In a similar manner, the humid tropical forests keep the climate cool through their capacity to transpire large volumes of water vapour. Their replacement by monocultures could precipitate uncontrolled climate fluctuations. Therefore, the fate of the tropical forests will be the major factor that determines the biological wealth of earth in the future. These dwindling ecosystems are the greatest single reservoirs of biotic diversity on the planet: approximately two thirds of the terrestrial species in the tropics occur in the rainforests.

Gaian responses to changes are covered by the rules of cybernetics Earth's climate and chemical composition are uniquely favourable for life and maintained by life. For example, the regulation of oxygen levels in the atmosphere has been constant for thousands of years. Such strongly homeostatic processes give the least warning of undesirable trends. By the time a malfunction is noticed inertial drag will bring things to a worse state before an equally slow improvement can set in. Population growth has inbuilt inertia, that is, population numbers will go up still further, for a time, before the effects of any controlling measures will become apparent. A major concern when dealing with equilibria is the possibility that systems that are homeostatically controlled may react violently and be destabilized by erratic or sustained oscillations if disturbed too far. If this were to occur then the extent of global warming or the depletion of the protective ozone shield may deviate greatly from present extrapolations. It is all very well for the politicians, having a short-sighted view extending only to the next election, to decree a change based on present projections of a reduction in CO_2 emissions by a specified amount by the year 2000. This is looking at the situation now, but the situation now is what will have an effect many decades from now. The more we know, the better we shall understand how far we can safely go in availing ourselves of Gaia's vital organs – the oceans and the earth's living surface.

THE PRESENT SITUATION

Everything depends on reducing our impact on the natural environment. The long-term consequences of abusing our present powers as the dominant species and recklessly plundering, exploiting and wasting the environment's most vital regions are not at all clear. The optimum number of people is certainly not as large as the maximum the earth can support. If the assumption is valid that the world's tropics and the seas close to the continental shores are Gaia's vital organs, then the principal dangers to our planet arise from human activities in those areas. Industrial society without any pollution control and the ever-growing world population are changing the land, sea and air so that the entire environment is under threat. For example, the damage caused to the fauna of the Pacific Ocean by the large-scale use of drift nets, many up to 65 kilometres in length, can be compared in gravity to the slaughter of Africa's big game and the destruction of the Amazonian rain forests. The drift nets operated by factory ships have helped increase the total worldwide catch of fish by more than five-fold since World War II and more boats are being recruited daily. They have also depopulated the seas of fish and threatened with extinction sea mammals such as dolphins, porpoises and small whales which drown in these walls of death. Many of those animals which have so far escaped the drift nets are faced with starvation because the humans are taking all their food. Areas hundreds of kilometres square have deteriorated into marine deserts bereft of fish, water birds, turtles and all kinds of sea mammals. The search to feed the ever-growing human population is denuding the planet – where will we end up when every non-human living creature is consumed? Population growth will be forcefully controlled if we cannot ourselves choose to control our natural birth rate, intelligently matching it with our power over the natural death rate. To take the stress analogy further, the planet is now at the stage of resistance but this stage is not static; either the situation improves or deteriorates, and we need to be very careful that our planet does not enter the stage of exhaustion.

B CONSERVATION BY MANAGEMENT

What can the individual do? Nature warns us that we must stop polluting the environment, felling the forests, depleting the soil and fishing and hunting species to a degree which exceeds their capacity for self-restoration. We can all contribute to reversing the destruction of life's genetic diversity by preserving the biosphere's still relatively intact gene pools.

Political aspects

In the section on the value of genetic diversity, I drew attention to some of Australia's polical/environmental triumphs. If human beings really do care about the environment, 'green politics' is an effective and powerful tool to bring about change. In the 1980s we have seen a consumer-driven revolution in the food industry which now, in all developed societies, has many checks and balances, not least those of honest labelling of ingredients and regulations enforced by food additive boards. National and international trade pressures, started by the individual, have brought this about. If enough of us want it, politicians will find it worth their while to place a value on our environment and to allocate funds for its restoration. Scientists either have or can develop technologies, often expensive, which can reverse or ameliorate most of our pollution problems; what is now needed is the strength of will to implement our collective expertise.

It is clear that major problems will be encountered in relation to legal aspects because different nations' development is uneven. Also the prevailing political situation is often crucial to environmental policy, with military governments outlaying most of their states' resources on methods of destruction, while some others demonstrate more successful environmental management. A similar argument to that proposed in Chapter 13 for linking emergency aid to population control may be advanced here. The idea is to tie aid to demonstrable progress in regard to environmental policies. The aim would be to wean developing countries from the aid trap and the debt cycle by encouraging free open markets with the removal of trade barriers and reduced pressures for protection. By encouraging efficiency, aid can lead to sustainable public good. The green revolution was largely based on aid. There is a need to actively reduce poverty through fecundity control and to rehabilitate the environment. These projects may be kick-started with carefully managed aid invested in sustainable productivity, and with none made available for military purposes or monuments to political self-agrandisment that often commemorate only the meaningless destruction of our planet with explosives.

Legal aspects

Illegal trade in wildlife is growing. Concurrently, the legal trade in endangered species is also increasing. For example, Brazil trades in endangered Amazonian animals to help alleviate its debt. Many of Australia's rare and unique parrot species are illegally traded. The illegal bird trade, in particular, is wasteful as at least 50% die in transport, a further 25% die subsequently in the markets and only 25% actually reach the buyer. This is despite the fact that the pet trade can be sustained from birds bred in captivity.

Another danger, particularly in Britain, is egg collecting. Eggs from endangered birds of prey are especially sought after. The British authorities have to build artificial osprey nests in secret locations and mount security guards to keep egg thieves away. Such environmental terrorism will not continue if the demand for such products ceases. Unfortunately, however, the wildlife trade yields enormous profits, with enforcement and penalties far lower than those for narcotic smuggling. A treaty based on national legislation in member countries, the Convention on International Trade in Endangered Species of Wild Fauna and Flora (CITES), has been set up for the control of this trade in animals and plants. CITES aims to control the legal trade of many hundreds of endangered species throughout the world via a special system of permits. However, international law is generally not very effective since local enforcement is weak or non-existent. It is not unusual, for instance, to see endangered animals being openly traded in markets in Indonesia and Thailand.

Linked to the trade in wildlife, there is also an illegal trade in wildlife products, for example ivory and rhinoceros horn. The black rhinoceros in particular (although the white rhinoceros is not safe from poachers either) is on the threshold of extinction because of the demand for its horn, either for use as dagger handles in the Yemen or in powdered form as a 'cure' or 'insurance' against male impotency in several Asian countries. International law has outlawed trade in rhinoceros products, but despite this it is still flourishing in China, South Korea, Taiwan and Thailand. These countries have been so successful in depleting the world's population of this beautiful and unique large mammal that in Java only 65 individuals are left alive. This illegal trade continues despite the availability of alternative preparations from non-endangered species. Animal tests have demonstrated that powdered buffalo and saga antelope horn is medically beneficial in aiding feverish rats (as is aspirin). Present research is aimed at the isolation of the active principle. Perhaps we should not expect too much from this scientific approach since most traditional cures for male impotency rely more on magic than pharmacological efficacy.

Education

Of course, the success of any conservation programme depends upon the effective education of all people. Education is powerful in alleviating a sense of helplessness and in formulating priorities based on a better understanding of the facts. Effective education needs to start at birth. Children respond positively to high quality literature, as witnessed by the popularity of Dr Seuss' book *The Lorax*. Suitable environmental education should not only be factual but also encourage the development of strong feelings for ecology, endangered species and the biological future of the planet through practical involvement. A collective responsibility toward our mutual problems can only raise our self-esteem and be for the general good – loving your neighbour means loving the environment.

Tourism and the environment

Tourism is the largest industry in the world. Between 200 and 600 million tourists turn over an estimated 20 billion US dollars each year, but this mass tourism is itself a threat to the environment due to ecosystem pollution, aesthetic pollution and cultural pollution.

Despite these disadvantages, the industry, because of its size, can dictate the development of tourist sites. It is the environmentally responsible tourist's duty to demand that the development companies get involved in conservation issues. Pressure must come from the consumer, as ultimately tourist companies will only adjust to market forces. A responsible tourist is one who demands to see sustainable tourism, and there are several strategies that can ensure success from this point of view.

Rejection Boycott tour operators who excessively destroy wilderness areas or open spaces with the development of environmentally insensitive buildings. Large parts of Australia's formerly unspoilt coastline, the asset that attracted the tourists in the first place, have not been sustained but covered with concrete. Countries like Tunisia, Gambia, Zimbabwe and

Madagascar have been improved by tourism because tourism reversed the process of land degradation and rejuvenated green areas. This has profited the people and, since foreigners go to those countries mostly to see wildlife in an unspoilt landscape, the wildlife has been marketed to its advantage. The money earned has gone back to sustaining the industry and further local prosperity. If an area, such as a forest system, is threatened, the tourist industry could be used to preserve that area. Care, however, must be taken that the nature reserves bring material benefit to the people living near them. It is the local people who hold the key to the survival of wildlife and wilderness and they must preserve a vested interest in it.

Restraint Limit the number of tourists per year allowed to visit certain vulnerable areas. This has been successful in, for example, the Himalayas of Nepal and the site of the prehistoric cave paintings at Lascaux in France. Entry to such places is by permit only. This principle may also work in areas similarly vulnerable to interference, such as Antarctica. Once having visited such places the overwhelming beauty leaves the tourist eager to preserve the environment.

Wisdom Select ecologically sound holidays. If more people insist on a minimum of environmental spoilage, then consumer pressure will favour companies that take green issues seriously. Recreation can then also be educational, especially if tour organizers are knowledgeable conservationists. For instance, Earth Watch combines travelling with the facility of tourism. These are holidays which actually are based on preservation such as Coral Keys Conservation in the USA. Divers can cause harm to coral reefs and volunteers can combine recreational diving, fishing and survey work. Marine parks, like that covering the Great Barrier Reef in Australia, profit from biologists giving their time gratuitously in teaching visitors. Areas such as these need to be allowed to grow and be thoughtfully exploited.

Forethought Spread the tourist season over more of the year. As an example, the staggering of industrial and school holidays would spread the burden of tourist numbers and increase the joy for the individual traveller.

Sustenance Promote exotic and vulnerable species as a food source for tourists which may help in their preservation because of their enhanced commercial value. 'Conservation cuisine' is popular in African countries where dishes made from crocodile, zebra and various species of ungulates sell for high prices. In Australia a switch from cloven-hoofed ungulates to kangeroo farming could rejuvenate grazing-induced deserts and provide consumers with game meat containing less than 1% fat, compared to the usual 40% found in mutton. Of course animals used as a food source must not come from the wild but be bred in captivity or culled from surplus animals on sustainably managed wildlife parks.

TACTICS OF CONSERVATION

Presently equity, justice, compassion, peace and comfort for all tend to take a subsidiary role to the drive to increase the gross national product and the acquisition of wealth. Well-being and wealth are not necessarily synonymous; people need a certain level of comfort beyond which there is no real further fulfilment, only a waste of resources. We are conditioned, particularly in the west, to the bizarre notion of linking fulfilment with wealth and material growth. Destruction of the environment for wealth acquisition may benefit one generation but the cost is paid by our children. Continual growth will eventually lead to an ecosystem collapse in which all species, including humanity, will fall victim. We desperately need a rationally planned transition from perpetual growth to a steady-state economic system, to stop using our capital and live on our interest. Our future has got to be moving towards a sustainable society, a society dedicated to living within environmental constraints. The World Conservation Strategy rests on three principles:

(1) maintaining essential ecological processes and life-support systems;
(2) preserving genetic diversity;
(3) utilizing species and ecosystems sustainably.

One answer to our problems is to preserve as much biological diversity as possible in zoos, botanical gardens, arboreta, national parks and game parks. This can never be the complete solution because of space limitations. Large-scale unmanaged reserves and changed human behaviour outside of reserves are the principal hope for slowing the extinction rate. New farming attitudes, as described previously, can be effective in preserving indigenous species in their habitat alongside human activities. The toleration of a variety of indigenous species, even if not farmed commercially, will in the long-term increase farming efficiency through the rejuvenation of the land. An additional bonus would be the beautification of the countryside by the development of pockets of natural bush among cultivated fields.

The 1980s has seen many developments in the technology associated with animal reproduction. Much of this technology had its origin in greater agricultural efficiency but was further stimulated in response to the need for human infertility treatments. Chapter 1 deals fully with such developments; however, the reproductive technologies of electro-ejaculation, artificial insemination, oestrus control, superovulation, *in vitro* fertilization, cell micromanipulation, embryo transfer and surrogacy are all tools used in the scientist's fight against species extinction. To this list can also be added recombinant DNA technology. It seems appropriate that the knowledge gained by human fertility research can now be used to the advantage of other species suffering from decline due to worldwide human population pressures.

Planned breeding programmes, aided by artificial insemination and embryo transfer technology, are used to amplify the international distribution of valuable genetic resources. These techniques do not improve fecundity but they are a solution to problems concerning the maintenance and spread of genetic diversity. Cryobiology enables the indefinite storage of advantageous genes from animals threatened with extinction or overcomes problems of inbreeding. For example, populations represented in several separate zoos by only a few individuals together may form the nucleus from which the species may be rejuvenated. Careful selection of donor males has already been shown to be effective in the maintenance of genetic diversity. The usual principle goal of such captive breeding is maintenance of 90% of the genetic variation in the source (the wild population) over a protracted period, for example 200 years. This period of 200 years is the amount of time population biologists have estimated that the human population will continue to grow before levelling out. For some species, tables are available that permit estimation of the minimum size of the captive group, given knowledge of the exponential growth rate of the group and the number of founders. In most cases, founder groups will have to comprise over 20 effective, that is breeding, individuals.

To make the most of the potential of cryobiology, it is necessary to create gene stocks or banks housing the gametes and embryos. Artificial insemination and embryo transfer greatly increase the intensity of selection if combined with other technologies, such as sex control. The separation of spermatozoa carrying the X and Y chromosome will enable the sex of the offspring to be chosen in accordance with a predetermined breeding programme. Embryo transfer can additionally be useful as a means of controlling or eliminating an infectious disease in a herd or flock, provided that the pathogen under consideration does not transfer to the embryo. Surrogacy can be of use for increasing the productivity of a given population and for the genetic management of populations. Interspecies and transgenus surrogacy has also been successful, leading to hopes for species without close relatives. In eutherian mammals the process is similar to that used in humans; however, for marsupials, surrogacy is a relatively new field with much more research required before it can be applied to saving endangered species. For instance, blastocysts collected after superovulation and natural mating can be transferred to synchronized surrogate mothers for gestation to full term. Alternatively, young can be pouch-fostered by matching up the lactating foster mothers to the stage of postnatal development. Progress in artificial breeding for marsupials in general has lagged behind that for eutherians because more research is required on individual species for traditional methods to be successful. However, Australia has considerable research strength in marsupial developmental biology and reproductive technology, and this knowledge and experience should soon develop effective artificial breeding techniques benefitting endangered marsupials.

Major breakthroughs in genetic engineering now make possible directed recombinations of individual gametes, interspecies transfer of genetic material and the creation of new geno- and phenotypes. The reconstruction of extinct species from preserved DNA, however, is not yet very realistic using to-day's technology, but reproductive science moves fast. A future possibility may be the reconstruction of a thylacine or woolly mammoth surrogated in a suitable mother!

Breeding strategies in zoos

Captive populations can serve as a last resort for species which have no immediate opportunity for survival in nature. The modern zoo plays an important role in education, conservation and research, as well as in the traditional roles of recreation and tourism. Zoos have been intimately involved in captive breeding and systematic re-introduction programmes for many species. For example, Przewalski's horse, *Equus przewalskii*, the last surviving species of wild horse once common in Central Asia, does well in captivity. However, to re-introduce it into the wild may pose difficulties because after only five to eight generations in the zoo environmental changes in reproductive characteristics have evolved. Most noticeable is that foals are often born outside the sharply defined foaling season typically seen in the wild, and this would make them vulnerable in a harsher, unprotected habitat. In contrast, the Arabian oryx, national symbol of Saudi Arabia, has successfully been returned to the wild. The

Bedouins were recruited into saving this beautiful white antelope and protecting it from poachers.

Unfortunately, not all threatened species can be successfully resurrected. Often, because of the stress of the artificial environment in zoos, animals have trouble breeding or do not breed at all (after all, zoos are analogous places to prisons). For example, giant pandas rarely breed in captivity and even artificial insemination has had limited success because frequently the few cubs born do not survive. The small wild population in China is continuously dwindling through habitat deterioration. More than 40% of the entire wild panda population is under threat from the patchy flowering and subsequent death of bamboo. In the past the pandas were able to migrate to other areas of food abundance, but such escape routes are no longer available because of encroaching human habitation.

The loss of a plant species often leads also to the loss of several animal species that are dependent on it. Plant conservation has lagged well behind concerns for animal conservation, despite the obvious interconnectedness of all ecosystems. Many Australian flowering plants, for instance, are endangered or have become extinct since European settlement. Only 23% of the endangered species are known to occur in a conservation reserve because 65% of Australia was cleared for agriculture without concern for the occurrence and distribution of the flora. An important task of the modern zoo emphasizes conservation and the relationships between plant, vertebrate and invertebrate species in simulated habitats. In this way the general public, often untrained in biology, can appreciate that the four major causes of species extinction are habitat destruction, environmental contamination, excessive exploitation and pressure from introduced species.

Breeding strategies in outside zoos

The primary problem of breeding in captivity is inbreeding, so that genetic diversity, essential in order for the animals to be equipped to readapt in the wild, cannot easily be preserved. Attempts to maintain critically endangered species in the biotic vacuums of zoos, laboratories and botanic gardens are of little use unless intact ecosystems are also preserved. Ultimately, conditions to improve the endangered species' chances for prolonged survival are related to the protection of the habitat, that is, the entire ecosystem which contains the endangered species. Species which can be saved in the wild before it becomes necessary to take measures involving captivity have their chances vastly improved. A drawback is that the fate of the species is closely linked to the fate of its limited habitat and to its accidental destruction by pollution or development.

Natural habitat reserves

In mature stable ecosystems, such as tropical rainforests, large reserves are needed in order to maintain stability. Many animals have a large range, and large areas are also essential for the reproduction of many plant species. Preservation of small patches of habitat is useless because if the animals (for example pollinators) die out so will the plants that depend on them, followed by the herbivores which feed on the plants. The management of wildlife parks has been very successful in those parts of Africa where the tourist's demands support their existence. Once reserves are set aside, however, problems of management, such as protection from poachers and squatters, usually arise. Management also involves brush clearance, control of water levels and control of undesirable plant and animal populations, all of which are costly. The beautiful Kakadu National Park, in the Northern Territory of Australia, suffers from all of the above problems. Feral buffalo churn up the billabongs and wetlands and introduced weeds choke up waterways. One of the worst is the giant sensitive plant (*Mimosa pigra*) introduced into the country in the late 19th century. Since then it has invaded vast stretches of open plains and the quiet backwaters of wetlands in tropical Australia. Neither chemical nor biological programmes can control its spread, and it has infested an estimated 30 000 hectares of wetland in the Northern Territory alone. It is now seriously threatening the wetlands of Kakadu National Park.

Despite their costs, national parks are vital to any community in serving its recreational and educational needs. If people are genuine in their concern for their parks, then short-term commercial demands for the immediate cash advantage from development or mining of the area will be resisted politically and ultimately outweighed by considerations in favour of long-term community benefits and financial gains.

Conservation beyond zoos and reserves

This, as highlighted before, involves radically changed attitudes towards the environment, individually, nationally and internationally. Some suggestions, which should not be too difficult to implement, are as follows.

- The areas between reserves need maintaining to ensure that they do not become biological deserts but continue to function as buffers supplementing the functions of the reserves by preserving biotic diversity at lower levels. Some possibilities are the development of greenbelt areas around cities and the growth of native vegetation along highways and railway lines.
- Agricultural land and natural resources need careful management: selective logging of forests and minimization of forest wastes; accepting some livestock losses to natural predators; and the maintenance of relatively undisturbed areas among agricultural land, especially near stream banks.
- The ubiquitous lawn, a biological wasteland surrounded by a border of exotic plants, may be put to better use so that these empty areas may come to resemble more complex natural communities which will attract associated animals. Much of this is already happening with the planting of more native trees and shrubs and less fencing in of Australian suburban gardens.
- The damage done by industrialization can also be consciously counteracted by efforts at restoration. A bonus is that shrubs and trees act, to some extent, as a shield or buffer against pollutants. Legislation enforcing regeneration after sandmining also exists.
- Agricultural methods can be restructured along more traditional lines, as practised prior to the industrial revolution. For example, permaculture, (agriculture based on permanence) is a traditional system which preserves biotic diversity and is at the same time suitable for the exploitation of private land. Spare open spaces can support a large variety of foods grown for private use with low energy expenditure. Farms, too, can improve productivity if fruit trees and vegetables are mixed with, for instance, wheat and other commercial crops. Greater genetic variability, that is, a 'cultivated ecology' will, as non-renewable resources run out, reverse the slow degeneration of the modern annual crop monoculture.

Fertility control of feral vertebrates using recombinant viral vectors

Uncontrolled breeding of introduced species is a major threat to indigenous fauna. Feral eutherian mammals are a particular problem in Australia where populations of rabbits, foxes, cats, dogs, brumbys (horses), donkeys, pigs, goats, buffalo, camels, rats and mice, just to name the most prominent, exterminate local species either by direct predation or by habitat destruction. Immunocontraceptives with crude preparations directed typically against zona pellucida or GnRH antigens and applied by trapping and injection or by darting have been successfully used in several countries including Australia. The principle of this technique is similar whether used in humans or other mammals and is fully described in Chapter 15. However, immunological and other methods of fertility control are labour intensive and expensive when applied to wild animals. What is required for pest species control is a cheap method of disseminating, irreversibly and specifically, the agent into the wider population. Research in the development of a species-specific contagious virus carrying foreign immunogens capable of targeting reproductive function has demonstrated good potential. Tyndale-Biscoe (CSIRO Division of Wildlife and Ecology) and a team of collaborators in Australia are developing an immunocontraceptive for the rabbit and the fox. For the rabbit, the myxoma virus (a highly infectious but not lethal mosquito-born virus of rabbits, which was artificially introduced into Australia to reduce the rabbit population) is used as the viral vector. The recombinant myxoma vector can transmit immunogens inducing an immune response in rabbits against zona pellucida and/or sperm surface antigens. This disrupts fertilization or implantation while having no effects on normal gonadal function and social behaviour. The approach is potentially so powerful, since the recombinant virus will be in the environment forever, that strict guarantees are necessary against fears concerning the safety of other species and cross infection to areas where the targeted species is not a 'pest'.

Preservation of human variants

The new-found consciousness toward other species has also highlighted problems in our own. Our evolution was characterized by the inbreeding of isolated subpopulations giving rise to human variants, occasionally rejuvenated by small gene migrations not exceeding a small percentage per generation. This ancient structure, except in isolates of indigenous people, has now been destroyed by intense global mobility and intermingling of distinct populations. Thus, the destruction or homogenization of separate gene pools will have the same effect and consequences as that seen in animals bred in zoos. It has been estimated that in a modern multimillion city population the mixing of large numbers of gene pools, each adapted to differing environmental conditions, can replace the populations' original gene pool in three to five generations.

In addition to global integration, indigenous cultures everywhere have to survive destructive advances of the more dominant (mainly western) cultures, indirectly through habitat destruction or directly by genocide. In this manner many of the native cultures of Amazonia, Australia and elsewhere have almost been lost.

General references

Australian Government (1990). *Ecologically Sustainable Development: A Commonwealth Discussion Paper*. Pirie, Canberra, Australia.

Ehrlich, P. & Ehrlich, A. (1982). *Extinction*. Victor Gollanz, London.

Goldsmith, E. (1989). Gaia and evolution. *Ecologist*, **19**, 147–153.

Gordon, A. & Suzuki, D. (1990). *It's a Matter of Survival*. Stoddart, Toronto.

Griffiths, R. & Seebee, T. (1992). Decline and fall of the amphibians. *New Scientist*, **1827** (June), 25–29.

Kennedy, M. (ed.) (1990). *Australia's Endangered Species*. Simon & Schuster, Sydney.

Lewin, R. (1993). Genes from a disappearing world. *New Scientist*, **1875** (May), 25–29.

Lovelock, J. E. (1987). *Gaia: A New Look at Life on Earth*. Oxford University Press, New York.

Lovelock, J. E. (1992). The Gaia hypothesis. In *Environmental Evolution: Effects of the Origin and Evolution of Life on Planet Earth*, eds. L. Margulis & L. Olendzenski, pp. 294–315. MIT Press, Boston, MA.

Mace, G. M. (1989). The application of reproductive technology to endangered species breeding programmes. *Zoological Journal of the Linnean Society*, **95**, 109–116.

Mollison, B. & Holmgren, O. (1987). *Perma-Culture One: A Perennial Agriculture for Human Settlements*. Tagari, Maryborough, Australia.

Rodger, J. C. (1990). Prospects for the artificial manipulation of marsupial reproduction and its application in research and conservation. *Australian Journal of Zoology*, **37**, 249–258.

Tyndale-Biscoe, C. H. (1991). Fertility control in wildlife. *Reproduction Fertility & Development*, **3**, 339–343.

Wilmut, I., Haley, C. S. & Woolliams, J. A. (1992). Impact of biotechnology on animal breeding. *Animal Reproduction Science*, **28**, 149–162.

Wilson, E. O. (ed.) (1988). *Biodiversity*. National Academy Press, Washington, DC.

15 Artificial control of fertility

HISTORICAL PERSPECTIVE

Contraception is not new: for thousands of years people have prevented pregnancy. In early and probably also prehistoric times, a naturally high mortality rate augmented by abortion, infanticide and deliberate pregnancy prevention controlled population size. Early human societies throughout the world developed measures to prevent conception, such as prepubertal coitus, prolonged lactation, delayed marriage, celibacy, withdrawal or various substitutes for 'natural' sexual intercourse.

Ancient documents from many different civilizations show a desire to prevent unwanted births. For example, a surviving Chinese medical text written around 2700 BC contains a prescription for an abortifacient. Prescriptions for contraceptives have also been discovered in ancient Egyptian medical papyri. The Kahun Papyrus, dating back to about 1850 BC, gives descriptions for varous vaginal pastes and menses-inducing tampons. The Ebers Papyrus (written about 1550 BC) is particularly interesting as it describes a medicated tampon. It makes use of ground acacia, a plant containing gum arabic that upon fermentation releases lactic acid, a substance recognized as a spermicide and still used today. Indian Sanskrit medical writings likewise refer to tampons, vaginal medications and abstinence. Coitus interruptus, or withdrawal, was probably the oldest effective and most universally used method of birth control in early times. This method is so effective that it is still practiced today, even in the many developed countries. The Talmud, the body of Jewish law and legend from the 2nd through to the 6th centuries AD, advises this particular practice whenever pregnancy would endanger life. This ancient Jewish contraceptive practice is also described in the Old Testament[1] 'And Onan knew that the seed would not be his; and it came to pass, when he went in unto his

[1] Genesis, Chapter 38, v 9. *The Holy Bible*, Authorized King James Version.

brother's wife, that he spilled it on the ground, lest that he should give seed to his brother'. The Talmud also recommended other methods besides coitus interruptus, such as the use of vaginal sponges, violent movements to expel sperm and various root potions designed to induce sterility in women. There is overwhelming evidence which indicates that men and women in the ancient world concurred that fertility should and could be controlled and that this control was desirable.

Classical Greece and Rome

Birth control was a subject of active discussion by men of learning such as Plato, Aristotle and followers of Hippocrates. In Crete, homosexuality was, according to Aristotle, officially supported as a population control tactic and anal intercourse with one's wife was not forbidden. Dedicated heterosexuals had their own options. It is said that Plato when asked how population stability could be assured, responded, 'There are many devices available, if too many children are being born, there are measures to check propagation; on the other hand, a high birth-rate can be encouraged and stimulated by conferring marks of distinction or disgrace'. Widespread resort to courtesans provided a sexual outlet for males who preferred not to impregnate their wives. Soranus (98–138 AD), considered the greatest gynaecologist of antiquity, wrote a comprehensive summary of contraceptive techniques in *Gynaecia*. In these writings, he clearly distinguished between contraceptives and abortifacients and described a number of preventive techniques including vaginal plugs and the use of astringent solutions and various fruit acids. Women naturally preferred to control their fertility by some form of contraception because recourse to abortion put them at physical and psychological risk. However, references to abortion are very common in ancient writings. Fertility control for many families also included practical solutions such as selling, killing or exposing of excess children. The Egyptians and Romans were opposed to the exposure of unwanted neonates, regarding this as an unpleasant Greek custom.

Like other peoples before them, the Romans used pessaries, tampons and plugs and wrote down recipes for pessaries such as sticky mixtures of peppermint, cedar gum, alum and axe-weed in honey. Caelius Aurelianus recommended a mixture of urine and vinegar. Remarkably similar solutions of vinegar or lemon juice were still being approved of by birth-control advocates in the early 20th century. If the Romans used them only after coition they would have a limited effect; if employed both before and after they would have been successful. A variety of herbal means was believed not only to prevent conception but also to destroy an already existing embryo. The herbalist Dioscorides described over 25 plants that had abortifacient properties, and many brews or potions prescribed from the ground-pine 'abiga', hemionion and bracken caused abortion. Later, the Romans contributed to contraceptive development through their use of goat and fish bladders as condoms.

It is of interest to note that Rome differed from the Greek city states in that at times it actively encouraged population growth. In good times, modest population growth could be expected; in bad times the intermittent scourges of war and famine wiped out entire generations, and concern for fertility was expressed in legislation to promote births (for example, in 59 BC Julius Caesar allotted land to fathers of three or more children). Heirs were so important that the Romans tolerated a variety of unconventional marriage arrangements in order to obtain them, such as the sharing of fertile wives. An interesting early surrogacy case was reported by Plutarch where he gives an account of Cato divorcing his wife Marcia so she could marry and bear an heir for a friend and later remarry him. It was also not exceptional for women to bear children for their infertile sisters.

Islam in the middle ages

Contraceptive practice spread to Europe through Islam, whose religious law at that time did not condemn either birth control or abortion performed for serious reasons before the 4th month of pregnancy. Popular contraceptive remedies, therefore, were innumerable and traditionally handed down by midwives, but several Islamic physicians also wrote comprehensive treatises on the subject. Physicians of the time considered contraception an integral part of medical practice and described a number of useful methods including various ointments, vaginal barriers and coitus interruptus. The last, known as azl, was probably the most frequently used and is even mentioned in the earliest Islamic teachings.

Although contraceptive knowledge and concern was widespread in the ancient world, only the privileged few had access to any of the medicated preparations. They were unknown to the average citizen who largely resorted to measures such as abstinence, prolonged breast-feeding and coitus interruptus.

Europe during the middle ages and the rennaissance

European physicians of the Middle Ages probably possessed contraceptive knowledge handed down from the Greeks, Romans and Arabs, but, because science and medicine were dominated by the church during this period, they applied it only in the most rigorously selected cases. The church condemned the practice of contraception as a vice against nature, an attitude which still influences personal behaviour and public policy in Christendom, especially in countries where there is a Roman Catholic majority. The position of Protestant denominations was similar to that of the Catholic Church until recent times when some denominations relaxed or abandoned their wish to control human sexual behaviour. The Christian abhorrence of sexual 'perversions' may have originated as a device to increase the numbers of the faithful but the logic for this may soon have been lost in the desire not to relinquish power over human destiny. In time the Church fell back on the argument that the need to procreate provided a rational reason for marriage and condemned any method of fertility control that permitted pleasure but prevented children. The idea that the use of a contraceptive might be offering protection to the mother and child from the dangers of unwanted pregnancies was not expressed. Meanwhile, legacies live on and as a result uncared-for children grow up in an atmosphere of contempt, suffering and violence; a particularly sickening example (given world coverage in 1991) concerns orphaned Brazilian street kids (numbering in their thousands) being murdered by privately organized exterminator squads in a futile attempt to clear them off the streets.

Because of society's abdication of its collective responsibility for reproduction, responsibility for procreation was placed primarily on women. The early medieval texts said little of the pessaries and suppositories mentioned by the ancients, and by 200 AD the Church forgot the teachings of one of our greatest social activists and emerged as a male-dominated, hierarchical institution intend on holding women to their 'natural' subordinate role.

Fertility control in early modern Europe and the fertility transition

In the early modern period, the discussion of fertility, once dominated by the church, was secularized. The state began to replace priests in policing motherhood and male obstetricians commenced their campaign to replace midwives in the delivery of babies. Childbearing was changing; by the 18th century Europeans not only postponed and spaced births, as they always had done, but they were preventing them altogether. In early 18th century England, condoms made of animal bladders or fine skins were advertised and sold by brothel owners; the English called them French letters and the French named them *la capote anglaise*.

In the 19th century, North America and western Europe entered a new demographic age. By World War I, family size was cut in half because fertility, particularly in the married, was brought to very low levels by the widespread adoption of birth-control practices. The invention of the diaphragm, for example, represented a significant innovation in fertility control. In the latter decades of the 19th century, contraceptives and abortifacients were freely advertised in newspapers and magazines, sold in barber shops and pharmacies and brought to villages and to working-class neighbourhoods by door-to-door salesmen.

The modern era

> Birth control is essentially an education for women
> *The Pivot of Civilization* Margaret Sanger (1922)

The restriction of family size that was achieved in western Europe and North America at the end of the 19th century continued into the 20th century and the modern era. What has characterized the modern era has been called the 'democratization of birth control', that is, the adoption of the practice by large segments of the population, poor as well as rich, illiterate as well as literate. The extension of family planning was carried by the general impetus surrounding enthusiastic human rights campaigns and the beginnings of the Women's Liberation Movement in the USA. Feminist Margaret Sanger (1879–1966) coined the term 'birth control' as a positive description of family limitation to replace the old economic term 'neo-Malthusianism' (after Malthus's message that contraception was the logical response to poverty posed by over-population). Marie Stopes (1880–1958) pursued similar work in England. Both successfully campaigned for legislative reform to permit the opening of medically supervised birth control clinics for the poor. Stopes and Sanger shared many of the same concerns; both were alarmed by the high maternal and infant mortality rates associated with large families. Needless abortion-related deaths in England increased from 10.5% to 20.0% of all maternal deaths between 1930 and 1934, and in the USA 8000 maternal deaths a year could have been avoided by legalizing abortion, at least until adequate contraceptives were made available. Kinsey's subsequent investigations revealed that one out of every five married American women had had an abortion. By the 1930s it was clear that medical abortion in the first trimester was safer (once antibacterial sulphonamides were available) than delivery at term if performed by trained doctors. The old argument in defence of the law, based on risk to the mother, was made irrelevant. In the west, the enforcement of the abortion laws was visibly class biased; a well-to-do women could usually find a doctor who would justify her need for a therapeutic abortion. The abortion issue became increasingly an issue of women's rights, and the campaign for decriminalization began to find advocates. Both Stopes and Sanger stressed the need for clinics supported by the government and directed by trained personnel to educate the public in contraceptive use. They believed that limitation of family size was not only economically necessary but morally acceptable. They developed the prevailing argument that contraception was not only compatible with pleasure but essential if the woman's full potential was to be expressed. Simone de Beauvoir, Sherry Ortner and Betty Friedan, likewise, re-enforced the idea that maternity had been used to keep women dependent.

The postwar democracies were less hostile to contraception; family planning was supported as part of the National Health Service in Britain. However, the proliferation of clinics occurred just as fertility rates rose due to the postwar 'baby boom' affecting, particularly, North America and Australia where birth rates remained high through the 1950s and into the 1960s. The domestic 'baby boom' was welcomed by the western governments because the rapid population growth of the developing world, brought about by a decline in mortality, was regarded as a threat to the global social order. For example, racial preoccupations in Australia, demonstrated by its 'White Australia Policy', blossomed in response to fears that the white nations could be submerged by the yellow and black. However, it soon became evident that there was a worldwide population problem of immense magnitude which could only be solved by the provision of cheap and effective contraceptives. Once fertility was seen as posing real dangers, scientists turned their full ingenuity and energies to tackling the problem.

NEW APPROACHES TO FERTILITY REGULATION

Humans now have come full circle by linking contemporary reproductive technologies with ancient wisdom and practices. Contraception by biological means involves the prevention of one or more of the following:

(a) Formation or release of gametes in the male or the female
(b) Fertilization
(c) Implantation of the fertilized. ovum, or development of the early embryo.

THE PREVENTION OF THE FORMATION OR RELEASE OF GAMETES

Contraception by the prevention of gamete formation or release can be accompanied in the female or in the male using synthetic steroids of immunological methods.

Oral hormonal contraception in the female

The ovulation-inhibiting effects of the oestrogens, androgens and progesterone were known since the early 20th century, but the full practical application of this knowledge was only possible when it became feasible, in the 1940s, to synthesize orally active steroidal compounds. Not until the mid-1950s, however, was it possible to synthesize a range of suitable compounds relatively cheaply. The use of oestrogen to inhibit ovulation was begun in the 1940s for the treatment of dysmenorrhoea, not for contraception. Several oestrogens, but particularly stilboestrol, were found to be beneficial. Two of the earliest orally active synthetic progestogens were norethynodrel and norethindrone. The progestogens which are related to either testosterone or 17α-hydroxyprogesterone do not possess some of the adverse side effects of the naturally occurring steroids, for example nausea, and are more effective orally. Medical scientist Pincus (Worchester Foundation for Experimental Biology) played a key role in pioneering work on the oral contraceptive pill. Ironically, his early research was directed at employing hormones to cure infertility. This was later reversed and after extensive clinical trials on norethynodrel, the first report of an effective oral contraceptive was published by Rock, Pincus and Garcia in 1956. In 1960, the Food and Drug Administration in the USA granted the Searle pharmaceutical company permission to market the synthetic anovulent as an oral contraceptive pill, and soon other drug companies produced similar products. The use of a synthetic progestogen provided a reliable control of ovulation, but the incidence of spotting, or 'breakthrough bleeding', was high. So it became an accepted practice to combine an oestrogen with the progestogen to maintain the endometrium. The orally powerful synthetic oestrogens are produced by chemical modification of oestradiol. A non-steroidal oestrogen, diethylstilboestrol, also proved to be highly effective when administered orally and was used extensively (Chapter 16).

The introduction of steroidal contraception was a major leap forward from the scientific, medical and social points of view. However, as always the times have to be right; this contraceptive revolution was both a cause and an effect of the dramatic social changes that were concurrently taking place in western societies. Now, with effective contraception, a mother who traditionally bore children for most of her reproductive life could cluster them and be free from childbearing whilst still relatively young. Accordingly, most women after World War II had the opportunity both to marry and to have a career. In Britain, for example, 10% of married mothers were employed in 1900, as opposed to 50% in 1976. Since its introduction in 1960, the history of hormonal contraception has involved a variety of synthetic compounds, differences in dose levels and differences in progestogen/oestrogen proportions. In general, there has been a progressive reduction in both the oestrogen and progestogen dose levels to reach the lowest dose compatible with the desired effect of inhibiting ovulation. There have also been efforts to deliver these hormones through a variety of routes.

Formulations

Fundamentally, there are only two categories that exist, the combined preparations and the progestogen-only pills (the mini-pill).

The combined method All of these preparations contain an oestrogen and progestogen. Combined formulations come in either 20- or 21-day packs or 28 everyday formulations with seven inert tablets containing iron and/or vitamins.

Biphasic and triphasic pills Biphasic and triphasic pills, a variation of the combined method, were introduced to imitate more closely the hormonal profile of the menstrual cycle. They differ from conventional combined pills in that the daily hormonal dose is not kept constant. By reducing the dose in the early part of the cycle the total dose per cycle is decreased and by using a slightly higher dose later in the cycle breakthrough bleeding and spotting may be diminished.

Low-dose progestogen (the mini-pill) The low dose progestogen-only pill is not as safe a contraceptive as the combined pill, but because of the inhibitory effect of the combined formulation on milk production, the mini-pill is useful during lactation. It is also useful in women who are unable to use non-hormonal forms of contraception and for whom oestrogen use is not advisable (that is, for women with hypertension, diabetes, a history of cardio-vascular disease and women who smoke).

Mechanism of action and efficacy

Combined oral contraceptives have a predominant action on the hypothalamus. Inhibition of GnRH reduces the secretion of LH and, to a lesser extent, FSH which in turn leads to the inhibition of the midcycle peak of LH. With the low dose formulation there is no significant inhibition of the background levels of either LH or FSH. The steroidal effects that are seen relate almost entirely to those in the pill because ovarian steroid biosynthesis is substantially inhibited; that is, the uterus, cervix and vagina are placed under the exogenous control of the imposed artificial cycle.

The mechanism of action of the low-dose progestogen-only pill is somewhat different. Inhibition of the hypothalamus is not complete, and many ovulatory cycles do occur. The mini-pill relies on other contraceptive actions. Most importantly, cervical mucus is thickened inhibiting sperm penetration.

The efficacy of contraceptive methods is often expressed in terms of failure rates. The combined oral contraceptive is extremely effective with failure rates of less than 1 pregnancy for 100 woman-years of use, if used correctly. The progestogen-only pill has a higher failure rate at about 1–4 per 100 woman-years.

Long-acting hormonal contraception in the female One goal of contraceptive research has been to develop effective, safe and reversible long-acting methods that do not require the user to remember a daily action. Progestogens are among the most promising drugs for the development of long-acting methods. At present, they are administered for this purpose in four different forms.

(a) Long-acting injectable contraceptives
(b) The subdermal implant
(c) The vaginal ring
(d) Hormone-releasing IUDs (discussed under intrauterine devices).

Injectable contraceptives

Progestogens This is usually an intramuscular injection of, probably, depomedroxyprogesterone acetate (Depo-Provera) and, less often, norethisterone enanthate (NET-EN). A single injection can provide highly effective contraception for 2 or 3 months. The two main advantages are that these agents do not contain oestrogens and are, therefore, free from adverse oestrogenic effects and that the need for repeated injections ensures that women will have periodic contact with medical or other health-trained personnel. With injectable progestogen-only contraceptives, ovulation is inhibited via inhibition of the hypothalamus. However, its effects on the oviducts, endometrium and cervical mucus may also play a role in reducing fertility.

Once-a-month injectables Monthly injectives deliver a lower dose of hormone per month than 3-month or 2-month injections and include both combined and progestogen-only formulations. The side effects, such as irregular bleeding and amenorrhoea, are less frequent but this advantage has to be weighed against the inconvenience and higher cost of more frequent injections.

Subdermal contraceptive implants: the Norplant system

Pioneering work began in 1967 on the concept of a steroid-filled silastic (silicone rubber in the form of polydimethylsiloxone) reservoir could be placed under the skin, allowing the gradual, continuous release of hormone as a protection against pregnancy. The advantage of a sustained-release system is that a long period of contraceptive effectiveness can be provided with a relatively small quantity of steroid. Implants which allow a continuous and constant release of hormone are different from oral contraceptives which produce high daily fluctuations of plasma levels of hormones and from injectables which produce high levels of hormone initially with the levels declining thereafter. After extensive testing the Norplant system was introduced by the Population Council and approved in 1983 by the National Drug Regulatory Agency in Finland, where the implant is manufactured. Since then, it has been used by more than 150 000 women in more than 37 countries.

The Norplant sustained-release system consists of six silastic capsules filled with the progestogen levonorgestrel. Norplant releases in the first year 50–80 μg and subsequently around 30 μg (approximately equivalent to the daily amount received when using the mini-pill) of the hormone per day through its polymer walls. Ovulation is inhibited with implants but not during all menstrual cycles. The important contraceptive action is that cervical mucus is thickened, inhibiting sperm penetration into the uterus. Other mechanisms that may contribute include effects on the endometrium that reduce the likelihood of implantation and depressed progestogen support during the luteal phase.

A level of protection sufficient to prevent conception is reached within 24 hours of insertion of the implant and one set of implants can provide protection for 5 years. Failure rates are generally below 1% at the end of 1 year and remain low for 5 or more years. Women who weigh more than 70 kg have a higher (up to 3.5%) probability of becoming pregnant than lighter women.

Other subdermal implants

Following the success of levonorgestrel-releasing implants, slow-release systems incorporating progestogens with other properties have been developed. The development of the progesterone implant is aimed at providing lactating women with contraception in the form of natural progesterone which is efficacious and will not affect lactation. The system provides contraception for 5 months following insertion.

Biodegradable implants Much research has gone into the development of a biodegradable implant which can be injected into the body. One advantage of such an implant is that it would not have to be removed when its hormone content is exhausted. Biodegradable microspheres containing small amounts of norethindrone and providing contraceptive protection for 3 to 6 months have undergone clinical trials and proved effective. The major disadvantage is that, once injected, they cannot be removed until their effect has worn off. Another variety combines the advantages of retrievable and biodegradable implants. Capronor is such a product. It is inserted subdermally and the capsule remains intact for 18 to 24 months. It is easily recoverable as the system is not completely absorbed for several years.

Efficacy

When properly used, injectables and implants give almost perfect security against conception – less than 1 pregnancy per 100 woman-years of use. In common with all progestogen-only methods, they can be used without an upper age limit in smokers and when oestrogen is contraindicated.

The implant method also combines a minimum of bioavailable steroid, released at a constant rate and with minimal user compliance. These advantages may help to make the implant a safer and more popular contraceptive method for many women in the future.

Transdermal delivery systems

Transdermal delivery systems for contraceptives have been designed to deliver progestogens only or oestrogen/progestogen combinations. This method of contraception is still in the experimental stage, since the method needs further refining. The contraceptive is delivered in combination with a potent penetration enhancer via a skin patch. The major concern, so far, has been associated with side effects produced by the enhancers because most are mild contact irritants.

Systemic contraception induced by vaginal administration

The vaginal ring Hormone-releasing vaginal rings, first produced in the late 1960s and early 1970s, are made of silastic, the same material used for contraceptive implants. They contain either a progestogen alone or a combination of an oestrogen and a progestogen and are designed to remain in the vagina for either 3 weeks or about 3 months depending on the type of ring. As for the implant, hormones are released from the ring at a constant rate and are absorbed through the vaginal wall into the circulation. The combined oestrogen-progestogen ring releases enough hormone to inhibit ovulation and is intended to be used intermittently; it can be left in the vagina for 3 weeks and be removed for 1 week to allow withdrawal bleeding to occur. It is about as effective in preventing pregnancies as the low-dose oral contraceptives. The progestogen-only type of ring can be worn continuously for 3 months and probably acts by rendering the vaginal mucus impenetrable to sperm in conjunction with local effects on the endometrium. However, studies of its efficacy have shown it to be less effective than implants or injectables.

The vaginal pill Despite the attractive features, the vaginal ring still carries significant drawbacks; most importantly, its continued presence may cause erosion and irritation of the vaginal wall and problems with hygiene. The vaginal pill containing a progestogen or a progestogen-oestrogen combination solves those concerns. The pill is inserted manually into the vagina, daily for 21 days, allowing 8 days for withdrawal bleeding before resumption. This method is almost 100% effective, with a lower risk of developing the common undesirable side effects of the oral contraceptive pill, such as nausea.

Health concerns

As contraceptive services have become more widely used throughout the world, it is important to weigh the benefits of contraception against the risks involved. Much of the focus has been on oral contraceptives, which are now among the most widely used systemic, preventive medications. Data have also been collected for injectables and implants but their evaluation has been over a shorter time-span. Modern methods of contraception have a large margin of safety; however, while the risk to the individual may be small, when drugs are used by large numbers of healthy people over a long period of time the potential long-term risks must not be overlooked. In 1988 over 60 million women around the world were using oral contraceptives. All hormonal preparations invented so far have short-term side effects and if taken for prolonged periods may pose health problems. In addition to the direct effect of the sex steroids on the reproductive tissues, there are indirect metabolic effects. These effects are anticipated because of the interrelationship that exists within the whole of the endocrine system and because of the ubiquitous actions of steroids, including the synthetic compounds. The systemic reactions are often physiologically similar to those of normal pregnancy, a time when the serum oestrogen and progesterone levels are also high. The most prominent effects are on liver function and carbohydrate and lipid metabolism. Other drug interactions with the contraceptives may also alter the overall physiological state. For example, drugs that speed the metabolism may lower the contraceptive's effectiveness through quicker elimination.

When evaluating the level of health risk attached to contraceptive use, consideration must be given to the alternative health risks attached to pregnancy. The health risks of unwanted pregnancies are not only physical but also social and mental and of immense significance to the mother and the fetus/child. Where the risk of maternal mortality and morbidity are high, as is the situation in most of the developing world, contraceptive failure may carry a risk of maternal death from a pregnancy that was neither planned nor wanted.

Minor side effects that often disappear after the initial few weeks include breakthrough bleeding or spotting, reduced menstrual bleeding, missed periods, nausea and weight gain. Established long-range health risks of combined oral contraceptives relate to an increased incidence of cardio-vascular diseases such as venous thrombosis and thromboembolism, acute myocardial infarction and stroke. It appears that the incidence of thrombo-embolic disease among oral contraceptive users is higher by 3 deaths per 100 000 users per year (US statistics). Another established risk is an increased risk of hepatocellular adenoma, an extremely rare liver tumour. Apart from the well-established health risks of combined oral contra-ceptives, concerns have been raised about other potential hazards. These may include a slightly increased incidence in the risk of breast cancer after prolonged early use (particularly of the 'old' high-dose formulations), preinvasive and invasive cancer of the cervix, malignant melanoma in association with high exposure to sunlight and impaired fertility. A small number of women taking oral contraceptives are affected psychologically and may develop a depressive syndrome characterized by despondency, tension and changes in libido. Fetal malformation if oral contraceptives are taken inadvertently during pregnancy comes in a special category and is discussed in Chapter 16.

A better perspective on the health risks involved is that they are confined almost entirely to women with other risk factors such as diabetes, hypertension, smoking or age. For a young woman who is a non-smoker, the cardio-vascular risks due to oral contraceptive use are almost negligible. For a heavy smoker (more than 15 cigarettes per day) aged 30 or over or for a woman 45 years of age or older the health risks may outweigh the benefits of contraception and may equal that of pregnancy. Stresses are cumulative and, in proportion to their number, intensity or extent, utilize the body's resources (Chapter 13). It has been estimated that the annual number of pill-related deaths in the USA would be brought down by 86% if no-one over 35 who smoked took the pill.

No established health risks have yet been reported for the injectibles and implants because they are associated with only minor changes in metabolic parameters. Short-term adverse effects may include disruption of the menstrual cycle (erratic bleeding pattern) and a slow return of fertility.

Progestogen medication during lactation may increase the milk volume and the duration of lactation. Although no problems have been detected during short-term observation of infants breast-fed by women using the mini-pill or implants, these drugs reach the milk and are orally active. There has been no long-term evaluation of their potential influence on the neuroendrocrine mechanisms that regulate the reproductive process in adulthood. The possibility of deferred effects, therefore, cannot be ruled out.

Non-contraceptive health benefits of hormonal contraception were discovered serendipitously during the search for adverse side effects, and they have been mostly documented for combined oral contraceptives. They include protection against heavy menstrual blood loss and iron deficiency anaemia (of great benefit to populations where anaemia is prevalent), benign breast disease, benign ovarian cysts, rheumatoid arthritis, pelvic inflammatory disease, ectopic pregnancy, ovarian cancer, endometrial cancer, dysmenorrhoea, premenstrual syndrome and the psychological wellbeing derived from certain protection against unwanted pregnancy.

Oral contraception in the male

The search for a male contraceptive has centred on substances that disrupt sperm with minimal undesirable side effects. Encouraged by the successes from manipulation of the female endocrine system, males have been exposed to various steroids, either alone or in combination, and to gonadotrophic hormone agonists or antagonists. Another approach to induce infertility has been to by-pass the endocrine system and directly disrupt either spermatogenesis or sperm maturation (spermiogenesis) with chemical substances. However, before a drug can be considered useful, the agent, or its metabolites, must first be able to cross the Sertoli-cell barrier to reach the germ cells and, secondly, it must interfere with their survival or differentiation without affecting libido. The capacity for sexual activity should be left unimpaired during suppression of male fertility; some of the commercially available anti-androgenic contraceptive pills, while suppressing spermatogenesis, have that defect.

Pharmacological agents The search for chemical agents acting specifically on spermiogenesis to render sperm incapable of fertilizing ova has been intensive. Despite this effort, however, an acceptable 'male-pill' remains elusive. The major problems associated with its development are the large number of sperm produced, the 3-month interval necessary to eliminate those already produced, the possibility of future sperm being abnormal but fertile and the problem of maintaining libido at the same time as facilitating contraception. To-date, direct and relatively selective actions on the seminiferous epithelium have been achieved experimentally with a range of compounds. For example, gossypol, a phenolic cottonseed derivative, has been successfully used in China as an oral contraceptive for men. Part of its mechanism of action seems to be a functional modification of the spermatocytes by the disruption of key mitochondrial enzyme systems which results in interrupted spermatogenesis and impaired motility. However, for many other compounds which suppress spermatogenesis, the mechanisms of action are unknown. Gossypol has much to recommend it, including its efficacy (100%) and the relative ease and low cost with which it can be extracted from the cotton plant, but pharmaceutical induction of oligospermia with a cytotoxin is always a concern.

Hormones Recently there has been considerable interest in exploring the use of inhibin as a potential male contraceptive. The logic behind this is that if FSH is essential for the initiation and maintenance of spermatogenesis, then selective suppression of FSH secretion by inhibin should render a man infertile without affecting his libido since LH and testosterone secretion would not be altered. Initial experiments in animals have been encouraging, and now that its full chemical structure has been determined many practical difficulties can be overcome. Regular injections of testosterone effectively suppress spermatogenesis. Only minor side effects such as increased greasiness of the skin and a small weight gain have been reported. Clinical trials, however, have demonstrated that even high (100 to 300 mg/week) doses of testosterone enanthate, despite consistently suppressing LH and FSH, achieved azoospermia in only 50–60% of the subjects. Recovery of sperm concentration after drug discontinuation, to at least 20 million/ml, took 3–7 months, demonstrating the reversibility of this mode of contraception. Although these results are encouraging for about two thirds of normal men, for the idea to become more widely acceptable the weekly injections will need replacing by longer acting androgen depot formulations and long-term health risks evaluated.

Temperature

Hyperthermia has been recognized as injurious to spermatozoa since early times; Hippocrates mentioned this in his reflections but it was not until this century that numerous clinical observations linked testicular hyperthermia with reduced spermatogenesis. Studies show that in the tropics conception is less frequent in the hot season. Men who are occupationally exposed to heat, sit for long periods of time or wear jock straps and tight pants may be risking sub- and infertility. In the rat, for example, protein synthesis and hexose transport by round spermatids is optimal at scrotal

(33–34 °C) temperature rather than at body (37 °C) temperature. Exposure of the testes to a bath temperature of 38–43 °C for approximately 15 minutes per day over a 2-week period leads to a significant reduction in sperm count. After an initial series of exposures, the sperm count can be maintained at a low level of one heat application every 3 weeks. The effect on fertility is reversible, but this method, although popular in parts of the USA and Japan, is definitely not recommended as physical interference may damage sperm without rendering them infertile. The physical properties of electromagnetic and ultrasound waves have synergistic effects with those of heat, and the combination has been investigated as a possible male contraceptive. Again, unknown risks in regard to potential congenital anomalies in the offspring render these methods suspect. It may be useful to identify specific biochemical events in spermatid metabolism which are highly sensitive to body temperature and then search for a pharmacological agent that can suppress those events directly.

Neuroendocrine regulation of fertility: the use of hormone analogues in females and males

With the knowledge of the structure of releasing hormones, an entirely novel approach to fertility control was possible. Synthetic compounds closely related in structure to the natural releasing hormones may act as antagonists of the natural hormone. Two such approaches for the control of fertility have been developed.

(a) Synthesis of competitive analogues of the target hormone of choice
(b) Synthesis of specific antisera capable of neutralizing the endogenous target hormone.

The development of potent, long-lived analogues to GnRH has received attention because the potential of a non-steroidal alternative in hormone contraception, free of side effects, is most attractive. By the replacement and/or deletion of some amino acids in GnRH, analogues have been made which can bind effectively to the native receptor but which lack those structural features necessary for physiological stimulation of the gonadotrophins (particularly LH release). In the female, the normal pulsatile GnRH and LH release is abolished by continuous administration of analogue, with oestradiol levels stabilizing at the low levels observed in the normal early follicular phase. After cessation of even prolonged therapy, a normal reproductive function is rapidly restored. In the male, the response to daily treatment with a GnRH agonist is suppression of circulating levels of testosterone into the castrate range and the absence of spermatogenesis. Impotency (and, interestingly, hot flushes) results from the drop in testosterone levels, so androgen supplementation is needed to maintain normal sexual activity. The GnRH analogues may be administered in a daily intranasal application or monthly intramuscularly in biodegradable microspheres.

Several synthetic antagonists of GnRH are potent enough to effectively suppress endogenous or GnRH-induced gonadotrophin release. However, antagonistic analogues are only active at considerably higher doses than agonists.

Immunocontraception

Immunocontraception represents an attractive mode of fertility regulation because it has the potential to be cheap, convenient, long-acting and reversible with few side effects – all attributes guaranteeing its wide usage and global acceptability. A search has been mounted for immune responses consistently associated with human infertility and to use the sera from patients with such responses for identification of the corresponding antigens. It is assumed that these antigens are clinically relevant as they could be suitable candidates for birth control vaccine development. With the advent of molecular biology and recombinant DNA technology (target antigens may now be produced in bulk), fertility control by immunological means is advancing rapidly. So far the major goal in immunocontraception has been the isolation and characterization of antigens of sperm and the zone pellucida that may be appropriate targets for immunological intervention in prefertilization or implantation events (see p. 295). Clinically, the presence of naturally occurring antisperm antibodies, especially those secreted locally in the reproductive tract, has been implicated in 10 to 15% of unexplained human infertility. The presence of sufficiently high concentrations of antisperm antibodies in the reproductive tract results in agglutation or immobilization of sperm, interference with fertilization or the blocking of implantation of early embryos. Individuals who suffer from sperm antibody-related infertility show no more health problems than do controls suggesting that this method of artificially inducing infertility would be safe.

The capacity to interfere with reproductive processes by immunizing against hormones has been recognized since the 1940s. In practical terms, gonadotrophin-based vaccines are still at an early stage of development;

the hormonal vaccines that are most advanced are directed against chorionic gonadotrophin. Encouraging progress has been made, however, in experiments using bonnet monkeys, where oligospermia was induced by active immunization with anti-FSH and resulted in infertility in all immunized animals. No long-term side effects were apparent. Similarly initial investigations in female baboons have demonstrated that active immunization against the LH receptor is possible and has potential for the development of a contraceptive vaccine suppressing ovarian function.

THE PREVENTION OF FERTILIZATION

Methods that prevent fertilization traditionally were barrier methods either alone or combined with spermicidal compounds. New approaches use immune methods to destroy the egg or sperm.

Barrier methods of contraception

Since the late 1970s there has been a swing back to the use of barrier methods, particularly amongst the single, because of the protection which sheaths offer against transmission of the AIDS virus and other sexually-transmitted diseases. Barrier methods are by definition based on various means of temporarily obstructing the entrance of the spermatozoa into the female tract or their ascent beyond the level of the external os of the cervix. The objectives of present research and new developments in vaginal barrier contraceptives are to increase the effectiveness of the methods while reducing the disadvantages associated with the currently available methods. For example, there has been a trend towards the use of medicated barriers that do not require an additional spermicidal preparation.

Male barrier techniques

This category includes the oldest of all methods (coitus interruptus or withdrawal), the simplest of mechanical devices (the condom) and sterilization by vasectomy. The most prevalent contraceptive device is the condom, a term derived from the Latin *condus* meaning receptacle. It was originally designed as a prophylactic against venereal disease associated with prostitution. When properly used the condom ranks relatively high in effectiveness; however, its actual failure rate, among the young especially, is often high (between 10 and 15 per 100 woman-years). Coitus interruptus is not very effective, and one reason for its high failure (apart for the need for considerable motivation and self-control) may be the deposition of glans fluid containing sperm into the vagina prior to ejaculation.

Female barrier techniques

This category includes some of the oldest and simplest forms of vaginal methods (spermicides, sponges, diaphragms, cervical caps) and sterilization by tubal ligation.

The diaphragm The diaphragm is a circular, shallow rubber dome with a firm but flexible outer rim. It fits between the posterior vaginal wall and the recess behind the pubic arch. When correctly fitted, a diaphragm blocks the upper vagina and cervix, but it may not fit completely in all coital positions and it may be displaced during female orgasm. Consequently, it is generally used with spermicidal jelly or cream. *In vitro* studies have shown that, in sufficiently high concentrations, some spermicides can inactivate the HIV virus, but frequent use of corrosive chemicals *in vivo* may damage the vaginal and cervical epithelium, which could in turn augment the transmission of HIV. Disposable non-fitted diaphragms are spermicide releasing and are used without spermicides.

The cervical cap The cervical cap is a thimble-shaped device made either of rubber, hard plastic or metal and is designed to be fitted tightly around the cervix and held in place by suction. Fitting by a physician is required. The insertion and removal of the cap is more complicated than for the diaphragm and needs some experience. However, once in place it can be left undisturbed from the end of one menstruation to just before the start of the next. The cervical gap is especially indicated when an anatomical defect, such as retroversion of the uterus, makes the diaphragm difficult to retain. Spermicides are used with the cap. The custom-fitted cervical cap has a one-way valve to release cervical secretions and menstrual blood and is intended to be worn constantly. To make this cap, an impression of the cervix is taken for use as a template. Both the diaphragm and the cervical cap can be highly effective contraception methods. A failure rate of 2–3 per 100 users seems usual; however, pregnancy rates of 13 to 23% have been observed in some studies. This discrepancy may be reflected in a low motivation to avoid pregnancy and faulty equipment. Care must be taken to ensure that caps and diaphragms do not degrade after a year of use, which in turn could decrease their contraceptive efficacy.

Chemical barriers: spermicides Spermicides have two basic components, namely, the active spermicidal agent and the carrier. They are contained in a contraceptive sponge and are also available in a variety of forms for use alone (creams, jellies, tablets, suppositories and foams). Almost all currently marketed spermicides are surfactants, that is, they are surface-acting compounds that destroy sperm cell membranes. They are safe but not very effective unless used in conjunction with some other means. Failure rates are between 17 to 22%. Currently, longer acting, more effective spermicides are being developed. Gossypol (a male oral contraceptive pill, see p. 288), is being investigated as a vaginal contraceptive. Also, other agents such as inhibitors of sperm enzymes which would result in inhibition of sperm motility or other metabolic functions are being developed.

Chemical barriers: the contraceptive sponge The spermicide-impregnated contraceptive sponge is made of polyurethane and is shaped like a mushroom cap with a loop attached for its retrieval. It is smaller than a diaphragm, is activated by water and can be placed in the vagina for up to 24 hours before intercourse. The sponge has three types of contraceptive action: it releases spermicide during coitus, it absorbs ejaculate and it blocks the entrance to the cervical canal. But because it can be pushed into the back of the vagina during intercourse, it is not as effective as the diaphragm, nor does it have the diaphragm's firm fit over the cervix. The range of failure (9–27%) is similar to that found for other spermicides.

The female condom The female condom or Femshield is simply a polyurethane vaginal sheath lining the whole vaginal surface. This development provides a barrier method that is female-based, gives protection against sexually-transmitted diseases including AIDS and is not dependent on male erection and intromission. The Femshield is stronger and lighter than the latex rubber used in male condoms, easy to insert and used in conjunction with a lubricant. Women find it very effective and are attracted to its protective function against disease.

Health concerns For all devices needing insertion and withdrawal there are health concerns relating to personal hygiene. Toxic shock syndrome may result from introduced infections and complications due to allergic reactions and vaginal irritation may also occur in a small percentage of users.

Female immunocontraception

Zona pellucida: target for a contraceptive vaccine The zona pellucida (ZP) is unique in the ovary as it is highly immunogenic with ZP3 being the species-specific receptor. It has been known since the 1970s that administration of zona pellucida antibodies results in long-term, reversible contraception but is often accompanied by interference with folliculogenesis. An ideal ZP-based contraceptive vaccine requires the production of zona pellucida antibodies that result in effective contraception without ovarian cytotoxicity. This means directing the immune response more specifically to ZP antigens of oocyte origin, which would not have deleterious effects on the ovarian follicles. So far a small number of non-human primate studies using more refined epitopes associated with the ZP3 α-component have been encouraging in that they demonstrated effective reversible contraception. The efficacy and safety of these vaccines will need further evaluation prior to human use.

Male immunocontraception: sperm antigens

Sperm antigen-based immunocontraceptive vaccines appear to be an ideal choice for pre- and postfertilization contraception in humans (and animals, Chapter 14). A sperm-specific protein vaccine has two theoretical advantages over anti-ovum or anti-fetus immunocontraception. First, it may be effective in both males and females, and, second, it would not raise autoimmunity problems in the female. It is hoped that male fertility would return with the decline of antibody titres such that a period of infertility is followed by a variable period of subfertility, then fertility. To-date a number of different sperm proteins have been used as the antigen.

Acrosomal antigens The acrosome and its membranes plays a critical role during sperm–egg binding. Only spermatozoa that have lost their apical membranes and acrosomal contents (as a result of an acrosome reaction) can penetrate the zona matrix and undergo fusion with the oolemma (Chapter 8). The World Health Organization Task Force on Contraceptive Vaccines, which met in 1986, designated the human sperm protein SP-10 as a primary vaccine candidate. SP-10 is a differentiation antigen that arises during spermiogenesis and is testis specific. It is an intra-acrosomal protein which remains associated with the sperm head following the acrosome reaction. The absence of cross-reactivity with somatic tissues coupled with its stage-specific expression during germ cell differentiation is relevant to the potential of SP-10 as a contraceptive vaccine immunogen. Problems of autoimmunity, which might be anticipated if common somatic antigens were utilized as vaccine immunogens, may not be found with SP-10. Trials of an SP-10-based vaccine may be carried out in primates as the distribution of SP-10 is similar in macaque, baboon and human spermatids.

Surface domain antigens Researchers have been involved in the generation of numerous monoclonal antibodies against specific sperm-surface antigens, and the interesting ones were further selected for detailed analysis. So far the success of the search has been variable: among the monoclonal antibodies that were produced, some reacted specifically with components of sperm-surface antigens but not with any other somatic tissues.

Epididymal antigens Because epididymal glycoproteins become associated with maturing sperm, antibodies to them may possess the requisite specificity for inactivating sperm in the female reproductive tract. An attraction of the immunization of females with sperm antigens is that the procedure would probably only interfere with some aspect of sperm function, while immunization of males with sperm proteins might compromise testicular function.

Lactate dehydrogenase isoenzyme (LDH-C_4) Lactate dehydrogenase is a glycolytic enzyme that exists in multiple molecular forms or isozymes, each with distinct tissue and functional specificities. The most rigid specificity is seen with the LDH-C_4 isozyme of mammalian testes and sperm and has been extensively studied for immunocontraceptive purposes. With the onset of puberty, this isozyme appears during prophase of the first meiotic division of the primary spermatocyte and is found on and in the mature sperm. Immunization of female baboons with LDH-C_4 resulted in antibody production and a reduction in pregnancies of up to 80% with no apparent immunopathological side effects. So far the data on immunized baboons suggest that immunization to LDH-C_4 does not impair embryonic development, does not damage the reproductive system and is reversible. Subsequent studies have been directed toward the development of a synthetic peptide that would elicit an immune response to the native proteins. There is strong support for the use of synthetic antigens to replace the natural product in immunocontraceptive vaccine development (see section below).

Immunological hazards in the use of fertility control vaccines

In theory vaccines to control human fertility run a higher risk of inducing anti-self reactions than do vaccines to control infectious diseases. This risk may be increased if powerful adjuvants are used and the vaccine is administered frequently. Safety, therefore, particularly the avoidance of autoimmune disease, is a dominant factor in their acceptability. There is a growing consensus that the main mechanism for induction of autoimmune disease is activation of T-cells, which are normally controlled by suppressor mechanisms. One approach to minimize the chances of activation of anti-self T-cells is to use a foreign carrier protein as the main or only source of T-cell epitopes in the vaccine.

Sterilization

As the success of the reversal operation is limited, sterilization should be regarded as primarily a means of terminating reproduction. Numerous surgical procedures for contraceptive sterilization have been devised but usually involve the bilateral cutting, tying or removal of a small portion of the vas deferens (male vasectomy) or of the fallopian tubes (female tubal ligation). Other methods for fallopian tube occlusion are electro- or thermocoagulation, mechanical (rings and clips) and chemical (sclerosing agents and tissue adhesives). The route to the tubes may be either via the abdomen or the vagina.

Voluntary female sterilization has become the most widely used and fastest growing contraceptive method in the world among women over 30 years of age. A recent population report estimated that, in 1990, 16% (138 million) of all married women of reproductive age worldwide were protected by this method. This was 45% higher than the 1984 figure. Both male and female sterilizations are routinely performed on an outpatient basis and their use is rapidly expanding in developing countries.

Health concerns

Complications are few in female sterilizations, with no identified long-term health effects due to the ligation. The most common complications are caused by infection and internal injury due to faulty surgery. Voluntary sterilization is a realistic acknowledgement that contraceptive needs change once the family is complete. The overriding advantage for the sterile woman is that she is free from the need to choose alternative methods of contraception with their attached inconvenience and/or health risks.

Vasectomy, to a variable degree, leads to the development of auto-antibodies against sperm but whether that, in itself, is a factor in the etiology of disease is not clear. Theoretically, sperm antibodies may elicit immunopathologic effects in cross-reacting tissues leading to autoimmune diseases and an increase in the chances of spontaneous neoplasm. In the human these antibodies have generally not been considered deleterious. Data from several large-scale human studies do not show any evidence of increased risk of cardio-vascular disease, despite some sub-human primate studies where sperm antibodies have been associated with atherosclerotic changes in the blood vessels. Other studies linked premature ageing and increased incidence of prostate and testicular cancer 10–15 years post-vasectomy, but none of these links has been convincingly established to-date. However, as for the oral contraceptive pill for women, a long-term manifestation of potential hazards may be confined to men with other stress-related risk factors. For healthy individuals, however, the benefits of sterilization probably outweigh the concerns, especially if counselling prior to the operation were to be made routine. Vasectomized men should be encouraged to examine their testes in the same fashion as women are encouraged to routinely examine their breasts.

THE PREVENTION OF IMPLANTATION OR DEVELOPMENT OF THE EARLY EMBRYO

The early stages of pregnancy are greatly dependent on the correct balance between the female sex hormones, and it is from among these compounds and their antagonists that useful methods of contraception have been developed. For example, in view of the importance of progesterone in pregnancy maintenance, methods which suppress corpus luteal function by the use of inhibitors or by direct progesterone antagonists are very effective (see RU 486 below). Artificially blocking the secretory activity of the endometrium initiates its erosion and prevents implantation or disrupts development. Without the hCG signal, the corpus luteum regresses, effecting a sharp decline in progesterone secretion and thus increasing uterine contractility and softening and dilating the cervix. Eventually, these changes cause the detachment and expulsion of the developing embryo and placenta.

Compounds which occupy receptors are more predictable in terms of effectiveness. They also pose fewer health risks because they are safer physiologically when compared to drugs which exert their effects by means of controlled toxicity. The term contragestion (for contra-gestation) is proposed to cover all modalities of fertility control interfering with the establishment of or continuation of pregnancy.

The antiprogesterone steroid RU 486 (Mifepristone)

RU 486 (Mifepristone or the controversial 'abortion pill'), as the first effective anti-progestogen ever brought to market, represents a major development in the field of hormone antagonists. RU 486 binds to the progesterone receptor with an affinity that is five times that of progesterone itself. It also binds to the glucocorticoid and androgen receptors with lower affinities. RU 486, a product of the Roussel-Uclaf company, was discovered in the late 1970s, ironically as a by-product of steroid research and not with the goal of pregnancy interruption. Etienne-Emile Baulieu of INSERM (the French institute for medical research) saw the potential importance of the discovery and convinced Roussel-Uclaf to pursue research into its anti-progesterone properties for fertility control. In September 1988, RU 486 was first marketed under the regulation by French law covering the termination of pregnancy.

RU 486 can be used at two stages as a fertility control drug: either preimplantation (contraceptive effect) or postimplantation (abortifacient effect). It has been successfully used to promote the onset of menstruation (in the manner of postcoital contraception) and to facilitate abortion up to the second trimester performed with a small amount of prostaglandin. Prostaglandin increases the frequency and strength of the uterine contractions used to expel an embryo. The 'once a month' approach is an invaluable form of contraception for the older or less fertile woman who can be saved from the need to take continuously active compounds that are physiologically effective. Based on 60 000 voluntary terminations, RU 486 is 96% effective as an abortifacient without any identified long-term ill effects, since endocrine function returns to normal soon afterwards. Short-term health concerns (as in a normal miscarriage) relate to 4–5% of cases, with heavy bleeding usually accompanied by strong and painful uterine cramps. In exceptional cases a blood transfusion is needed and, because of the risk of haemorrhage, the prostaglandin given at the end of RU 486 treatment must be administered in a medical facility where women can be monitored and, if necessary, treated. As with all potent medications such as prostaglandin, caution should be taken to avoid complications in women who are heavy smokers or are suffering from illness. In countries where RU 486 is available, between a quarter and a third of women now choose this chemical approach over standard surgical procedures to interrupt an early pregnancy. The advantages are that it does not involve mechanical intervention, avoiding the risks associated with anaesthesia, infection and scarring. This is of overwhelming significance in the developing countries where conditions of hygiene may not be high and where WHO officials estimate that more than 125 000 women die every year as the result of abortions. Of these deaths, 95% are the result of simple hygiene problems.

A major spin-off from RU 486 research has been the increased understanding of the basic physiology of the reproductive and menstrual processes. These insights would not have been possible without the development of a means of switching off progesterone secretion. Outside the realm of pregnancy, RU 486 may become a useful drug through its role as a progesterone antagonist in the treatment of cancers that bear progesterone receptors (including breast cancers) and also through its role as a glucucorticoid antagonist in the treatment of over-production of cortisol which can lead to hypertension and osteoporosis.

Prostaglandins

Prostaglandins have been known since the mid 1930s, but their role as labour-inducing agents was not fully exploited until the 1970s. They are very efficient in terminating unwanted pregnancies but are physiologically powerful in very low concentrations and can cause violent uterine contractions. The use of prostaglandin analogues offers important advantages over the naturally occurring substances. Some of the new prostaglandin analogues, in suppository and injection form, are approaching vacuum aspiration in effectiveness in the termination of very early pregnancies, although their side effects, especially nausea, can still be quite severe. Combination procedures are also frequently used for second-trimester terminations; for example, lower doses of intra-amniotic prostaglandin administered together with other agents such as urea and, more recently, RU 486. Prostaglandin is also used as an intravaginal application to bring on a late period, thus inducing menstruation without the woman knowing whether she had conceived.

Postcoital contraception

These methods, sometimes called the postcoital pill or the postovulation pill, are aimed at preventing implantation in women who are exposed to unpredicted intercourse during the fertile time of the cycle. Diethylstilboestrol (DES) alone was used initially, but it is now more common to take two tablets, repeated within 12 hours, containing a high dose of oestrogen/progestogen within 72 hours of intercourse. The pregnancy rate is about 1% and nausea is a common side effect. This method is not suitable for recurrent use because of the relatively high dose of steroids given. Since the method is not 100% effective a back-up strategy is required in cases of ensuing pregnancy.

Menstrual regulation

Menstrual regulation (menstrual induction, menstrual planning, endometrial aspiration, preemptive abortion or atraumatic termination of pregnancy) is a method which moved into the gap between foresight contraception and hindsight abortion and involves vacuum aspiration of the uterine lining within a few weeks after failure to commence menstruation (usually within 2 weeks). The technique is simple and safe and is performed on an outpatient basis and requires only a local anaesthetic, but less than 40% of the world's population have access to this service. Social conscience should be alerted when a woman in one country may visit a clinic shortly after a missed menses and have, on request, a safe, private and dignified abortion, whilst in another country a woman is forced to resort to a life-threatening action or continue a pregnancy considered hazardous.

From the historical perspective it is seen that folklore is rich in methods by which a woman might induce abortion. Abortion is once again a most divisive issue and opinion in western societies ranges from disapproval under all circumstances (even to save the mother's life), through qualified approval, to it being a human right which should be available on demand. The abortion debate among institutions and activists has the power of dictating political policy without consideration of equality, individual needs, global survival strategies and population realities. During the 1980s, political attitudes had become more conservative, with many attempts being made to ban or at least restrict abortion.

Non-steroidal oral contraception in the female

In 1991 a new type of oral contraceptive pill, developed in India, was released by India's National Family Planning Programme. Centchroman (also known as Saheli or 'female friend') deviates from the usual steroid-based formula in that its active compound is a coumarin. It is believed that its mechanism of action is directed at the endometrial oestrogen receptors without disturbing the hypothalamic–pituitary–ovarian axis. As a result, the drug interferes with implantation and the menstrual cycle can proceed normally. Centchroman's efficacy is better than that of the condom or diaphragm but it has a higher failure rate than the combined oral contraceptive pill.

The intrauterine device or IUD

The origin of the intrauterine device (IUD) dates back to antiquity. The ancient Arabs (who still employ the method) used to prevent conception in their saddle camels on long journeys by introducing a round smooth stone into the uterus; the camel then repulsed the advances of the male as if she were pregnant. The IUD was introduced as a contraceptive method in the early part of this century but did not achieve popular acceptance till the 1960s. In 1928 Grafenberg from Germany introduced intrauterine rings made of an alloy of copper and silver, while at about the same time Ota from Japan worked with a flexible ring. The first generation of modern IUDs, designed and produced in the late 1950s, were unmedicated devices produced following the development of the biologically inert plastic polyethylene. During the 1970s medicated or bioactive IUDs were developed which carried substances such as metallic ions (copper acting as a spermicide) or hormones. These medicated devices were developed to reduce the incidence of side effects and to increase their contraceptive effectiveness.

Mechanism of action All IUDs stimulate a foreign-body reaction in the endometrium. The addition of metallic copper wire to a polyethylene intrauterine device potentiates the efficacy of that device. Copper metal incubated in biological solutions produces Cu^{2+}, which at critical concentration are inhibitory to steroid hormone–receptor interactions. Cupric ions also inactivate spermatozoa. It is unlikely that any single mechanism of action accounts for the anti-fertility effect of IUDs. Alteration or inhibition of sperm migration, fertilization, ovum transport and failure to implant may all be implicated. Implantation may be prevented by biochemical changes in the endometrium as a result of local inflammation with histamine and prostaglandin release which, in turn, increase uterine motility.

Efficacy Generally IUDs are safe, effective, reversible and practical forms of fertility control. However, the IUD offers slightly less protection against unwanted pregnancy than do injectables, implants or oral contraceptives with pregnancy rates ranging from less than 1 to about 5 per 100 women years. The reliability of the IUD varies with its retention rate, which is generally inversely related to the ease of insertion; large devices are less likely to be expelled than are smaller ones. The major adverse side effects are heavy bleeding and pain which are reasonably common in the early months of use. The levonorgestrel-releasing IUD has the advantage that it strongly reduces menstrual blood loss, a health benefit in countries with a high incidence of anaemia and dietary iron deficiency. There is also the less common possibility of the complications of ectopic pregnancy, pelvic inflammatory disease (PID) and even perforation. PID is a prime cause of tubal infertility (Chapter 1); however, the association of multiple sexual partners and IUD use is an important additional prerequisite, since other risk factors, including prevalence of pathogens, are significant in the causation of PID. The users of the Dalkon Shield IUD had a five times greater chance of developing serious complications, which resulted in its removal from the market in 1975.

Vaccine against pregnancy

A birth control vaccine is different from almost all other vaccines in that it requires the long-term presence of a sufficiently high antibody titre in the female reproductive tract to give protection. A sufficiently high titre to give conveniently long infertile periods may be achievable by initial immunization or might require a method for long-term administration of the antigen; for example, a subdermal implant that would slowly release a protein antigen.

Human chorionic gonadotrophin

Of all the hormones which have been researched as immunogens over the last few years, hCG is the most promising. A birth control vaccine consisting of a synthetic peptide corresponding to the C-terminal 109–145 amino acid region of the β-subunit of hCG and conjugated to diphtheria toxoid as a carrier is now available for human evaluation after its safety and efficacy has been established in monkeys. The vaccine's anti-fertility effect may be mediated through ablation of the luteotrophic signal of CG or other events occurring at the embryonic level. With one protocol, immunity against pregnancy was achieved for 1 year by two initial injections and one booster every 12 months for continued protection.

Trophoblast vaccine

Development of a trophoblast birth control vaccine is dependent on the identification of one or more antigens that are expressed on the very early trophoblast and whose recognition by immunological means will prevent or disrupt pregnancy. Monoclonal antibodies recognizing specific antigens on the human syncytiotrophoblast and cytotrophoblast have been produced.

Possible teratogenic effects of vaccines

Before conducting clinical trials with any immunocontraceptive it is essential that careful studies are carried out to determine if fetal anomalies occur in immunized individuals that become pregnant in the presence of sub-threshold levels of immunity. Potential hazards can be divided into two types.

Chemical teratology Chemical teratology might be manifested through, firstly, components of a fertility-regulating vaccine which may have a direct or indirect effect on the gametes, in both males or females, and result in subsequent sublethal defects in the fetus. Secondly, fetal exposure to these same compounds may occur as the result of pregnancy occurring during the waning phase of the immune response.

Immuno-teratology The material immune response elicited to a fertility-regulating vaccine might have a teratological effect following alterations in maternal physiology on which the fetus is dependent. It might also result from antibodies passing across the placenta and being deposited in, or reacting with, fetal tissues. Many sperm antigens, for example, are shared with antigens of the developing embryo/fetus. The consequences of such events may not be detectable until the offspring reach sexual maturity, when fertility may already be compromised.

OVERVIEW

According to United Nations estimates, the global human population will continue to increase for another 100 years. Most of this future growth (95%) will occur in the developing countries, with the next 15–20 years being crucial. The continued momentum of population growth will lead to an estimated world population of 8.2 billion by the year 2025, with 55% of the total located in Asia. The connection between the stresses of severe crowding, poverty, inequality of resource distribution and political instability has been made in Chapter 13. The present chapter has described the variety of ways in which high rates of population growth can be reduced. New, effective and acceptable options are discovered with astonishing speed. Implicit in the scientists' urgency of discovery is the understanding that global health and survival depends on ecological integrity brought about by effective fertility control. Global peace depends on the collective and individual demand for contraception as a basic human and ecological right, free from government intrusiveness. No person has the indisputable right to have a baby if that baby's human rights are later denied; every child must be wanted in order to fully express its genetic potential. The inevitable result of unchecked population increase is seen in developing countries where millions of people spend their lives uneducated, unemployed, ill-housed and without access to even the most elementary health and welfare services. The challenge is not the availability of the technology for contraception, rather it is one of management and delivery of the services. The urgent need for fertility control in developing countries can be met only after social and economic reforms have taken place. History shows us that voluntary fertility decline has not preceded improved literacy and infant/child mortality in any society. For example, much of the fertility decline that occurred in Europe, the USA and Japan did not follow the availability of modern contraception but was the result of the population's enjoyment of better health and economic status. Thus strategies that enhance a child's survival through increased maternal and child health will increase the demand for contraception and improve the effectiveness of the management and delivery of family planning services. Given the western experience over the last 20 years, women are likely to remain the principal users of contraceptive services (although the role of the future male will certainly be more active). Because of this discrepancy in reproductive responsibilities, the education of women, particularly in developing countries but also amongst minority groups in developed countries, is of primary importance if replacement or zero population growth is to be a viable option. Therefore many more women in influencial and decision-making positions are required to ensure the success of the evolution of a new world order. A joint, more just future requires the integration of women's unique talents, perspectives and life experiences.

General references

Alexander, N. J., Griffin, D., Spieler, J. M. & Waites, G. M. H. (eds.) (1990). *Gamete Interaction: Prospects for Immunocontraception.* Wiley-Liss, New York.

Bromham, D. R. (1991). New developments with IUDs. *British Journal of Family Planning*, **15**, 34–41.

Davies, G. C. & Newton, J. R. (1991). Subdermal contraceptive implants – a review: with special reference to Norplant. *British Journal of Family Planning*, **17**, 4–8.

Eden, J. A. (1991). Progestogens: an occasional review. *Asian-Oceania Journal of Obstetrics & Gynecology*, **17**, 289–295.

Friend, D. R. (1990). Transdermal delivery of contraceptives. *Critical Review in Therapeutic Drug Carrier Systems*, **7**, 149–186.

Harlap, S. (1991). Oral contraceptives and breast cancer. *Journal of Reproductive Medicine*, **36**, 374–395.

Mauck, C. P. & Thau, R. B. (1990). Safety of anti-fertility vaccine. *Current Opinion in Immunology*, **2**, 728–732.

McLaren, A. (1990). *A History of Contraception – From Antiquity to the Present Day*. Blackwell, Oxford.

Rock, J., Pincus, G. & Garcia, C. R. (1956). Effects of certain 19-*n* steroids on the normal human menstrual cycle. *Science*, **124**, 891–893.

Sanger, W. G. & Friman, P. C. (1990). Fit of underwear and male spermatogenesis: a pilot investigation. *Reproductive Toxicology*, **4**, 229–232.

Sivanesaratnam, V. (1990). Vasectomy – an assessment of various techniques and the immediate and long-term problems; review. *British Journal of Family Planning*, **16**, 97–100.

Stafa, J. A., Newschaffer, C. J., Jones, J. K. & Miller, V. (1992). Progestins and breast cancer: an epidemiologic review. *Fertility & Sterility*, **57**, 473–494.

Swahn, M. L. & Bygdeman, M. (1990). Medical methods to terminate early pregnancy. *Baillière's Clinical Obstetrics & Gynaecology*, **4**, 293–306.

Talwar, G. P. (ed.) (1988). Contraception research for today and the nineties. *Progress in Vaccinology*, Vol. 1. Springer-Verlag, New York.

Wall, D. M. & Roos, M. P. (1990). Update on combination oral contraceptives. *American Family Physician*, **42**, 1037–1050.

Woolley, R. J. (1991). Contraception – a look forward, Part II: mifepristone and gossypol. *Journal of the American Board of Family Practice*, **4**, 103–113.

16 Hormonal contributions to errors of sexual differentiation

Errors of sexual differentiation were among the first human birth defects to be recognized. Since antiquity, individuals with genital anomalies were either deified or ridiculed (Hermaphródîtos was the son of Hermes and Aphrodite who became united in body with the nymph Salmacis while bathing in a fountain). Fortunately, scientific understanding of the etiology, pathogenesis and management of errors of sexual differentiation has improved in the last 2000 or so years. Congenital anomalies, which result in failure to achieve normal sexual dimorphism, may be due to intrinsic defects in gene expression and/or epigenetic modifications of normal gene expression. Genetic defects of sexual differentiation are inherited by Mendelian patterns of inheritance, that is, autosomal dominant, autosomal recessive or sex-linked. Examples of genetic errors causing pseudohermaphroditism can be found in Chapter 2 alongside a description of the orderly process of normal sexual differentiation. This chapter is concerned with potential teratogenic derangements in sexual differentiation resulting from therapeutically introduced or environmentally derived steroid-like compounds and highlights the sensitivity of the reproductive system to early stimulation by hormonally active xenobiotics. Epigenetic factors can operate at any time during sexual differentiation and can cause developmental abnormalities which may not be evident until puberty or old age. Mechanisms underlying biochemical teratology are further discussed in Chapter 19.

The term Dice syndrome (the roll of the dice) describes the random exposure of any individual to unsuspected disturbing influences of environmental origin. A special form of Dice syndrome is iatrogenic disease, meaning originating from the physician and, by analogy, any intervention inadvertently deviating a developmental process. In many instances, the teratogenic results of early gonadal steroid exposure are induced either by a hormonal environment inappropriate for the sex of the fetus or by effects which seriously impair the development of the target organs of both sexes. The action of steroids on target tissues can be divided into a short-term organizational phase during embryogenesis, which leads to an essentially irreversible differentiation of the tissues, and a long-term activational phase, leading to the potentiation of pre-established functions and characteristics. Compared with the differentiation of the phenotype, hormone-dependent functional differences in the CNS are complex and not well understood in the human. It is clear, however, that androgens and oestrogens play an essential role in the development of the intrinsically bipotential CNS leading to the control of tonic (male) or cyclic (female) secretions of GnRH, FSH and LH (Chapter 2).

EXPOSURE TO SYNTHETIC HORMONALLY ACTIVE AGENTS

There is a clear case for the careful evaluation of the potential risks associated with disturbances of the normal hormonal milieu during pregnancy and the early postnatal period. Risk assessment, however, is complicated by a lack of information relating to the properties and mechanisms of action of compounds to which the mother and the fetus may be exposed. Although the principal activity of a drug appears risk-free to the embryo, side effects of the drug may well possess other hormonal teratogenic activities. There are many man-made substances either introduced therapeutically or derived from the environment which, due to their androgenic or oestrogenic qualities, are embryotoxic and/or interfere with normal sexual development.

Diethylstilboestrol

The synthetic non-steroidal oestrogen, diethylstilboestrol (DES) (Fig. 16.1), was first described in 1938 by Dodds and subsequently promoted

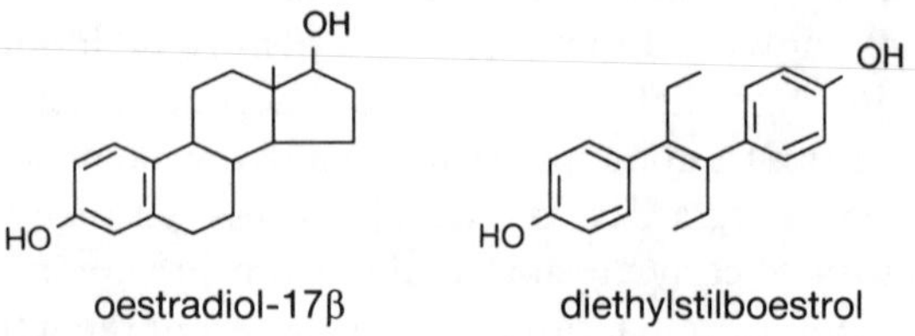

OH HO oestradiol-17β

OH HO diethylstilboestrol

Fig. 16.1 A comparison of the chemical structure of diethylstilboestrol (DES) with oestradiol-17β.

for the treatment of women at risk of miscarriage, intrauterine fetal death, toxaemia and preterm birth. In 1971, a statistical association was found between maternal ingestion of DES and the occurrence in the daughters of a rare form of clear cell adenocarcinoma of the vagina and cervix. Even after DES exposure, this form of cancer is still a rare occurrence in women younger than 24 years (0.14 to 1.9 per 1000 exposed individuals). The risk is higher in women who were exposed to DES earlier (before 9.2 weeks) than later in gestation and the survival rate is better for women who are older than 19 years at its onset, since the carcinoma is more aggressive in the younger patients. Subsequently, a link between prenatal DES exposure and a variety of other abnormalities in the reproductive tract of both the male and female offspring was established. Between the late 1940s and its prohibition in the early 1970s, an estimated 6 million persons (the majority in the USA) were affected by prenatal exposure to DES. In addition, the complete range of consequences of DES exposure may not be known for many years since affected individuals are now in their twenties and thirties and further problems may emerge as women enter the menopause and both sexes approach old age.

The major abnormalities identified, in addition to vaginal cancer, included benign adenosis of the cervical and vaginal epithelium, distortion in the shape of the cervix and uterus, abnormally small uterine cavity, obstruction and distortion of the fallopian tubes, ovarian cysts, pelvic adhesions and disturbances of fertility, such as an- (oligo)ovulation, increased rates of spontaneous abortion, risk of ectopic pregnancy and premature and complicated labour. Structural abnormalities are found in 40–50% of exposed women. A retrospective study analysing the effects of intrauterine DES exposure in fertile women, demonstrated that the incidence of complications during the third stage of labour was significantly higher in drug-exposed women when compared with controls. The DES-exposed group was more likely to have a greater blood loss after vaginal delivery, suffer postpartum haemorrhage and need manual removal of the placenta. It was suggested that the smaller intrauterine dimensions may impair placental separation and be responsible for the complications of labour. There is, however, a remarkable resemblance between *in utero* DES-exposure and the effects observed in rats exposed to caffeine *in utero*; these experience significant delays in the progression of labour and parturition (Chapter 11). Steroids transported in the plasma are mostly bound to plasma proteins (such as CBG and SHBG) with a small free fraction being the active component. Certain synthetic oestrogenic compounds, including DES, are not bound which may in part explain the high potency of DES. Prenatal exposure to DES may, in both sexes, affect the development of other CNS functions that affect psychological and behavioural parameters with subsequent long-term consequences. Although many aspects of existing behavioural studies are inconclusive, the overall picture which emerges is that prenatal exposure to synthetic oestrogens, particularly DES, feminimizes/demasculinizes male-oriented behaviour (manifested as, for example, decreased interest in contact sport and decreased scores in visuospatial tasks) and defeminizes/masculinizes certain aspects of female CNS function (manifested as, for example, decreased verbal skills). Conclusions from this area of research must be regarded as tentative since they were drawn from studies of selected average behaviour patterns considered significant in an Anglo-Saxon population living in middle-class America. More substantial data, based on the examination of DES-exposed sons and daughters between the ages of 24 and 30 years, relate to the prevalence of psychiatric disease. Depression and anxiety, especially, was found to be twice as frequent in the DES-exposed individuals when compared to controls. The difference was consistent in both sexes and the findings particularly notable because neither the DES subjects nor the general practitioners were previously informed of the history of DES exposure.

Structural abnormalities of the reproductive tract of male offspring from DES-exposed mothers have also been found. These include increased incidence of hypotrophic testes, cryptorchidism (maldescended testes), epididymal cysts, testicular lesions and primary infertility due to low ejaculate volume and low sperm density. The relative risk of malignancy and germ cell tumours in maldeveloped and cryptorchid testes has been reported as 10–40 times that in normally descended testes. The association between cryptorchidism, the early presence of abnormal germ cells later developing in a premalignant histological pattern (called carcinoma-in-situ, CIS) of the seminiferous epithelium and the subsequent development of germ cell neoplasia is established. CIS cells probably develop during sexual differentiation but are mostly not transformed into malignant cells until sexual maturity has been reached. The structurally related defects are the result of DES-induced teratogenic changes in the differentiation of the organs derived from the Müllerian and Wolffian ducts but the teratogenic biochemical changes are less obvious. Sex hormones regulate growth by stimulating cell proliferation and differentiation and their involvement in the modulation of T-cell-mediated immune functions may explain the low resistance to different types of neoplasm and increased autoimmune disease associated with *in utero* DES exposure. A common finding in women was the detection of human papilloma virus DNA in cervical lesions. In mice, neonatal administration of DES inhibits $CD4^+$

T-helper cell activity and reduces the ability to mount a defence against cells infected with mammary tumour virus.

The DES experience makes it obvious that, before any treatment is introduced to the population at large, exhaustive clinical trials are essential in the assessment of both drug efficacy and prolonged safety. It may, of course, be argued that bias in evaluating the data has arisen because the drug was used in the prevention of threatened miscarriage or in problem pregnancies. Thus exposure to DES would serve as a prognostic marker to identify children born from problem pregnancies and this might increase the risk of their developing disease in later life. This interpretation would, however, contradict the experimental data from animals which clearly demonstrate that early maternal exposure to DES leads to reproductive tract abnormalities and cancer of the genital tract in offspring of both sexes.

Recognition of the adverse effects of DES on sexual differentiation coincided with increasing concern over the release of oestrogenic compounds into the environment. DES was also widely used in the farming industry to increase meat yields in cattle and poultry with the result that meat was contaminated with residual DES. In 1977 it was estimated that almost 30 000 kg of DES was being used per annum for agricultural purposes. It was established in the 1980s that oestrogens present in poultry excreta caused hyperoestrogenism when fed to sheep or cattle and were responsible for lactation in non-pregnant ewes and heifers. One study found that heifers eating 6 kg/day of chicken manure silage received a dose of more than 1 mg/day of oestrogen (10–30 μg of DES/day is sufficient to cause lactation in the non-pregnant ewe). It is unlikely that such grand-scale contamination of the environment presents a long-term hazard, since DES is rapidly excreted by livestock and broken down in the soil. It is worth noting, however, that the number of operations performed in England and Wales to correct cryptorchidism doubled in the period from 1962 to 1981. Since the approval for the use of DES has been withdrawn, other types of hormonal treatments have been introduced in agriculture to take its place.

Synthetic progestogens and other agonists, antagonists and inhibitors of steroid biosynthesis

In addition to DES, women at risk of spontaneous abortion were also treated with synthetic progestogens, some of which possessed androgenic and progestational activities. The effects of treatment varied depending on the type of steroid used and the duration of therapy. In some cases treatment *in utero* with androgenic progestogen caused clinical masculinization of the female external genitalia. Fetal exposure to exogenous androgens may, in addition to clinical applications, also arise accidentally from a number of other sources, such as exposure to contraceptive steroids or herbal medicines. Many progestogens contained in birth control pills/injectables/implants are 19-nortestosterone derivatives and such agents may induce a degree of masculinization of female fetuses. Other progestogens, including depomedroxyprogesterone acetate (Depo-Provera), have also been implicated. For instance, exposure to Depo-Provera during pregnancy has been associated with an increased risk of IUGR, low birth weight and neonatal complications accompanying suboptimal fetal growth rate.

Although no problems have been detected during short-term observation of infants breast-fed by women using the mini-pill or implants, these drugs reach the milk and are orally active. There has been no long-term evaluation of the effects of fetal exposure to steroids on the neuroendocrine mechanisms that regulate the reproductive process in adulthood and the possibility of deferred effects cannot, therefore, be ruled out.

Tumours derived from steroid hormone-producing cells are rare, but are important because fetal exposure to abnormal levels of maternally produced androgens of ovarian and/or adrenal origin may be sufficient to masculinize, or otherwise damage, a female infant as well as the mother. The luteoma of pregnancy produces large amounts of weak androgens but other rarer types of ovarian tumours, such as arrhenoblastoma which produces high levels of testosterone are much more serious. The fetus may also produce excessive quantities of androgens. This situation most often arises when a female infant has the adrenogenital syndrome (congenital adrenal hyperplasia) caused by various autosomal recessive enzyme defects in steroid biosynthesis which, in turn, cause lack of cortisol synthesis with resultant increased ACTH production. The increased ACTH causes increased production of adrenal androgens which virilizes, in variable degree, the female external genitalia. Males with congenital adrenal hyperplasia show no evidence of genital ambiguity but may have an enlarged phallus and a hyperpigmented scrotum. It is important to recognize this condition in the newborn because a lack of a specific adrenal steroid may be life threatening.

Cimetidine and fertility

The H_2 histamine antagonist cimetidine effectively reduces gastric hypersecretion and is widely used in the treatment of ulcers. In the early 1980s, an increased frequency of infertility during long-term cimetidine treatment of adult men became apparent. Fertility was restored on cessation of cimetidine therapy and it was established that cimetidine had significant anti-androgenic activity. Not surprisingly, cimetidine treatment of rats during gestation induced serious abnormalities of the sexual differentiation of the external genitalia in male fetuses and impaired adult masculine sexual behaviour. These observations assumed a particular significance when it was disclosed that cimetidine was considered for use in pregnant women for the treatment of nausea. Clearly, such use would constitute a hazard to the developing male fetus and could have been another example of disease at the random fall of the iatrogenic dice.

Polychlorinated biphenyls (PCBs) and organochlorine insecticides

In addition to synthetic oestrogens such as DES, a number of other laboratory-made chemicals released into the environment have been shown to exhibit significant oestrogenic activity. These include the polychlorinated biphenyls (PCBs) and the organochlorine insecticides methoxychlor, kepone and dichloro-diphenyl-trichloroethane (DDT) as well as a number of closely related chlorinated hydrocarbons (chlordane, dieldrin, endin, heptachlor) used in aerial spraying. Although the oestrogenic potencies of these agents are relatively weak, they are stable, fat-soluble compounds, and thus continued exposure leads to their accumulation in the tissues, resulting in reproductive dysfunction. For example, in animal tests methoxychlor is metabolized to a oestrogenically active material with fetotoxic and teratogenic effects. This pesticide also reduces fertility through a significant reduction in implantation leading to embryo loss, possibly due to changed oviductal embryo transport and/or reduced circulating progesterone levels. Reported increases in the incidence of precocious sexual development in Puerto Rico appear, likewise, to be related to environmental contamination by man-made compounds with oestrogenic activity. It is well known that many species of birds are now endangered or extinct due to high concentrations of DDT in the gonads and eggs. DDT, by interfering with the oestrogen-dependent mobilization of calcium in the oviduct, results in the production of eggs with exceedingly thin or no shell which make successful incubation impossible. Birds of prey, such as the bald eagle and the osprey, that are at the top of the food-chain are particularly vulnerable.

Another class of compounds with significant reproductive side effects is the imidazole antimycotics used for treatment of fungal infections. Similar preparations are also widely used in agriculture. Male rats exposed *in utero* to the agricultural fungicide fenarimol (α[2-chlorophenyl]-α-[4-chlorophenyl]-5-pyrimidine-methanol) fail to copulate. This compound inhibits aromatase activity and so reduces oestrogen biosynthesis in the CNS, blocking the normal masculinizing effects of testosterone.

A more attractive alternative may be the use of microbes (virus, bacteria or fungus) as biopesticides. The ideal bioinsecticide causes acute symptoms at low doses, has high specificity and can potentially be made more potent by the genetic introduction of toxin genes. However, special care with genetically modified organisms is essential. To-date bacterial (*Bacillus thuringiensis* and *Xenorhabdu* spp.) and fungal (*Verticillium lecanii* and *Metarhizium anisopliae*) insecticides have been successfully used in Europe and Australia.

EXPOSURE TO NATURALLY OCCURRING, BIOLOGICALLY ACTIVE COMPOUNDS

Compared to the attention that hormonally-active xenobiotics have received in relation to reproduction, relatively little information is available for naturally occurring compounds with steroid-like effects. Plant products have been used in folk medicine since prehistoric times, as aphrodisiacs, aids to childbirth, abortifacients and promoters of fertility. In European traditional herbal medicine, fennel and anise have been used to increase milk production, promote menstruation, facilitate labour and increase libido.

The phytooestrogens

The phytooestrogens are naturally occurring plant substances which have weak oestrogenic or anti-oestrogenic activity and are normal constituents of the diet. Very few plants synthesize animal oestrogens; the earliest oestrogens to be isolated from plants were oestrone (from palm kernel) and oestriol (from willow catkins). In the early 1950s Djerassi and Pincus realized the potential application for steroids derived from Mexican yams (from the Dioscurea family) in suppressing ovulation. Many more plant substances structurally resemble oestrogens or DES and, therefore, have oestrogenic properties. Because they assist the plant's resistance to patho-

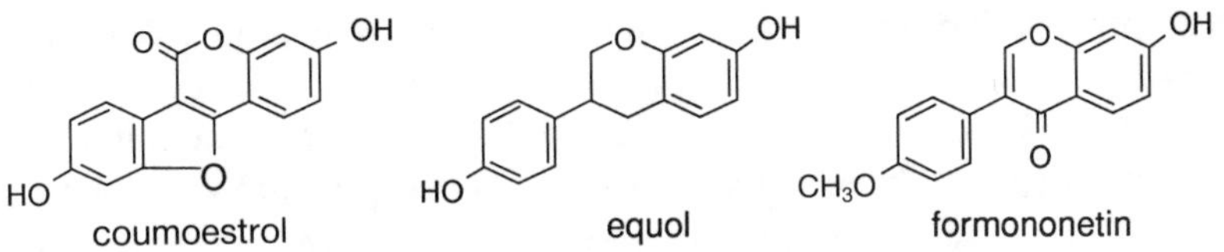

HO, O, O, OH, O — coumoestrol
HO, O, OH — equol
CH_3O, O, OH, O — formononetin

Fig. 16.2 The chemical structure of some common phytooestrogens.

gens, the synthesis of phytooestrogens usually increases following trauma, for example mould infection. It has also been suggested that plant oestrogens evolved as an ecological device to control the fertility of populations of grazing herbivores. The major source of phytooestrogens found in the human diet are the isoflavones (in soy beans, chick peas, cherries), the coumestans (in alfalfa, peas, beans) and the resorcyclic acid lactones (in peas and a variety of grains). Coumoestrol is an important coumestan and the most widespread isoflavones are formononetin and equol (Fig. 16.2). The resorcyclic acid lactones, for example zearalenone (a mycotoxin produced by *Fusarium* spp.), are not intrinsic to the plant in which they are found but are produced as a fungal metabolite which accumulates on poorly stored foodstuffs, such as grain. Mouldy feed is a problem in the pig and poultry industries because zearalenone frequently causes hyperoestrogenism in swine and oviductal prolapse in hens.

A link between ingestion of dietary phytooestrogens and their effects on the reproductive system has not been established in humans, despite the fact that the phytooestrogens are ubiquitous in the environment and their metabolites have been identified in the urine of many species, including humans. Phytooestrogens are orally active and, in common with other oestrogens, stimulate the synthesis of SHBG, thereby attenuating hormonal effects of the endogenously released steroids. Zearalenone also exhibits considerable anabolic activity which has been exploited commercially in the use of the closely related synthetic compound, α-zearalenol as an anabolic agent in beef caffle and sheep. The isoflavonoid content of soy meal is high, which may prove of particular concern because of the increasing human use of soy protein, especially in infant formulae. However, there have not been any documented cases of oestrogenization resulting from soya-rich diets. The age of puberty is relatively constant in children raised on vegetarian or omnivorous diets and infants raised on soy protein do not exhibit abnormal sexual development or disrupted fertility. It seems probable that since mammals evolved together with the plants they eat they can cope with their phytooestrogen. Nevertheless, the potential for subclinical impairment exists, even if it is not measurable, since reproductive function in laboratory or domestic animals on phyto-oestrogen-rich diets is affected. Commercially prepared rat chow containing soy beans produces a significant uterotrophic response in laboratory rats, an effect which has been correlated with the isoflavonoid content of soy meal. Sheep grazing on pastures rich in *Trifolium subterraneum*, a species of clover that is widely distributed in Australia, become infertile. It is estimated that over 1 million ewes fail to lamb each year but, since the loss is so widespread, Australian farmers accept their diminished lambing rates as 'normal'. Clover disease is the result of the presence in this species of clover of the isoflavone formononetin. Formononetin is demethylated to equol and coumoestrol by bacterial metabolism in the sheep rumen (Fig. 16.2).

On the positive side, however, it has been suggested that the lower level of breast cancer in primarily vegetarian cultures may occur because the higher phytooestrogen intake antagonizes the cancer promoting effects of endogenous oestrogen. Dietary phytooestrogens may inhibit oestrogen and androgen biosynthesis and partially antagonize their actions by reducing the effective free concentrations of circulating steroids through the stimulation of SHBG synthesis. Women have higher endogenous oestrogen levels than most other mammals, so the plant anti-oestrogenic impact may be beneficial. Chapter 18 compares the menstrual profiles and corresponding gonadotrophin levels under vegetarian and high-meat diets.

Panax ginseng and its relatives

As touched on in the previous section, phytotherapy (medical treatment using plants) dates back to ancient times and is still preserved in folk medicine. New discoveries concerning the therapeutic value of substances contained in plants, together with an increased understanding about the functioning of the human body, have led to a reassessment of many ancient traditions in phytotherapy. The use of ginseng (*Panax ginseng*) originated more than 5000 years ago in the mountain provinces of Manchuria and is steeped in traditional Chinese medical lore and mythology. The Chinese pharmacopoeia, *Chen Nung Pen Ts'ao*, has Manchuria's sacred Jen-shen (meaning shaped like a man) as a prominent ingredient in many herbal prescriptions. Originally it was consumed for its restorative qualities and promoted as an anti-stress, anti-ageing drug. Its botanical name *Panax* comes from the Greek *panacea* meaning universal remedy. Now it is regaining popularity for many undergoing physical or mental exertion who view ginseng's use as an alternative to 'pep' pills or caffeine. Ginseng is

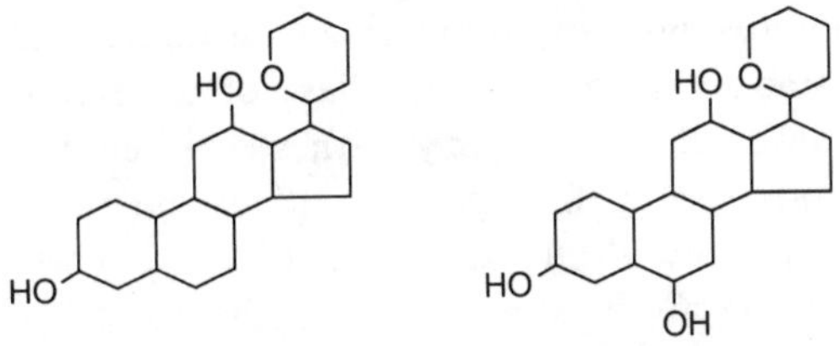

panaxadiol panaxatriol

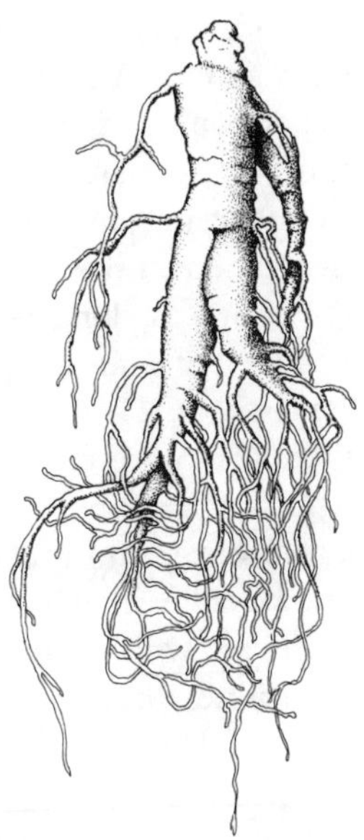

Ginseng (*Panax ginseng*)

Fig. 16.3 The chemical structure of panaxadiol and panaxatriol glycosides found in *Panax ginseng* and the ginseng root.

used either as a short-term stimulant to restore vitality or as a long-term tonic to delay the effects of fatigue.

Within the family Araliaceae, the important medicinal species of ginseng, apart from *Panax ginseng* (Chinese or Korean ginseng), are *Panax pseudoginseng* (Japanese ginseng), *Panax quinquefolium* (America ginseng) and *Eleutherococcus senticosus* (Siberian or Russian ginseng). The principal active constituents of *Panax* spp. are saponin glycosides named panaxosides or ginenosides which are, significantly, steroids (Fig. 16.3). Ginseng saponins stimulate the secretion of ACTH and adrenal hormones, in particular cortisol and sex steroids. These, it is thought, assist the body's normal adaptive response by inducing a well-directed alarm reaction and increased resistance to stress. Regardless of its efficacy, however, it is unwise to continue in self-therapy with steroid-containing preparations without also considering their possible effects on reproductive capacity. Since ginseng has the ability to stimulate the hypothalamic–pituitary–adrenal axis, concern does exist about the safety of its use during pregnancy. An estimated 5 million people in North American alone consume ginseng regularly; yet, no systematic reproductive studies have been conducted in regard to its safety during pregnancy and lactation. Hyperstimulation of maternal adrenals with concomitant androgen secretion may cause fetal androgenization. A recent case of clinical masculinization of a male infant (hypertrophy of external genitalia including the presence of pubic hair and an excessive growth rate) whose mother consumed large amounts of Siberian ginseng throughout pregnancy and lactation generated lively discussion in the medical literature.

Little is known about the short- or long-term reproductive effects of herbal medicines beyond the fact that some are potent, yet the preparations are freely available across the counter. The commercial marketing of herbal medicines must be accountable and product quality control should include consumer information concerning possible side effects on reproductive function. Aside from fetal exposure, continued use of the same herbal preparation may also be inadvisable in adults because the active constituents, which may transiently improve symptoms, can, in the long-term, be aggravating and even cause illness due to their physiological effects or to toxin accumulation.

General references

Chang, H. M., Yeung, H. W., Tso, W. W. & Koo, A. (eds.) (1985). *Advances in Chinese Medicinal Material's Research.* World Scientific, Singapore.

Cheeke, P. R. (ed.) (1989). *Toxicants of Plant Origin.* IV. *Phenolics.* CRC Press, Boca Raton, FL.

Field, B., Selub, M. & Hughes, C. L. (1990). Reproductive effects of environmental agents. *Seminars in Reproductive Endocrinology*, **8**, 44–54.

Folb, P. J. & Dukes, M. N. G. (eds.) (1990). *Drug Safety in Pregnancy.* Elsevier, Amsterdam.

Gondos, B. (1991). Gonadal disorders in infancy and early childhood. *Annals of Clinical & Laboratory Science*, **21**, 62–69.

Koren, G., Randor, S., Martin, S. & Danneman, D. (1990). Maternal ginseng use associated with neonatal androgenization. *Journal of the American Medical Association*, **264**, 2866.

McClamrock, H. D. & Adashi, E. Y. (1992). Gestational hyperandrogenism. *Fertility & Sterility*, **57**, 257–274.

McLachlan, J. A. (ed.) (1985). *Estrogens in the Environment* II: *Influences on Development*. Elsevier, New York.

Meyer-Bahlburg, H. F. L. & Ehrhardt, A. A. (1987). A prenatal hormonal hypothesis for depression in adults with a history of fetal DES exposure. In *Hormones and Depression*, Raven Press, New York.

Müller, J. & Skakkebaek, N. E. (1991). Gonadal malignancy in individuals with sex chromosome anomalies. *Birth Defects*, **26**, 247–255.

Pardthaisong, T. & Gray, R. H. (1991). In utero exposure to steroid contraceptives and outcome of pregnancy. *American Journal of Epidemiology*, **134**, 795–803.

Reinisch, J. M. & Sanders, S. A. (1992). Effects of prenatal exposure to diethylstilbestrol (DES) on hemispheric laterality and spatial ability in human males. *Hormones & Behavior*, **26**, 62–75.

Thorp, J. M., Fowler, W. C., Donehoo, R., Sawicki, C. & Bowes, W. A.. (1990). Antepartum and intrapartum events in women exposed in utero to diethylstilbestrol. *Obstetrics & Gynecology*, **76**, 828–832.

17 The acquired immunodeficiency syndrome (AIDS) epidemic: a global emergency

The conditions under which people live and the ways in which they behave have a significant bearing on their reproductive health. That medicine has to be practised in a social context was emphasized by Hippocrates as long ago as 400 BC, when he wrote 'Whoever wishes to investigate medicine properly should proceed thus: in the first place to consider the seasons of the year . . . Then the winds . . . In the same manner, when one comes into a city in which he is a stranger, he should consider its situation, . . . the water which the inhabitants use . . . , and the mode in which the inhabitants live, and what are their pursuits . . . '. (From the *FIGO Manual of Human Reproduction*.) This good sense is as relevant to reproductive health today as it always has been to health in general. Reproductive health reflects the stresses of poverty and environmental pollution and is especially relevant in the context of infectious diseases. Reproductive health, particularly in developing countries, is lamentable and will deteriorate further unless all nations take responsibility for improving their environments.

Sexually-transmitted diseases (STDs) are an extremely broad subject area and are discussed here only in relation to AIDS, which now represents our greatest human concern. Much of the experience with AIDS can be generalized because there are close parallels between the current AIDS epidemic and previous epidemics of other sexually-transmitted diseases. Syphilis which, like AIDS, spreads most rapidly among the poor and mostly in developing countries, is such an example. Syphilis and AIDS are very different diseases medically, but if they are juxtaposed then their similarities, especially in relation to their initial incurability, popular reactions provoked and the sense of emergency, become evident.

HISTORICAL BACKGROUND

Acquired immunodeficiency syndrome (AIDS) is the clinical end stage of human immunodeficiency virus (HIV) infection which results in severe, irreversible immune suppression. After sufficient immune system damage has occurred, the individual becomes susceptible to infection by many opportunistic diseases including cancers. These then become the indirect markers of AIDS: AIDS is a syndrome or collection of many diseases resulting from HIV infection. AIDS was first recognized as recently as 1981 but since then has become a global phenomenon. Ten years later (in 1991) over 170 000 AIDS cases, from over 130 countries, had been officially reported to the Centres for Disease Control. Figures on reported AIDS cases, however, grossly underestimate the true incidence of disease associated with HIV infection since, in geographic terms, HIV was already an epidemic by the late 1970s. The global dimensions of HIV infection and AIDS are now clear, although their extent remains an estimate as there is considerable uncertainty about past and current HIV infection rates and thus future trends in AIDS incidence. The growth of present estimates should galvanize the global development of effective strategies in the prevention of further spread of HIV infection.

A typical timespan between HIV infection and the development of the disease appears to be 5 to 8 years in 15–35% of cases (studies using different methodologies have produced fairly consistent estimates of the incubation period/time distribution). From probability studies, the progression to AIDS within 2 years of seroconversion to HIV positivity is less than 2%, 25–35% will progress to AIDS within 7 years and 50% within 10 years. The HIV incubation period varies according to the general resistance of the individual. As stresses are cumulative, the additional presence of existing disease (in particular other sexually-transmitted diseases), malnutrition or drug use compromises immune function and increases the chances of immunosuppression by HIV. Accordingly, in developing countries the time it takes for HIV infection to manifest itself as AIDS is much shorter (44 months on average compared to 10 years in affluent countries). For perinatally infected infants the mean timespan is between 1 and 2 years. The highly variable incubation period may, in part, be due to unpredictable HIV mutation with the emergence of new viral forms or quasispecies. In this case immune breakdown follows at the point at which different viral forms exceed the immune system's capacity to control the virus. As a result, the use of estimates relating to the current prevalence of HIV and projections of AIDS-related morbidity/mortality into the future may be misleading. The 1993 WHO estimate of the number of people who may be HIV-positive is 14 million worldwide with up to 60% of those infected living in Africa. In some of the worst afflicted regions, such as sub-Saharan Africa, AIDS is now one of the main causes of infant mortality and the leading cause of mortality in adults. In the general adult population there is a seroprevalence of the order of 30% and the levels of infection in female prostitutes are estimated to be 60–80% in the worst afflicted urban areas. In Uganda the disease is out of control; little is done about this as the country cannot afford either to care for the people already infected and sick, or to promote, fund and encourage community-based awareness in health issues. A significant factor in the disease's rapid spread is war which moves large numbers of men through a countryside where poverty and ignorance forces women, who do not necessarily consider themselves as prostitutes, to sell unprotected sex. In countries such as Thailand, AIDS poses a double threat because of the thriving 'sex tour' tourist industry. Even in areas far removed from the capital such as Chiang Mai, near the Burmese border, and the Golden Triangle, 40–70% of all women working in brothels are HIV positive. Despite cultural acceptance of brothels, protected sex is not generally popular; for example, the African use of condoms is 0–2%. There is also a need for counselling in general contraceptive use since the intrauterine device (IUD), which causes inflammation, and perhaps also oral contraceptives, which cause cellular changes in the cervix, facilitate the transmission of the virus. One study demonstrated that the spermicide nonoxinol-9, used for contraceptive purposes, does not prevent HIV infection but is associated with a significantly increased incidence of genital ulcers and fungal vulvitis. Observations such as these indicate that the use, in high-risk women, of genital irritants increases the risk of HIV infection by increasing the prevalence of genital ulcerative disease. Viral transmission is also facilitated during menstruation and tearing during the loss of virginity.

For developing countries in Africa, Asia and Latin America, most horizontal transmission of AIDS occurs through heterosexual contact, with vertical transmission from mother to infant. Epidemiological data collected from those countries also predict large continuing increases in HIV seroprevalence. For example, in sub-Saharan Africa, a cumulative total of 10 million HIV infections are expected by 1995, in Latin America 2 million, and in south and southeast Asia 2.5 million. The WHO projection is for 15 million HIV-infected people worldwide and an annual adult progression to AIDS of about 1 million by 1995, with 90% of the cases occurring in developing countries. India, for example, which has one-sixth of the world's population, has to urgently focus on AIDS education and prevention if an epidemic of unprecedented magnitude is to be averted. By contrast, the incidence of HIV infections among homosexual men and intravenous drug users has, since the mid-1980s, markedly slowed in the more developed countries. AIDS kills people during the most productive period of their lives, which spells disaster for developing countries where poverty and disadvantage prevent the development of the necessary infrastructure, together with community based activism, to educate and protect the citizens.

The western AIDS experience was characteristic in that it showed an initial exponential growth of infection rates up to 1981 among homosexual men, intravenous drug users and, shortly thereafter, in haemophiliacs and other blood transfusion recipients. Since 1981, infection rates among these groups grew sub-exponentially due to the rapid reduction in high-risk behaviours (frequency of unprotected sex, use of contaminated needles, numbers of new sexual partners) and the screening of blood and its products. The diffusion of the epidemic among heterosexuals was generally initiated via sexual contact with either bisexual males or intravenous drug users. In 1992 an estimated 70% of all new AIDS infections were through heterosexual contacts, with an estimated 80% by the year 2000. In the west, heterosexual spread of HIV is expected to increase in accordance with the characteristic pattern experienced for all new sexually-transmitted diseases at first appearance, that is, infection first spreads rapidly among people whose behaviour puts them at greatest risk, then diffuses into the population at large. Heterosexuals seem to lag behind the high-risk groups in facing up to their responsibility to change their behaviour: a 1990 survey reported that German men travelling to Thailand were highly likely to buy sex without use of a condom in cities where an estimated 60–80% of the prostitutes are infected. Hidden and separated from adult statistics is the impact of the disease on infected children, highlighting the socio-economic and political aspects of AIDS as much as the pathology of the virus itself. The spread of AIDS must be stopped by educating children and adolescents to avoid becoming infected themselves and transmitting the virus to their offspring.

Because past estimates were based on an AIDS definition linked to AIDS-related diseases, the figures obtained failed to reflect the true magnitude of the epidemic. A revised definition of AIDS now includes anyone infected with the HIV virus and having a $CD4^+$ lymphocyte count (CD4 is the molecule on the cell surface that acts as the main receptor for HIV) under 200 per mm^3, compared with the normal lymphocyte count range of between 1000 and 3300 per mm^3. Therapy is phased into the population of individuals with advanced stage HIV ($CD4^+$ cells/mm^3 of less than 200).

Recently, researchers from the Pasteur Institute, Paris have reported that CD26 may be a second cellular receptor for the HIV virus. In this scenario, CD4 is the outer membrane receptor with CD26 transmitting the virus into the cytoplasm nucleus.

HIV AND THE IMMUNE SYSTEM: PATHOGENIC PROPERTIES AND EVOLUTIONARY CONSIDERATIONS

Certain families of viruses, like herpes, wart and hepatitis B viruses, contain DNA as their genetic material and the evolutionary success of these viruses depends on their ability to persist for the life of their host. Other families of viruses contain RNA as their genetic material. The mutation rate of RNA is much greater than that of DNA giving RNA viruses, such as the influenza virus, a great propensity for change. HIV is a RNA virus that belongs to the third sub-family of human retroviruses, the Lentivirinae. The HIV virus has an external lipoprotein envelope composed of regions of surface glycoproteins involved in binding surrounding a central core that contains a helical nucleocapsid. The HIV infection cycle is initiated by binding of the envelope glycoprotein to the cellular ligand, the CD4 receptor, present on the surface of cells such as the T-helper lymphocytes that are essential for the proper functioning of the immunological defence.

What sets the retroviruses apart among the RNA viruses is their unique method of reproducing involving an enzyme, reverse transcriptase, which 'retroverts' RNA into DNA. The proviral DNA then becomes integrated into the host's DNA, forming a virogene and establishing permanent infection. Each time an infected host cell divides, viral copies are produced along with more host cells. However, the virogene can now exist in a latent state when neither viral RNA nor protein are made. This can cause a problem in diagnosing the disease because the virus, in effect, can 'hide' in cells, infecting them without triggering the production of antibodies. This can result in a seronegative test when, in fact, the patient is carrying the HIV virus. Consequently the HIV virus poses a double-threat: it persists, like many DNA viruses, and it readily mutates, like many RNA viruses. Genetic variability in viral populations, both in and between infected individuals, is a major influence on the likelihood of their transmission and the subsequent development of disease. A retrovirus infection may, by this method, not cause any ill effects for years. Alternatively, the proviral DNA can make viral RNA and proteins with the eventual production of infectious viruses. This process of virus release does not necessarily lead to immediate cell death but infected T-helper cells will die in time. In many cases the infected T-helper cell fuses with uninfected cells and these multinucleated cells continue releasing virus for some time. The first clinical manifestation of advanced HIV infection is a profound immunosuppression and depletion of helper T-cells as the infected cells progressively die. The consequent loss of immunity, however, is selective and affects primarily those parts of the immune system involved in the defence against parasitic, viral and fungal organisms. People with AIDS often develop certain unusual infections but can resist other, more common, illnesses. Additionally HIV infection not only depletes T-helper cells but also may prevent surviving cells from functioning normally. For example, they may mount a HIV-induced autoantibody reaction against their own modified antigens thus further eliminating themselves from the body. There is some experimental evidence supporting the suggestion that AIDS is essentially an autoimmune disease in which the T-helper cells have lost the normal anti-self inhibition.

A general concept that has developed from the study of the evolution and possible origins of viruses is that ancient viruses have, over time, adapted to live with their host without causing them undue harm. As they are essentially parasites, any new forms of viruses which were too virulent would die out if their hosts were rapidly killed. For example, the DNA herpesviruses (members of which can be found in many lower species, such as fish and birds) cause severe disease or death only under special circumstances. Among the exceptions are when a virus 'crosses species'; for example, when a monkey herpesvirus (the macaque B virus) infects humans accidentally, the ensuing disease in humans is almost invariably fatal. Retroviruses are also members of an old family of viruses with many representatives infecting non-human animal species.

There are many different strains of HIV, but they can mostly be classified within two major types of human immunodeficiency virus, namely HIV type-1, first found in central Africa and which has spread most widely throughout the world and type-2, mostly found in west Africa with limited spread elsewhere as yet. In mid-1988, a third variant, HIV-3, was reported. The close genetic similarities, particularly between HIV-2 and a simian (monkey) retrovirus (SIV), strongly support a monkey source for at least this HIV type, as well as an African origin. The high prevalence of HIV in heterosexual populations in Africa, by comparison with developed countries, has been linked to the longer period of occurrence there and the higher rates of sexual partner change in African societies. The lethal nature of HIV in humans suggests a relatively recent origin, which is often observed with viruses which 'cross' species. Even though the virulence of a virus in the new host will decline with time it will still take a large number of generations for a benign form to evolve. If HIV is not of recent origin it may be that an attenuation of the virus may have occurred many centuries ago in highly separated 'islands' of people who only recently began to intermingle with other humans. Measles virus, for instance, which probably originated from domestic animals, became adapted in Europeans over time but when introduced into virginal populations, such as South American indians and Australian aborigines, proved to be very virulent and often fatal. A historical view point of the likely spread of HIV-1 indicates initial heterosexual transmission from Zaire to Haiti, thence the first recognition of the virus in western homosexual men and a secondary spread among intravenous drug users and blood recipients.

BODY FLUIDS AND INFECTION

Cells in the haematopoietic system are particularly sensitive to HIV infection. B-Lymphocytes, macrophages and promyelocytic cells can be infected in addition to T-helper lymphocytes. Generally, the T-helper cells release the highest titres of virus and undergo cell death. The other cells release smaller amounts of virus and continue to act in the body as reservoirs of the virus. HIV has been isolated from virtually all body tissues, fluids and secretions, but it is found in the greatest concentrations in body fluids which harbour leucocytes, particularly $CD4^+$ lymphocytes. Sperm cells carry CD4A and CD4F antigens. These have been postulated to possess HIV receptor activity, thus making the sperm cell the primary cellular element in the transmission of HIV by semen. Leucocytes can be found in a variety of body fluids, including blood, menses, lymph, cerebrospinal fluid, breast milk, vaginal/cervical secretions and semen. In HIV-infected individuals, a proportion of the lymphocytes present in genital secretions will carry the virus. A high cell-free virus titre is also present in blood and lymph but only low levels of free virus can be found in saliva, tears and urine. Fluids with a low titre could be a minor source of HIV contagion. In contrast, however, the virus-infected cells found in blood and genital secretions are a major source of transmission. Infection can be readily transmitted into the circulation through breaks in the mucosal lining of the anal or vaginal canal. HIV could also pass, via infected sperm or blood cells, through the cervical os and directly infect macrophages and lymphocytes in the endometrium, or be transferred from cell to cell through fusion via the $CD4^+$ receptor interacting with the viral envelope glycoprotein. Immediate penetration of lymphocytes by infected cells is an efficient method of passing virus unaffected by neutralizing antibody. Therefore, any approach at the control of HIV spread must involve cytotoxic methods. Since HIV infects a wide range of host cells, approaches to its control must be made early in infection before HIV spreads to many tissues in the body.

AIDS is the end stage of HIV infection and, at best, all that drugs can do is control the pace of bodily deterioration. The spectrum of opportunistic infections and other conditions in individuals with AIDS varies in different regions, usually reflecting the infections prevalent in these regions, although some differences recorded may also be due to lack of diagnostic facilities in some countries. In Africa, the most common opportunistic infections are tuberculosis, cryptococcal meningitis, herpes simplex, oral or oesophageal candidiasis, cryptosporidiosis, CNS toxoplasmosis and skin rashes. Chronic diarrhoea and weight loss are very common. In the USA, in contrast, 63% of people with AIDS are diagnosed with *Pneumocystis carinii* pneumonia. Cryptococcal infections, cryptosporidiosis and toxoplasmosis have also been reported. Kaposi's sarcoma is the most common malignancy in people with AIDS. It has been diagnosed, mostly among homosexual men, in 33% of AIDS cases in Europe and 24% of cases in the USA. Other types of cancer, especially Burkitt's lymphoma and non-Hodgkin's lymphoma of the CNS, also occur in people with AIDS. Severe neurological disorders, including progressive memory loss, dementia, psychiatric symptoms, encephalitis and meningitis, can occur in adults with HIV infection.

In addition to the infection of cells derived from the haematopoietic system, HIV can infect other cells bearing the CD4 receptor including cells of the macrophage/monocyte stem cell lineage, such as dendritic cells, Langerhans cells of the skin, glial cells of the brain and glomerular cells of the kidney. These cells may, since they divide infrequently, become latent reservoirs of the virus in body tissues. HIV has also been recovered from cerebrospinal fluid in individuals who, in the absence of other clinical findings, present with persistent headache and/or fever.

In summary, the major biological features of HIV include its replicating properties in a variety of different cells, its ability to induce cytopathology and its existence in a latent state in some cell types. After virus infection, the host mounts both humoral and cellular immune responses against HIV and the antibodies to the virus can be readily detected by a variety of procedures including cytologic (fusion, lysis, plaque formation), immunologic (immunofluorescence, enzyme/radioimmunoassays) and molecular (Western/Northern blots, reverse transcriptase and nucleic acid hybridization) methods. The polymerase chain reaction is another useful technique in investigating early or latent HIV infection, since the reaction amplifies integrated HIV DNA into amounts large enough to be detected by standard molecular techniques. The presence of antibodies (seropositivity) can usually be detected 2–8 weeks after initial infection, although in a small minority of cases, 6 months or more may elapse before their development. The majority of individuals who become infected develop antibodies without any immediate symptoms, others experience a short-term illness similar to mononucleosis 2–5 weeks after the initial infection or, more rarely, exhibit acute neurological symptoms such as seizures and temporary motor impairment. There are rare individuals who do not develop detectable antibodies to HIV but who are acutely infected. It is also possible to have a virus-negative/antibody-positive state suggesting that if the immune system of an individual is prepared it can mount a successful response against HIV, suppressing it by cellular immunity. Compromises to this immune reaction by other bacterial, viral or parasitic infections or lifestyle stresses could then permit HIV progression.

ANTIVIRAL DRUGS AND CHEMOTHERAPY

At present AIDS is a fatal disease: there are no vaccines to protect against infection and no antiviral drugs that can cure by permanently clearing the body of HIV. The two major factors that prevent the development of such drugs are, firstly, that HIV 'hides' in the body that it infects (to effectively kill the virus a drug would have also to destroy host tissues) and, secondly, that most antiviral drugs cannot cross the blood–brain barrier so the brain still remains a safe haven for HIV. Thus strategies for prophylactic and therapeutic intervention require the understanding of the biological and molecular properties of HIV. Progress in both areas has been rapid, as the genomes of several variants of the retrovirus have been sequenced and every effort is being applied to develop a vaccine. However, a vaccine is preventitive so, even if a safe and effective vaccine were available, many millions of already infected individuals are going to develop AIDS. A two-pronged effort is essential, and more direct approaches, such as antiviral chemotherapy or chemoprophylaxis, are also being developed. It is also essential to develop safe and effective antiviral drugs for the prophylactic use of individuals who are still antibody negative but who are at risk of infection with HIV, for example the partner or parent of a haemophiliac. Because patients will be taking the drug on an outpatient basis for a prolonged period of time (perhaps more than 10 years), it is also important to find an effective oral formulation, or a formulation that can be implanted intramuscularly as is done for long-term contraceptives (Chapter 15). Other delivery systems that may decrease the toxicity of the drug and allow physiologically effective levels of the drug to be maintained in the blood (skin patches, liposomes or prodrugs) will also be useful once suitable drugs are developed.

To-date much research effort has been directed toward antiviral agents which prevent viral reproduction and the spread to more body tissues. While some are at different stages of clinical development, others (the first-generation drugs) are available but produce serious side effects even in the short term. The most promising drugs so far are azidothymidine (3′-azido-3′-deoxythymidine) or AZT/retrovir and ribavirin (2′,3′-dideoxy-cytidine or ddC), but many other 2′,3′-dideoxy analogues are also being developed. Both AZT and ribavirin cross the blood–brain barrier (at levels up to 70% of the plasma level within 4 hours) and can be taken orally (oral bioavailability is about 63%). Various combinations of drugs have also been evaluated, almost all in combination with AZT; for example, a combination of recombinant alpha interferon and AZT. Interferons and other immunotherapies alone are not useful in cases of overwhelming infections, but they are important in helping to control the development of drug-resistant viruses because virus mutation and selection are unlikely to occur in the presence of interferon. Patients must be treated with antiviral drugs not just to eliminate HIV and decelerate the progression to AIDS but also to eliminate other opportunistic infections that may frequently occur. In 1989, the Public Health Service of USA recommended AZT therapy for asymptomatic infected individuals with less than 500 $CD4^+$ cells/mm^3. However its high cost (US $10 000 per year in 1991) prohibits its widespread use, especially for the many AIDS victims in the developing world. Since the treatment must be continued for the lifetime of the patient, only countries with the most well-financed health systems can afford it.

Mechanism of action of AZT and other analogues

AZT resembles the nucleotide thymidine except that it has a 3′-azido instead of a 3′-hydroxy group on the sugar moiety (Fig. 17.1). It acts as a thymidine analogue and accumulates in HIV-infected cells. Thus AZT and ddC, by blocking the action of reverse transcriptase and DNA synthesis, slow viral replication. There are, however, several problems related to the metabolism and toxicity of these drugs. Since AZT undergoes almost complete hepatic biodegradation, its half-life (about 1.1 hour) is short, with the production of a metabolite possessing no antiviral properties. The drug must, therefore, be administered fequently. AZT's major toxic side effect relates to the dysfunction of the rapidly dividing bone marrow stem cells and, consequently, patients undergoing AZT therapy will eventually require blood transfusions. The need for repeated

thymidine 3'-azido-3'-deoxythymidine (AZT) 2',3'-dideoxycytidine (ddC)

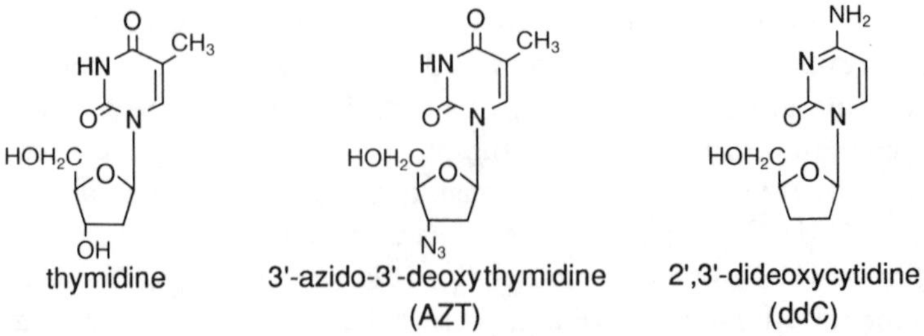

Fig. 17.1 Structure of AZT and ddC showing the 2′,3′-dideoxynucleoside nature of the drugs.

blood transfusions over extended periods would rule out any large-scale use of AZT for AIDS patents in developing countries. Nausea, insomnia and muscle pain are also side effects suffered by patients receiving AZT. Despite toxic effects, however, the data from placebo-controlled studies reveal that the clinical status of AZT-treated patients deteriorates less markedly and they live longer when compared with placebo-treated patients. In one trial of 56 patients treated over 1 year with a combination of AZT and ddC, an increased count in the $CD4^+$ lymphocytes was reported.

The second generation of pharmaceuticals developed were non-nucleoside reverse transcriptase inhibitors. Several of these have performed well in clinical trials and the good news is that they appear to lack some of the toxic side effects of the first generation drugs. A third promising approach to therapy involves the HIV protease which is crucial to the assemblage of mature viral particles. Several pharmaceutical manufacturers have focussed on anti-protease compounds. Trials have also taken place with a recombinant soluble CD4 molecule which competes with the CD4 receptor for the binding of HIV, thus protecting cells from infection and eliminating the bound virus from the bloodstream. Initial trials indicate that the body does not produce antibodies against this synthetic chimeric molecule. Boosting of the immune response, such as intermittent therapy with cytokines, in already infected individuals has also been tried as a measure to prevent progression of the diseases.

THE SEARCH FOR A VACCINE

The search for a vaccine, to protect those who have not yet become infected, has been intensive. There are a number of approaches for developing a vaccine each directed against a different target, such as the glycoprotein envelope or the core genes of HIV. Human trials on infected patients, designed to test safety and immunogenicity of AIDS vaccines, have already been completed; however, no-one is ready to test uninfected individuals to determine whether a vaccine protects against future viral contact. Therefore, scientists have based most of their research about AIDS vaccines on studies in macaque monkeys. Barring a totally unprecedented breakthrough, most virologists believe that even limited marketing of an effective HIV vaccine is several years away because, once it enters the body, HIV mutates at a rate hundreds of times faster than most other viruses. The availability of a vaccine is also of utmost urgency from the fetal point of view. The currently available chemotherapies, such as AZT, are contraindicated in pregnancy because of their teratogenic effects. This makes any treatment of HIV infection in pregnant women difficult and hazardous in those who may become pregnant.

HIV INFECTION DURING PREGNANCY

The most important route of HIV infection in infants is maternal via perinatal transmission, where the fetus/newborn is infected during pregnancy (transplacental), delivery (contact with maternal blood) or during breast-feeding. An estimated 30–60% of live births to infected mothers acquire infection and fewer than 5% will remain asymptomatic: a highly unfavourable prognosis. There are differences between estimates because mothers with advanced HIV disease are more likely to transmit the infection and this may be related to a higher titre or a more pathogenic strain of virus. Pregnancy also adversely influences the course and progression of the disease in infected women, raising the problem of who will care for the infants in the cases of those pregnant women who die quickly from AIDS. At the antenatal clinics in Port-au-Prince, Haiti, 6% of women attending for their first pregnancy, from 1986–88 were HIV seropositive. The global estimate of 12–15 million AIDS orphans places a horrendous burden on the elderly, usually grandparents, who are left to care for the mostly sick offspring of their own dead children. Healthy orphans often live alone, or have to look after their sick siblings, and are deprived of human love, care and comfort. This epidemic is thus unique in that the providers, as opposed to the usual vulnerable young and old, are most affected.

HIV transmission to the offspring can also be paternal at the time of fertilization with an infected sperm or during development via fetal exposure to infected semen. The diagnosis of HIV seropositivity in the newborn is difficult since the maternal HIV antibodies pass the placental barrier leading to misleading results; the diagnosis cannot be certain until a minimum 9–12 months of age when the maternal antibodies will have completely disappeared from the infant's blood. An infected child is, however, very likely to die, or be severely ill, by 2 years of age. Infected infants fail to thrive and suffer from recurrent coughs, diarrhoea, fever, pneumonia, candidiasis, lymphodenopathy and neurological impairment. Transmission through lactation was thought to be low until researchers in Africa reported that breast-feeding women were passing the virus to their newborns with a frequency of greater than 50%. These observations follow campaigns which have helped reverse the effects of the notorious effects, in the 1970s by companies like Nestlé, to convert women in developing countries to bottle-feeding. Mothers and public-health experts find themselves in an insoluble dilemma. Bottle-feeding remains perilous because the use of dirty water to mix the formula in unclean containers has been shown to increase infant mortality by up to 500%, primarily by triggering diarrhoea and pulmonary infections. Yet allowing HIV-infected mothers to breast-feed is also unethical. Continued nursing is also threatening to the mother, causing aggravated illness due to the physiological burden of lactation.

Pregnant HIV-positive women pass the virus to the fetus in many pregnancies yet, how can and should one restrain an HIV-positive person from spreading the virus? The balancing of individual's rights and society's needs raises many legal and judicial issues regarding testing and whether pregnancy should be interrupted in an individual found to be seropositive. The majority of women with AIDS are 13–39 years of age and thus have the potential for transmission to a sexual partner or, since virtually all women with HIV infection in developing countries are heterosexual, to an offspring. Perinatal transmission of HIV and paediatric AIDS must be reduced. One possibility is by screening young women and counselling those positive for HIV against future pregnancies. However, one of the most depressing problems that the health profession has encountered is that, despite behavioural counselling and the knowledge of being HIV positive, there are still persons who will not consider abortion, or, if they already have had a HIV-positive child, will subsequently have additional children that are also infected. Reports show that less than 50% of infected women who know early enough to abort choose to do so. In one study, only one out of three infected women who could have aborted chose to do so, and another study reported 16 mothers with infected children who subsequently had other babies despite the knowledge that they were infected. These facts are hard to rationalize; for many women childbearing is a cultural expectation and may be the anchor of the individual's womanhood. If the infection is kept secret, the woman will feel isolated and deprived of desperately needed support from family, friends, community and, particularly, from other infected women. In general, heterosexual women do not have the sense of community that is enjoyed by the gay culture. In the absence of a vaccine or effective therapy for AIDS, education, changes in sexual behaviour and a caring responsibility for the unborn are the only defences against the rapid spread of HIV.

STRATEGIES FOR THE PREVENTION AND CONTROL OF AIDS

AIDS represents the worst human communicable disease threat to arise for hundreds of years. In view of the exponential rise in the prevalence of seropositivity, the spread to the heterosexual population and the increasing incidences of babies born with the HIV virus, an AIDS strategy is urgently needed. The overall strategy for AIDS prevention and control is divided into long-term and short-term components. The long-term strategy is to develop vaccines and prophylactic and therapeutic drugs (as already discussed) whereas the short-term strategy is to modify the environment and behaviour. Simple effective strategies such as needle-exchange programmes and distribution of condoms where men are congregated (prisons and military institutions, for example) make a large positive impact in reducing infection. The basic educational emphasis is on the young in local communities, schools, colleges and universities, with the government's role being to facilitate and stimulate the teaching of AIDS prevention. While all who are concerned with AIDS management agree on the urgent necessity of preventing its spread and providing the best scientific and humane management available to those infected, there is disagreement on the strategies that should be adopted.

Mandatory testing, for example, is one such area. Mandatory testing has well-established precedents in tuberculosis and sexually-transmitted disease control and HIV could be listed as a venereal disease under the Communicable Disease Law. To be effective, legislation must co-ordinate activities against the spread of HIV infection and dispel discrimination against afflicted groups. About 40 countries have some kind of immigration restriction on HIV-infected individuals or compulsory testing for incoming foreigners. Many governments recommend that all surgeons and hospital patients should be routinely tested for HIV. The alternative of voluntary testing, even if adequate resources were provided, would be likely to identify only an estimated 10–15% of the total infected. For prenatal or premarital programmes, an estimated 1–2% of the infected population would be identified. In the light of these figures, legislation for compulsory testing could be justified by the courts, overriding the fundamental constitutional right of privacy in favour of the fundamental human right to be protected from infection and by the urgent public health necessity to prevent the spread of AIDS. An essential concurrent requirement is to make legal provision for the strict protection and confidentiality of test results and to provide counselling to seropositive individuals on safe sex practices, the danger of sharing needles and the risk of transmitting HIV infection to a fetus during pregnancy. Discrimination against persons with HIV disease and social stigmatization are two of the most worrying social challenges; the moral obstacles of some prevent the free access to others of information, condoms and choice of continuing a pregnancy. The community has a moral right to give its members information which dispels ignorance and allows for informed decision making. To develop a deep trust in a partner can be difficult, especially in sexual matters, and HIV involves discussions about medical facts, honesty, trust and fidelity. The more such matters are openly discussed in a community context the easier it will become to discuss them in a relationship context.

Supporters of mandatory testing contend that it would be effective in controlling the spread of AIDS and would reach high-risk groups. Persons with knowledge of their condition would be highly motivated and amenable to education for changed behaviour and to notification of their sexual partners. Women exposed to the AIDS virus would be counselled against pregnancy and transmission of the virus to the fetus. Mandatory testing would contribute to stopping the transmission of the virus before it becomes so prevalent that it is unmanageable. Past experience with tuberculosis and sexually-transmitted disease provides a successful model. AIDS should be treated like any other communicable disease, with strict provision for confidentiality of test results and criminal penalties for disclosure and for discrimination in employment, housing, or other entitlements. HIV-antibody testing need not be made more dramatic than, for instance, the routine perinatal screening for the rubella virus.

Opponents of this view state that compulsory testing would not be effective since this approach would undermine the acceptance of voluntary testing of high-risk individuals and discourage behavioural change and the notification of their sexual partners. Opponents also fear that the protection of privacy will not be secure. Making routine testing a standard and widely available medical procedure which requires informed consent and is attached to strict protection of confidentiality, may be an effective, non-intrusive strategy preferable to voluntary testing at designated test sites.

ADOLESCENT SEXUALITY

Adolescents are not yet grown up; they are neither child nor adult. Adolescents can be impulsive, with a desire for immediate gratification and a tendency to question authority. Because of their youth and relative freedom from health problems, adolescents may also feel that they are invulnerable to catastrophic events, which only happen to others. They are willing to experiment with drugs and sex which, along with strong peer pressure to conform, may place them at special risk from HIV infection.

To-date two general approaches for controlling the spread of HIV infection among the young have been utilized. One depends on encouraging morality and abstinence, the other on encouraging behaviour modification by exposure to sex education programmes which include specific information about the prevention and consequences of infection. Simplistic efforts to control sexually-transmitted diseases (and unwanted pregnancies) by abstinence have not been successful in the past so there seems no reason to suspect that they will be in the future. However, a co-ordinated programme of sex education as part of the school curriculum should modify sexual behaviour and help control the spread of HIV infection. Alongside knowledge, it is also important to teach adolescents to care for each other and for the potential of the unborn. Outreach programmes will have to provide this information to school dropouts. A responsible media (newspapers, radio, television) can do much in disseminating information regarding HIV infection and methods of prevention. In conjunction with education it is important that suitable condoms are easily obtainable by sexually active adolescents. While condoms are not foolproof, they do provide the safest penetrative sex possible with an HIV-infected partner. In the laboratory, latex condoms have been shown to block passage of the virus but natural skin condoms are not always effective. The effectiveness of condoms in preventing pregnancy is about 90%; the failure rate is most often among the young and inexperienced. In a study of couples who used condoms after knowledge of one of the partners' HIV seropositivity, 17% seroconverted despite the use of condoms over a mean follow-up period of 18 months. The usual cause of condom failure in contraception is that the condom comes off or ruptures; the risk of rupture and consequent infection can be minimized by proper placement, retention and removal of the condom, and by use of water soluble (rather than petroleum based) lubricants in conjunction with a viricidal spermicide, such as nonoxynol-9. The female condom is increasing in popularity among women in the west as it gives protection against sexually-transmitted disease and is not dependent on male co-operation (Chapter 15).

Unprotected sexual activity, often signalling sexual insecurity, is only one of a number of interrelated risk behaviours of adolescents, another being intravenous substance abuse. It is important for parents and teachers to understand and accept the interactive roles of biology, psychology and the environment of adolescence in order to develop the most appropriate and effective education and intervention strategies for the prevention of AIDS in the adolescent population. The goals of education are, firstly, to provide the young individual with correct data to rectify misinformation; secondly, to change behaviour; and thirdly, to nurture a caring, thoughtful attitude towards others. The last goal should help prevent Draconian measures being introduced which may seriously threaten civil liberties in an attempt to control the widespread dissemination of HIV related disease.

General references

Anderson, R. M., May, R. M., Boily, M. C., Garnett, G. P. & Rowley, J. T. (1991). The spread of HIV-1 in Africa: sexual contact patterns and the predicted demographic impact of AIDS, a review. *Nature*, **352**, 581–589.

Benditt, J. (feature ed.) (1993). AIDS the unanswered questions. *Science*, **260**, 1253–1293.

Brookmeyer, R. (1991). Reconstruction and future trends of the AIDS epidemic in the United States. *Science*, **253**, 37–42.

Hamer, M. & Brown, P. (1991). Those old jelly roll blues: what can we learn about AIDS from the moral panic created when syphilis spread through the west? *New Scientist*, **21/28** (Dec.), 54–57.

Hu, S. L., Abrams, K., Barber, G. N. *et al.* (1991). Protection of macaques against SIV infection by subunit vaccines of SIV envelope glycoprotein gp160. *Science*, **255**, 456–459.

Johnson, P. M., Lyden, T. W. & Mwenda, J. M. (1990). Endogenous retroviral expression in the human placenta. *American Journal of Reproductive Immunology*, **23**, 115–120.

Naz, R. & Ellaurie, M. (1990). Reproductive immunology of human immunodeficiency virus (HIV-1) infection. *American Journal of Reproductive Immunology*, **23**, 107–114.

Pantaleo, G., Graziosi, C., Demarest, J. F. *et al.* (1993). HIV infection is active and progressive in the lymphoid stage during the clinically latent stage of disease. *Nature*, **362**, 355–358.

Sattaur, O. (1991). India wakes up to AIDS. *New Scientist*, **2** (Nov), 19–23.

Schinazi, R. F. & Nahmias, A. J. (eds.) (1988). *AIDS in Children, Adolescents & Heterosexual Adults: an Interdisciplinary Approach to Prevention*. Elsevier, New York.

Smith, J. R., Kitchen, V. S., Botcherby, M. *et al.* (1993). Is HIV infection associated with an increase in the prevalence of cervical neoplasia? *British Journal of Obstetrics & Gynaecology*, **100**, 149–153.

Tindall, B., Cotton, R., Swanson, C., Perdices, M., Bodsworth, N., Imrie, A. & Cooper, D. A. (1990). Fifth international conference on the acquired immunodeficiency syndrome. *Medical Journal of Australia*, **153**, 204–214.

18 The effect of nutrition and exercise on the hypothalamic–pituitary–gonadal axis

In this book the close connection between stress, of varying etiology, and reproductive dysfunction has been repeatedly pointed out. Reproductive dysfunction caused by environmental factors, such as excessive physical activity, food restriction and abnormal eating behaviour (as in obesity and anorexia nervosa) has been well documented. Since the 1970s, a striking increase in eating disorders occurring most frequently, but not exclusively, in young women living in affluent societies has been reported. It has been hypothesized that a primary cause for this increased prevalence has been an obsession with thinness and obesity. It is interesting to note that the fattest of all mammals are humans whose adipose tissue to total body mass varies from 8% to over 35%. The regulation of fat distribution is complex, involving hormones, neurotransmitters and neuropeptides which integrate environmental, behavioural, psychogenic and physiological mechanisms.

NUTRITION AND THE MENSTRUAL CYCLE

Ovulation is readily suppressed by food shortage. It can also be suppressed by excessive exercise or cold exposure if these metabolic drains are not fully compensated for by an increase in food consumption. The evolutionary basis for these associated effects relate to the female mammal's need to reproduce within good foraging conditions in order to maximize her and her offsprings' survival. Generally, humans no longer need to obtain sustenance by extensive foraging, but they still possess these ancient evolutionary responses. Pubertal ovulation is particularly sensitive to energetic and nutritional constraints (Chapter 3). Facets of ordinary life such as the food eaten and the effort expended in acquiring it also affect the menstrual cycle. It has long been known that there are differences in steroid hormone metabolism between vegetarians and non-vegetarians, although the underlying reasons are still conjectural. A calorie-sufficient vegetarian diet can decrease plasma LH levels and shorten the menstrual cycle by several days. An interesting study on three groups of women, rural black South African, mixed race and white, all eating their customary diets, illustrates the strong link between diet and menstrual periodicities. The duration of menses and the menstrual cycle length were found to be significantly different between black South African/mixed race women (whose staple diet was corn with seasonal vegetables and fruits) and white women (whose staple diet also contained a generous meat component). The women of mixed race and black South Africans had significantly shorter menstrual periods and duration of menses than did the white women. In a follow-up study, the black South African women were given, in addition to their staple diet, a daily supplement of meat for 10 weeks. The addition of meat, (but not an isocaloric quantity of soybean) significantly increased the length of the menstrual cycle and the episodic release of LH and FSH but did not alter the LH peak. It is possible that the higher dietary phytooestrogen intake inhibited endogenous oestrogen and androgen biosynthesis via the suppression of the hypothalamic– pituitary–ovarian axis. Furthermore, phytooestrogens can partially antagonize endogenous steroid action by reducing the effective free concentrations of circulating steroids through the stimulation of SHBG synthesis (Chapter 16). The significant response following a changed eating habit, even after only 10 weeks, dramatically demonstrates a potential pitfall, that of attributing racial differences to what are environmentally induced functional characteristics.

FOOD RESTRICTION IN THE FEMALE RAT

Much research on food restriction has been done on experimental animals, particularly rodents. Studies with adult rats have shown that reduced food intake inhibits the secretion of all the glycoprotein hormones of the anterior pituitary except ACTH, which is increased, and FSH, the secretion of which is little affected unless food restriction is severe. These experiments in essence demonstrated that maintaining the rat at approximately 75 g bodyweight largely, but not completely, inhibited the pulsatile release of LH. Conversely, if weanling females are fed *ad lib* and allowed to grow normally until they reach about 75 g in bodyweight before having their food restricted, this effectively halts their reproductive development at a prepubertal stage. If, after several weeks, the rats are returned to *ad lib* feeding, 'catch-up' growth and reproductive development results in ovulation soon afterward. Food restriction does not, however, alter adult pituitary LH concentration or its half-life in the circulation, because normal pubertal development can be immediately initiated by GnRH administration whilst the animals are still under food restriction.

The natural mediation of food restriction involves both the adrenergic and opionergic pathways, since the opioid peptide β-endorphin stimulates prolactin and inhibits gonadotrophin release, probably by inhibiting the release of dopamine and GnRH from the hypothalamus. GnRH is under tonic opioid inhibition which varies during the menstrual cycle and is dependent on gonadal steroids exerting their negative feedback effects via an opioid link (Chapter 4). Opiate abuse has long been known to be associated with amenorrhoea in women and impotence in men. CNS-peptide hormones, including β-endorphin, stimulate appetite and, concomitantly with food ingestion, a number of satiety hormones (cholecystokinin, bombesin and neurotensin among others) are released from the gastrointestinal tract. These satiety peptides increase the opportunity for absorption by delaying gastric emptying and/or intestinal transit. They are also directly involved in modulating the hypothalamic feeding centres to suit environmental variables. The significant increase in the basal metabolic rate during the luteal phase of the menstrual cycle can be attributed to progesterone, which has an inhibitory effect on intestinal smooth muscle consistent with increased absorption. Under β-endorphin stimulation, progesterone levels are also elevated, which further acts in the suppression of GnRH. Progesterone, prolactin and β-endorphin are released by stress and function as stress hormones (Chapter 13).

It must also be mentioned that the sympathetic nervous system and catecholamines (elevated during food restriction) are physiologically involved in the control of ovarian function. The intraovarian distribution of sympathetic fibres is similar in all species of mammals and is associated with the vasculature, the fibres travelling along the interstitial tissue and surrounding developing follicles. The catecholamines, which control ovarian function, may originate from the extrinsic nerve supply, local cellular elements and circulating adrenaline from the adrenal medulla.

In the final analysis, experiments with protein, fat and carbohydrate deficiencies lead to the conclusion that the signal causing suppression of reproductive function in adult rats is not deficiency of a particular dietary nutrient but is the result of a deficiency of calories. For instance, food-restricted females can grow and achieve a normal pubertal cycle when their restricted daily level of rat chow is supplemented by access to drinking water containing glucose.

FOOD RESTRICTION IN THE HUMAN

In humans, the normal reproductive function in both males and females can be profoundly influenced by changes in the body's nutritional status. In women, especially, the reasons for a reproductively significant state of food deprivation are not esoteric and can be attributed to:

(a) Protein/calorie malnutrition in the underprivileged and poor; this is often further aggravated by residual metabolic deficiencies from previous pregnancies in women suffering dietary insufficiency (Chapter 9)
(b) Diets for weight loss in the privileged but nutritionally ignorant
(c) Eating disorders such as anorexia nervosa and bulimia nervosa which have, in recent years in the west, become disorders of major concern.

Each of these is usually associated with an increased incidence of disturbed and anovulatory menstrual cycles and amenorrhoea.[1] In general, malnourished women with reproductive dysfunction have low circulating levels of oestrogen, LH, FSH and decreased pituitary responsiveness to GnRH due to the suppression of GnRH neuronal activity. In men, undernourishment is accompanied by decreased circulating LH and testosterone levels resulting in depressed sexual drive and libido.

In response to prolonged food restriction, thyroxine (T_4), triiodothyronine (T_3) and insulin are also suppressed and cortisol elevated. Variations in stress levels and corresponding endocrine adjustments have profound effects on an individual's reproductive capacity. Chronic hypersecretion of CRH reduces GnRH secretion whilst the body's resources are directed in supporting the hormonal backup of the general adaptation syndrome. Ovulatory women spend more energy during the luteal phase and, therefore, an energy saving can easily be made from a shift from fertility to infertility. The caloric saving from suppressing ovulation and menses has been estimated to be 9% in the 24-hour energy expenditure during the luteal phase (Chapter 13). It must be remembered that suppression of the reproductive axis is a normal response to short periods of decreased food intake which occurred frequently in our ancestors' environment, but it is the prolonged activation of the stress response which leads eventually to dysfunction and diseases of adaptation. The subsequent section describes some of these as they relate to eating disorders.

[1] Amenorrhoea is a clinical term denoting cycles longer than 90 days duration with probable anovulation. The term oligomenorrhoea refers to irregular menstrual cycles of 39–90 days duration. The terms eumenorrhoeic, regular and cyclic are used interchangeably to refer to normal menstrual cycles which recur consistently at intervals of 25–39 days duration.

EATING DISORDERS

Food deprivation for reasons other than poverty is increasingly prevalent particularly in more affluent societies.

Dieting

Delayed menarche, secondary amenorrhoea and infertility have all been linked to excessive dieting for weight control because fasting causes an almost total suppression of LH secretion in otherwise healthy women. Semistarvation or weight loss of approximately 1 kg/week over a 4–6-week period brought about by an energy intake of less than 1000 kcal (4200 kJ) per day causes a significant proportion of luteal-phase defects and anovulatory cycles in women who previously showed normal menstrual cycles. Dieting can induce either disturbances of the luteal phase (progesterone values smaller than 6 μg/ml, luteal phase length less than 8 days) or disturbances of the follicular phase (with low oestradiol plasma levels and an absence of ovulation) (Fig. 18.1). Secondary amenorrhoea (Fig. 18.1*c*) is common when bodyweight falls to more than 10% below the ideal; this is seen in an extreme form in anorexia nervosa.

Dieting, ironically, is extremely ineffective in permanently controlling weight. The induction of a homeostatic emergency conservation strategy, linked to the suppression of thyroid activity and a drop in the metabolic rate, renders crash diets useless unless they are complemented with increased physical activity. This, of course, makes adaptive sense because, at times of food shortage, it is wanton to squander the body's precious fat reserves on non-productive physical activity.

Anorexia nervosa

Anorexia nervosa is a complicated syndrome, found mostly in women, and is characterized by very low food intake (particularly of carbohydrates) leading to extreme emaciation of the body. It is typically an affluent malady with, in the United Kingdom, an estimated 90–95% of afflicted individuals being women of college age and 1% girls of high-school age. Anorexia nervosa is a maturational problem in adolescents who, in the majority of cases, come from homes where family relationships have ceased to exist. One survey reported that 70% of patients with eating disorders were sexually molested by a family member or friend during childhood, or who

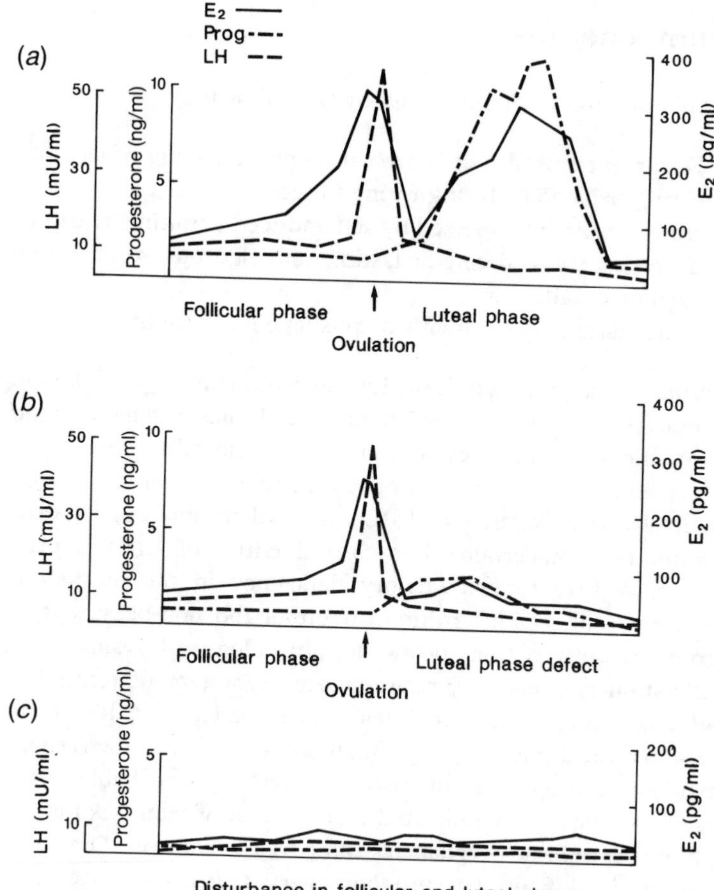

Fig. 18.1 Hormone levels in common menstrual disturbances as a result of food restriction, eating disorders or excessive energy expenditure. (*a*) The normal menstrual profile; (*b*) luteal-phase deficiency; (*c*) disturbance of both follicular and luteal phases. E_2 is oestradiol-17β. (Note that the hormone concentrations are drawn to different scales so are not directly comparable.)

associated eating with sexual involvement and anxiety. For these patients treatment is often difficult, but a multidisciplinary approach involving contact with a physician, psychologist and nutritionist is advantageous.

The reproductive effects in anorexia are as described for dieting but are more severe. Following increased calorific intake and weight increase, the infantile LH secretion pattern mostly returns to the pubertal or adult mode. Recovery takes longer in older patients and ones who have been ill for a long time.

Bulimia nervosa

Bulimia nervosa is characterized by the following.

(a) Recurrent episodes of binge eating and a feeling of lack of control of eating behaviour during eating binges
(b) Binge eating interspersed by self-induced vomiting, use of laxatives or diuretics, strict dieting or fasting, and rigorous exercise all to prevent further weight gain
(c) A persistent concern with body shape and weight.

Bulimia is recognized clinically by a minimum average of two binge-eating episodes per week for at least 3 months. Bulimics may have disturbed luteal phases and anovulatory cycles not readily detectable from their menstrual history alone, and they often have a history of anorexia nervosa as well.

Luteal-phase deficiency (LPD) is a silent disease; its extent in the community is uncertain. The clinical entity of LPD is present when there is recurrent postovulatory deficiency in the production of progesterone leading to infertility or habitual abortion because, typically, the corpus luteum functions below the physiological threshold necessary for fertilization or pregnancy maintenance. Synonymous terms for LPD are inadequate corpus luteum, luteal-phase inadequacy, luteal insufficiency and short luteal phase. It is estimated that LPD may be responsible for a considerable proportion of cases of infertility and habitual miscarriage. Results of one survey indicated that the risk of fetal loss may be twice as high in first bulimic pregnancies compared to controls. Diagnosis of LPD, however, is difficult because not every menstrual cycle is necessarily deficient in progesterone.

Healthy women at the extremes of reproductive age (immediately postmenarchal or premenopausal) often experience physiological LPD. Psychological or physical stress can also precipitate LPD; particularly sad are reports of LPD in women who have lost children to sudden infant death syndrome.

THE RELATIONSHIP BETWEEN HYPERACTIVITY AND FOOD SUPPLY

There is some overlap between anorexia and bulimia eating disorders and increased body activity, sometimes to the extent of exercise addiction. During therapy directed toward weight gain, patients with anorexia nervosa often report that they exercise in an attempt to 'distribute the newly acquired fat over the body'. The physiological consequences of excessive energy expenditure are a lack of growth and reproductive development due to the absence of pulsatile release of LH. The relationship between energy intake and energy outlay is at the core of the evolution of all mammalian foraging strategies: wild animals spend costly energy on locomotion only when they must find food/homes, fulfill social needs or avoid danger. Forced and prolonged exercise (swimming, running, dancing) may be an evolutionarily inappropriate means of becoming 'fit'. It has been suggested that a cycle of increasing activity and decreasing food intake is the basis of activity anorexia (runners' amenorrhoea), characterized by GnRH suppression and reproductive dysfunction.

There is evidence that reproductive dysfunctions attributed to exercise may be related to eating disorders and dieting because exercise may alter the tendency to eat meals when they are available. Experiments on rats demonstrated that physical activity interfered with the amount eaten during a single daily meal, even though there was time to consume more food. Sufficient food was consumed when the food was presented as four 15-minute meals rather than as a single 60-minute meal because in exercised animals satiety quickly sets in. Human studies, likewise, demonstrated that high-mileage runners who were amenorrhoeic became more energy efficient by significantly lowering resting metabolic rate, when compared with eumenorrhoeic runners and sedentary controls. Although both groups of runners were similar in age, weight, percentage of fat and training pace, the amenorrhoeic group displayed a more aberrant eating pattern then did the eumenorrhoeic and sedentary groups. The amenorrhoeic women maintained their energy balance and compensated for their less adequate diet, while still maintaining a large energy output, by lowering caloric needs for other bodily functions such as fertility maintenance. It could be that strenuous exercise decreases the value of food reinforcement resulting in reduced food intake. Reduction in food intake increases the reinforcing value of physical exercise. This produces an escalation of activity that further suppresses appetite. In the human context, anorexias may have resulted from natural selection favouring those individuals who become active in times of food scarcity. When activity increases, wild animals become anorexic, since stopping to eat is negatively balanced against reaching an area where food may be abundant. The nomadic !Kung bushwomen of the African Kalahari Desert show an adaptive seasonal suppression of ovulation. Women of this tribe ovulate only at certain times of the year when food is plentiful and hunting activities are restricted. At other times they are on the move and, while carrying their children on their backs, may walk up to 60 km each day in search of food. Therefore, in the human female, food restriction or excessive physical exercise is preferentially compensated for by physiological mechanisms which conserve energy by cessation of ovulation and menses and also by a drop in the resting metabolic rate, a strategy which preserves the body fat. The endocrine control involves decreased thyroid function and increased cortisol secretion with the endogenous opiate system mediating between eating and running. It may be that anorexia athletica is the result of an ancient interplay between biology, physiology and behaviour occurring in a modern setting.

WOMEN AND EXERCISE: MODERN TRENDS

Physical exercise and athletic training have become an important part of many women's lifestyles. This is in contrast to earlier this century when girls and young women were discouraged from participating in physically active or competitive sports, especially during menstruation. On the other hand, boys were expected to participate in strenuous athletic training. The beneficial effects of physical fitness on the cardio-vascular, musculo-skeletal and metabolic systems are now well recognized for both men and women, and regular exercise has become an important component of a healthy existence. There is also a positive correlation between exercise, improved fitness and enhanced sexuality. Consequently, a great amount of interest and research has been generated by the numerous observations of menstrual irregularity (athletic amenorrhoea or oligomenorrhoea) seen in women runners and ballet dancers who undergo frequent strenuous physical training and who are required to maintain a low bodyweight. Swimmers, by contrast, exhibit less reproductive dysfunction because low bodyweight is not so critical, making dietary behaviour less necessary. The incidence of amenorrhoea among dancers, gymnasts and athletes varies but may afflict up to 78% of adults and 50% of adolescents. Several factors contribute to the problem, including a particular athlete's body size and composition, the physical stress of the specific sport or activity, the severity of the training regimen and the age and reproductive maturity of the athlete. The dietary intake and the degree of psychological stress associated with athletic performance is also of immense importance.

As discussed above, there is excessive athleticism in individuals prone to anorexia nervosa. The connection between physiology and behaviour has been investigated by numerous neurological studies. One model attempts to demonstrate how anorexia nervosa might develop within a vicious cycle of starvation and hyperactivity. A proportion of young women, mostly from western societies are not satisfied with their bodyweight and shape. This may be due to sociocultural conditioning; advertising exploits feelings of dissatisfaction and insecurity by promoting special diets reputed to be effective in the maintenance of the desired low bodyweight. In terms of body shape, only 5% of the population has the body structure depicted in fashions, yet the model is approximately 23% less in bodyweight than average (25 years ago she was 8% below average). Women have also, to their disadvantage, internalized the external pressures equating thinness with career prospects, success in relationships and other achievements to the extent of emulating the anorexic Twiggy, a famous model of the 1960s. The fashion environment does not tolerate diversity, but women must resist being psychologically imprisoned. Some responsible organizations have recognised that a problem exists; for example, the fashion magazine *Elle* has a policy not to use painfully thin models. Since diets alone are notoriously unreliable as a means of weight loss, many women on diets also spend their leisure time burning calories by zealously jogging or participating in some other sport. A proportion of these women with, perhaps, a genetic predisposition then develop an addiction to exaggerated dieting and excessive exercise. Starvation produces internal stimuli for hyperactivity and, conversely, hyperactivity ameliorates the aversive sensations associated with starvation. For example, the semi-starvation-induced alterations in neurotransmitter turnover can be reversed by hyperactivity; noradrenergic, dopaminergic and endorphinergic neuronal activities are stimulated by vigorous exercise and might thus act as some kind of reward for exercise. It is also, of course, these same neurotransmitters which control reproductive function and affect, through neuroendocrine changes, the immune system.

The combination of vigorous training regimens, dieting and the stress of competition can lead to a state of immune vulnerability to a range of viral and bacterial infections. The release into the circulation of ACTH and, in turn, cortisol, in response to the stress of daily high intensity exercise, may override the positive effects of general fitness and lead to a degree of chronic immunosuppression.

MEN AND EXERCISE

Evaluating the effect of exercise on male puberty is more complex than it is in women because early sexual maturation has greater potential benefit to athletic performance in boys. The onset of puberty in boys is influenced by anabolic energy balance, and, in prepubertal boys, the significant strength and muscle mass increase gained during resistance training may accelerate the onset of puberty. Conversely, prolonged endurance training, such as long-distance running, may lead to a catabolic influence sufficient to delay maturation and the onset of puberty. It seems that conventional sports, removed from the pressures of competitive stress, have little effect on the timing of puberty in the majority of adolescent boys. However, boys and young men with an eating disorder are often involved in competitive track events, wrestling or gymnastics. The driving force seems to be the coach's requirements to improve performance; for example, a lower bodyweight in runners or reduced body fat by 1% or 2% in gymnasts.

In postpubertal men a parallel exists between the sexes in the reproductive effects of chronic exercise. Endurance training and other forms of strenuous physical activity are associated with an initial increase in testosterone followed by suppression; lower mean testosterone levels have been observed in runners with a training distance of more than 65 km but not in those running less than this per week. Thus, physical activity has a range of effects on male reproductive function depending upon the intensity and duration of the activity and the fitness of the individual. In general, it appears that relatively short, intense exercise increases serum testosterone levels, but there is debate as to what degree haemoconcentration, decreased clearance and/or increased synthesis are involved. It is clear from the immediacy of the testosterone increment that its mechanism does not involve gonadotrophin stimulation of the testis as there is suppression of circulating testosterone levels during and subsequent to more prolonged exercise (and to some extent in the hours following intense short-term exercise). A variety of systems could contribute towards decreases in testosterone synthesis, including decreased gonadotrophin; increased cortisol, catecholamine, β-endorphin and prolactin levels; or perhaps even an accumulation of metabolic waste materials. In the final analysis, endurance training induces changes in the function of the reproductive axis in men in a similar manner to that in women.

Despite physiological reduction in testosterone, symptomatic changes in spermatogenesis are slower and less clearly defined. There is evidence that demanding physical activity does impair fertility. In an artificial insemination programme, it was observed that sperm from donors with a high physical activity profile and low semen volume scored significantly lower pregnancy rates than sperm from those with normal activity and low semen volumes. One interesting study was of soccer players from Australia and those from European countries where football is the most popular form of male physical exercise. The results suggested that long-term trained, non-professional soccer players suffered some derangement in the control of LH secretion, possibly due to repeated increases in adrenal steroids and other stress-related hormones in response to competitive play. A small but significant decrease in sperm motility was also noted. However, contradictory observations in other sportsmen may indicate that, once again, 'stress is where you find it'; as with the baboons (Chapter 13), different individuals handle competitive (life-threatening) situations differently.

PREGNANCY AND EXERCISE

The fact that pregnant women can and do perform feats requiring a high level of physical fitness is often (perhaps incorrectly) used as proof of their physical capabilities, and the compatibility of strenuous sport and pregnancy is only occasionally questioned (participation in, for example, the Olympics by pregnant women is often mentioned with approval in the press). Pregnancy-induced physiological alterations may not be necessarily disadvantageous for physical performance. Some changes are advantageous: the most obvious are increases in cardiac output and blood and plasma volumes as well as in ventilation and oxygen transport capacity. In essence, pregnancy changes many variables in the same direction as does regular training in response to the extra demands placed on the body. However, the changes which work positively on the performance capacity might be counter-balanced by other pregnancy-related features such as weight increase and the shift of the body's centre of gravity. Very few data are available on the postpartum recovery, quality of lactation and resumption of menstruation in mothers who trained heavily during pregnancy, or who start training within a short time after labour. More information is needed to be able to counsel women about the effects of strenuous training during pregnancy and the puerperium. The most important concern related to pregnancy and exercise is whether exercise influences the course of pregnancy and fetal development.

Since exercise is an integral part of normal life for many women, a proportion do not want to interrupt their usual routine even when this involves high performance and competitive sport during pregnancy. The available evidence indicates that moderate exercise, on a regular basis during pregnancy, leads to an improved course of pregnancy when compared with that of a sedentary lifestyle. Such exercise is also without notable health hazards to the woman or her fetus, providing she is in good general health and has no pregnancy-related disease. Exercises and sports which are especially beneficial involve rhythmic motion of large groups of muscles, are aerobic and do not require delicate balance. These activities improve physical fitness. Being able to maintain good bodily control, together with the pleasure derived from activity, also works positively on the emotional wellbeing. An added bonus is that exercise aids in the prevention of certain pregnancy-related disorders such as backache, leg cramps, swollen legs and varicose veins. On the debit side, the risks of sport injuries might be increased during pregnancy due to changes in the centre of gravity, shape of the body posture and softening of ligaments. For these reasons, along with the increase in weight, certain activities such as riding, surfing, water-skiing, team and contact sports should be discouraged.

Strenuous physical activity results in many changes that can affect the well-being of the fetus. Some of these changes are obvious; for example, exercise stimulates increased production of noradrenaline causing immediate contractions of the uterus which may, if severe, negatively affect the uteroplacental circulation and the fetus. There are indirect observations which show that the safety limits of exercise may have been exceeded in some situations. Fetal tachycardia has been observed during strenuous exercise, and reductions in birthweight, dependent on the intensity of the physical activity during pregnancy, have also been recorded.

The most serious argument against endurance sport and also prolonged immersion in saunas, during pregnancy concerns the consequences of maternal hyperthermia. Animal studies have shown a consistent relationship between birth defects and exposure to high temperatures, although the data on human pregnancy and elevated temperature are not as consistent. Hyperthermia is a natural result of certain types of sport: values of up to 41 °C have been recorded during long-distance runs, depending on their duration and the ambient temperature. In such cases there may be a risk to the fetus, especially since retrospective studies have reported CNS (microphthalmia, seizures, abnormal EEGs and faulty neural tube closure) defects in offspring where core temperatures had been elevated above 38.9 °C. Resting fetal core temperature mirrors maternal core temperature but remains 0.4–0.6 °C higher due to the fetal–maternal gradient. During exercise, the redistribution of blood away from the uterus to the working muscles and the skin in order to thermoregulate has a significant effect on fetal heat dissipation. Hyperthermia has at least four deleterious effects.

(a) Teratogenic effects of core temperatures of greater than 40 °C in early pregnancy have been conclusively demonstrated in animal studies.
(b) Effects on oxygen transport caused by the maternal and fetal haemoglobin-binding curves shifting to the right at elevated temperatures.

Although this shift results in an improved oxygen supply to the placenta, oxygen uptake from the fetal blood becomes more difficult (Chapter 9).

(c) Effects on O_2 consumption which increases due to temperature-induced elevated metabolic rate.

(d) Shifts in maternal blood flow with increased flow to the skin for thermoregulation. Several animal studies have indicated that uterine blood flow decreases by 20–60% during exercise. On the other hand, within 20 minutes after exercise the blood flow returns to normal, sometimes with a slight overshoot. Furthermore, other changes might fully or partially compensate for the blood flow reduction. For example, blood flow redistribution within the uteroplacental unit favouring intervillous space flow at the expense of myometrial flow, increased O_2 extraction from the blood and haemoconcentration at the beginning of exercise with an accompanying increase in O_2 transport capacity.

Since hyperthermia is a potential consideration for pregnant exercising women, non-weight bearing exercise in the water may be better as it is less strenuous and provides for greater heat loss. A study comparing land and water exercise found that pregnant women had significantly lower heat rates and systolic blood pressures during immersion exercise when compared to their land controls. Fetal heart rates also showed a tendency toward being higher after land exercise compared to water exercise, which may reflect increased core temperature and alterations in uterine blood flow during and after land exercise.

General references

Clapp, J. F. (1991). Exercise and fetal health. *Journal of Developmental Physiology*, **15**, 9–14.

Cook, C. L. (1991). Luteal-phase defect. *Clinical Obstetrics & Gynecology*, **34**, 198–210.

Cumming, D. C. (1990). Physical activity and control of the hypothalamic–pituitary–gonadal axis. *Seminars in Reproductive Endocrinology*, **8**, 15–24.

Garner, P. R. (1990). The impact of obesity on reproductive function. *Seminars in Reproductive Endocrinology*, **8**, 32–43.

Jones, G. S. (1991). Luteal phase defect: a review of pathophysiology. *Current Opinion in Obstetrics & Gynecology*, **3**, 641–648.

Keizer, H. A. & Rogol, A. D. (1990). Physical exercise and menstrual cycle alterations, what are the mechanisms? *Sports Medicine*, **10**, 218–235.

McMurray, R. G. & Katz, V. L. (1990). Thermoregulation in pregnancy, implications for exercise. *Sports Medicine*, **10**, 146–158.

Mitchell, J. E., Seim, H. C., Glotter, D., Soll, E. A. & Pyle, R. L. (1991). A retrospective study of pregnancy in bulimia nervosa. *International Journal of Eating Disorders*, **10**, 209–214.

Vander Walt, L. A., Wilmsen, E. N. & Jenkins, T. (1978). Unusual sex hormone patterns among desert dwelling hunter gatherers. *Journal of Clinical Endocrinology & Metabolism*, **46**, 658–663.

Wade, G. N. & Schneider, J. E. (1992). Metabolic fuels and reproduction in female mammals. *Neuroscience & Biobehavioural Reviews*, **16**, 235–272.

Weidemann, M. J., Smith, J. A., Gray, A. B. *et al.* (1992). Exercise and the immune system. *Today's Life Science*, **4**, 24–33.

White, J. R., Case, D. A., McWhirter, D. & Mattison, A. M. (1990). Enhanced sexual behaviour in exercising men. *Archives of Sexual Behavior*, **19**, 193–209.

Wolf, N. (1991). *The Beauty Myth*. Vintage, London.

Xie, Q. W. (1991). Experimental studies on changes of neuroendocrine functions during starvation and refeeding. *Neuroendocrinology*, **53** (Suppl. 1), 52–59.

18 Principles of teratology and an update on nicotine, ethanol and caffeine abuse

DEVELOPMENTAL TOXICOLOGY

This book contains many examples of abnormal development due to either intrinsic defects in gene expression or epigenetic modifications of normal gene expression. Chapter 16, in particular, is concerned with recognized teratogenic derangements of sexual differentiation resulting from exposure to hormonally active xenobiotics. The present chapter outlines the principles underlying mutagenesis, teratogenesis and delayed development. Specific biochemical and physiological mechanisms and common developmental disturbances associated with nicotine, ethanol and caffeine consumption are also examined. It has been repeatedly emphasized that epigenetic factors can, by influencing the activity of genes, impinge on the differentiating oocyte or sperm and modify the identity of the future conceptus. Genomic or parental imprinting during gametogenesis which, at syngamy, begins to direct the normal development of the conceptus' unique potential is a special case of epigenetic influence (Chapter 8). Teratology, on the other hand, is the science which deals with biological, genetic, biochemical, and behavioural aspects of maldirected development and spans the period from germ cell differentiation to the termination of functional development in the postpartum individual. Behavioural teratologogy refers to long-term effects on behaviour and psychological development. For example, infants prenatally exposed to substances like ethanol, opiates, various catecholamines and caffeine may experience neonatal withdrawal symptoms and neurobehavioural impairment such as hyperactivity, shortened attention span, co-ordination problems and specific learning difficulties.

The early stages of mammalian differentiation are times of greatest development flexibility. A robust genetic constitution may be able to repair mutational damage, although major defects are likely to cause spontaneous abortion or remain unrestored. Minor variations in developmental programmes can also influence biochemical function in subtle, non-visible ways and adversely affect the emerging identify of the new offspring. In human populations, it is difficult to distinguish between a developmental divergence beyond the usual range and an anomaly because of the existence of a continuum of variations ranging from the normal to the abnormal. Nevertheless, since each of us is already a rare survivor of natural selection (approximately 20% of conceptuses survive to the blastocyst stage and of those only about 75% establish clinically recognized pregnancies), it seems natural justice that those which do survive, against all odds, should be permitted to express their full genetic potential. The term 'congenital anomaly' describes the results of disturbed prenatal development and includes all forms of defects present at birth but assigns no cause. The inclusion of visible anomalies that are detectable at birth or in the early neonatal period only has led to a gross underestimation of the incidence and importance of the problem. On the basis of numerous surveys, a consensus has been reached that approximately 3% of newborn children are affected with major congenital malformations. Numerous congenital abnormalities (such as pyloric stenosis, delayed development or absence of teeth, carcinoma in infants and young children, mental retardation and behavioural anomalies) are undetectable at birth and become evident only after the neonatal period. A realistic estimate raises the risk of major malformations in childhood from the popularly quoted 3% at birth to 8–9% at 7 years of age. Major congenital defects have increased as deaths due to other causes have declined and may now account for up to 22% of perinatal deaths in western countries. Chromosomal abnormalities are believed to be associated with 50% of spontaneous abortions and 8–10% of stillbirths.

TERATOGENESIS

A teratogen can be defined as any agent introduced during the period spanning germ cell differentiation and the end of prenatal life that causes morphological or functional change in the offspring. The earliest written expression of attempts to understand the significance of malformations in humans dates back to the Babylonian era, about 2000–1000 BC. In the ancient world it was generally felt that malformations were manifestations of divine intervention and served as an omen. The words monster (from the Latin *monstrum*, meaning warning or ill omen) and teratology (from the Greek *téras* meaning abnormal form and *logos* meaning the study of) were interpreted as 'to point out' or 'to show'. The ancients believed that malformations were caused by noxious occurrences during pregnancy and warned of something deleterious occurring in the environment. Malformed children were seen to be privileged in some way because they were 'chosen' to direct attention to something that needed correcting. Jews and Christians, on the other hand, taught that the malformed child was God's punishment for a grave sin committed by the parents. This interpretation resulted in disinterest in the role of environmental factors in the etiology of birth defects. It took the thalidomide tragedy of the late 1950s and early 1960s for the average modern physician to be convinced that not all congenital malformations have a genetic cause. The accidental introduction of the powerful teratogen thalidomide (a supposedly harmless sedative hypnotic agent effective in treating nausea in pregnancy) 'showed' or 'pointed' to the hazards of exposing the developing embryo/fetus to foreign substances. However, not until thalidomide was shown to be directly responsible for the birth of up to 10 000 deformed infants with absent limbs or flipper-like stumps was the drug removed from the market. The thalidomide-associated pattern of malformations included, in addition to the well known limb reduction and/or deformation, facial capillary haemangiomas, hydrocephalus, intestinal, cardio-vascular and renal anomalies, and eye and ear defects. Intelligence is not affected. The 1950s was also the decade marked by the DES tragedy, where an estimated 6 million persons were affected (Chapter 16). Not many medications consumed during pregnancy can be trusted to be completely safe, and care in drug consumption should also be exercised by the breast-feeding mother. Several drugs not usually noted for their teratogenic potential, including ampicillin, acetylsalicylic acid and codeine, have reportedly been taken more frequently than the average by mothers of children affected with congenital heart disease.

Infectious diseases, nutritional stresses, hyperthermia, radiation, pharmaceuticals, drugs of abuse, environmental pollutants and emotional/psychological stresses can all cause disruption of normal embryonic development. The classic results of hazardous epigenetic influences on the developing mammal are, in descending order of severity: embryonic death, gross congenital malformation and intrauterine growth retardation. These may manifest themselves postnatally as reduced viability, impaired growth and/or CNS effects and may be permanent, not only for the affected offspring but also for the next and subsequent generations if the germ line is implicated. Whether an agent is teratogenic depends upon several factors.

The stage of development at disturbance The time in gestation during which the embryonic/fetal development is disturbed determines its susceptibility to teratogenic agents. Little cell differentiation takes place during the early embryonic period so all blastomeres are equally vulnerable to injury and exposure to toxic substances often results in death of the embryo rather than specific malformations. The sensitivity of an embryo to external disturbances is greatest immediately preceding or during the differentiation of a given organ field. In the human the limb buds, for example, can be seen at about week 5, but the limb field can be delineated much earlier (at approximately 3–4 weeks of embryonic life); at this stage the embryo is extremely sensitive to any toxic agent which may adversely affect the limb field.

The specific agent causing the disturbance The physiological properties of agents affect development according to specific mechanisms. For example, the pharmacokinetic properties of many drugs can change following placentation and the establishment of exchange systems between mother and embryo.

The dose and duration of exposure to the agent Teratogenicity is described by a dose–response curve. In addition, the simultaneous presence of other stresses may compound the potential toxic effect of an individual teratogen. Stresses are cumulative, and incidental or occupational exposure to chemicals, pesticides and ionizing radiation may interact, synergistically, with the primary agent.

The genetic susceptibility of the developing offspring and the physiological state of the mother The importance of the genotype of the embryo or fetus in determining its response to xenobiotics is apparent in the variation of defective development that may occur between dizygotic twins exposed to a teratogen during the critical period (see next section). Chronic maternal vascular diseases, such as essential hypertension, heart disease or diabetes mellitus, are likely to contribute to uteroplacental insufficiency and exacerbate the deleterious effects of teratogens.

The absorption routes Absorption routes are via the gastrointestinal tract, the lungs or the skin. Because of the large surface area of and efficient blood flow in the lungs, pulmonary absorption of environmental pollutants, such as gases, volatile compounds and small particulates, is rapid. Metal absorption, lead for example, is many times greater in the lungs than in the intestine.

In summary (Fig. 19.1), a major teratogenic insult during the first 2 weeks of embryonic life (or the first 4 weeks following menstruation) is usually embryotoxic, but, if the embryo survives, no organ-specific anomalies result. Organ differentiation spans embryonic weeks 3–8 (or gestional weeks 5–10), and during this time the human embryo is especially

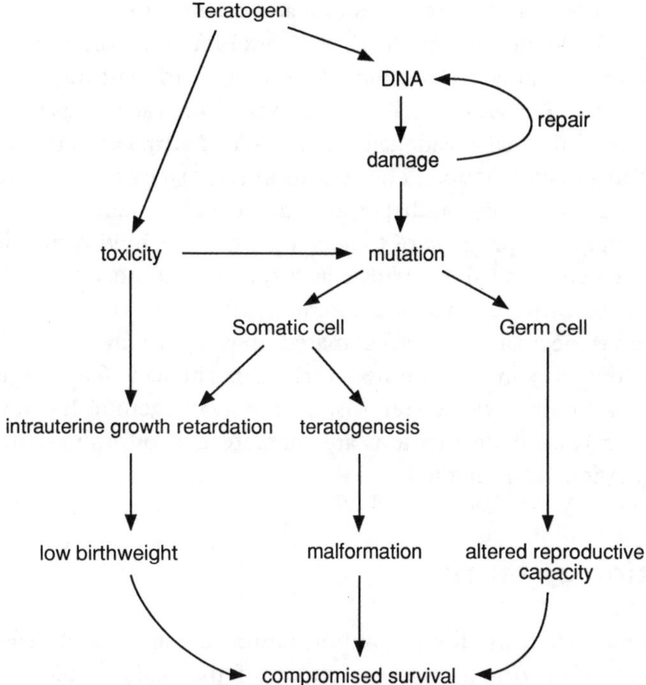

Fig. 19.1 A prenatal insult may impair and/or deform growth, resulting in compromised viability of the offspring and, if the germ line is implicated, of the next and subsequent generations.

susceptible to teratogens which may cause germinal and/or somatic mutations. Organogenesis is followed by organ growth, when a teratogen can deleteriously affect growth of the entire fetus or of a particular organ or organ system but usually does not produce a specific malformation. Specific effects cannot be excluded, however, since the development of eyes, ears, teeth, external genitalia and the CNS extends well into the fetal period. The final severity of the effect resulting from a certain teratogen depends on individual genetic predisposition, reflected in cell and tissue-specific repair processes, and whether the defect can be transmitted to a subsequent generation.

THE CRITICAL PHASE OF INTRAUTERINE DEVELOPMENT

The critical period of intrauterine development is that time during development at which the embryo or fetus has the greatest sensitivity to a particular stressful influence. Because each organ, or organ system, has a particular critical period, exposure to xenobiotics at different times of development is likely to produce lesions in different organ systems, reflecting these individual critical periods. For each organ system, a threshold for abnormal organogenesis exists, and surpassing this threshold will result in a malformation. The actions of teratogens require the embryo or fetus to be genetically predisposed to a particular malformation and the teratogen triggers expression of the defect. The threshold is thus locked to one or more genes and the extrinsic agent; as the number of predisposing genes increases, the severity of the teratogenic insult required to surpass the threshold decreases. It is estimated that approximately 50% of all malformations may involve multifactorial inheritances. Cleft lip and cleft palate, spina bifida, pyloric stenosis, clubfoot, congenital hip dislocation and certain cardiac malformations are characteristic of multifactorial types of malformations in humans.

Definition of terms

Malformations and disruptions Malformations are structural defects of an organ, parts of an organ or larger region of the body resulting from an intrinsically (genetic) or extrinsically (epigenetic) abnormal developmental process. If the development is abnormal due to intrinsic causes, the resultant primary developmental defects are called malformations and are often carried by a single mutant gene. In the case of extrinsic causes, the resultant developmental defects are called disruptions or secondary malformations. The anomalies resulting from teratogens are disruptions because of the extrinsic interference with an initially normal developmental process. As mentioned previously, genes may predispose to and influence the final outcome of the etiological agent. Disruptions are causally heterogeneous and may be anatomically identical to malformations. In any given case, a distinction may be made on the basis of the associated malformation together with the history of known gestational exposure to a teratogen.

Deformations A deformation is an abnormal form, shape, or position of a part of the body caused by mechanical forces. Environmental causes of deformation are usually limited to the gestational sac and the uterus. For example, restricted fetal movement occurs in oligohydramnios (amnion rupture causing intrauterine constraint), multiple pregnancies (twins and multiple fetuses may interlock), malformed uteri and uteri with fibroid cysts. Fetal restriction can result in clubfoot (talipe equinovarus), syndactyly, limb deficiency, body wall defects, neural tube defects and congenital dislocation of the hip. Deformation may also be secondary to a dysfunction in the fetus itself and may be referred to as autoteratogenesis. For example, defective CNS function resulting in fetal hypomobility may cause congenital limb deformities. Rare complications causing deformation have also been reported following amniocentesis.

Dysplasias Dysplasias are an abnormal organization of cells into tissue(s) and represent defects of tissue differentiation which may or may not be metabolically induced. Minor dysplasias are very common in the normal human population and include freckling, café-au-lait spots, moles and naevi. Dysplasias may be induced environmentally through a variety of influences, including radiation, viruses and carcinogens.

It might be helpful if the terminology of developmental anomalies of differing etiologies were defined. The arbitrary categorization of the continuous spectrum of congenital malformation, genetic metabolic disease and tissue dysplasia is, however, too restricting because human dysmorphology often arises at the interfaces.

TERATOGENICITY: ETHICAL AND LEGAL CONSIDERATIONS

Because the etiologies of approximately 65% of spontaneous malformations are unknown, the contribution of unidentified environmental agents to the causation of malformations is uncertain. It is certain, however that exposure to an increasing number of pharmaceutical, industrial, agricultural and recreational chemicals lowers the susceptibility threshold to each individual agent. Notwithstanding growing public concern, every year hundreds of new drugs and chemicals are introduced into the market and the environment before their effects on human reproduction are known. According to various studies, between 40–90% of pregnant woman consume one or more medications during gestation, and in the USA the average is about four drugs (but may be as high as ten) per patient. Pregnant woman also use over-the-counter drugs, such as vitamins, iron, aspirin, antacids, analgesics, etc., perhaps in combination with licit and/or illicit recreational drugs. Despite protests from toxicologists, geneticists, obstetricians, neonatologists, occupational, addiction and drug-information specialists, psychologists, sonographers, epidemiologists and ethicists, little has been done to stem this flood-tide of ingested toxicants. Abortion debates, on the other hand, are taken so seriously that they have the power to polarize communities and cut across political, religious and social groups; yet the unborn's 'right to life', often enforced by restrictive laws, is not qualified by the notion that a basic human fetal right is protection from serious prenatal abuse. Such an illogical state of affairs only strengthens the cynical view that the birth of a child has political rather than biological value; the unborn is the victim of a power-acquiring strategy to regulate reproduction according to specific social ideologies. Therefore, setting a limit on what may be negligently done to gametes, embryos and fetuses needs urgent attention.

Health care professionals and the scientific community, mostly in western societies, have for some time now faced important issues related to drug exposure in pregnancy, teratogenicity and resulting litigation. With increased certainty that a wide range of human malformations are caused by non-genetic environmental factors, an accompanying philosophy is emerging that someone must be responsible for the fetal damage incurred and should therefore provide compensation. This has resulted in an increase in litigation by parents on behalf of their malformed infants. On the other hand, establishing a causal relationship in a human population is difficult because there are no appropriate controls. Frequently the dosage of medication(s) is not quantified and patient characteristics such as their smoking and drinking history, caffeine use, illnesses during pregnancy and/or the presence of chronic disorders such as diabetes mellitus are often missing. It is on account of the above variables that the medico-legal issues related to teratology are so complex. On the one hand the legal rights of the unborn fetus have to be protected, and on the other justice requires the unequivocal demonstration of negligence by conclusively proving causality. The major problem in apportioning causality is the element of chance, or the 'dice syndrome'. A good example of this conflict involves Bendectin, a drug used in the treatment of nausea and vomiting in pregnancy. In 1983 the Merrel Pharmaceutical Company announced that it had decided to stop production of a formulation (composed of substances with antihistamine, antinauseant and antispasmodic properties) marketed as Bendectin in North America, Debendox in Great Britain and Lenotan in West Germany. The decision to cease production originated from widespread negative publicity and financial concerns based on numerous lawsuits launched in the USA. A suspicion about possible teratogenic effects of Bendectin was first raised in 1969 by Paterson, who described congenital limb deformities in an infant whose mother had allegedly taken the drug during pregnancy. Description of other similar cases soon followed, and on the basis of a total of 86 cases Paterson concluded that

Bendectin was not safe in pregnancy. However, among the estimated 30 million infants exposed to Bendectin early in gestation (given a background 'normal' malformation rate of 3%), chance could account for up to 900 000 malformed infants. Equally, if the rate of spontaneous occurrence of limb reduction defects is considered to be 1 in 3000, then 10 000 such defects would have occurred by chance in the same group. Thus litigation cases alleging teratogenic effects of Bendectin must be concluded in favour of the defendent, Merrell-Dow.

As related above, any existing legal rights that children may have are, typically, taken up by the parents on their child's behalf and usually relate to the nature of medical treatment. Questions of whether a drug-abusing parent should be legally responsible for compensating a child born with anomalies remain unresolved. However, a small number of test cases involving parents and children have initiated discussion, heightened awareness and been educational in promoting changed behaviour and improved lifestyles. Interestingly, a practice termed 'corporate exclusionary policies', meaning the exclusion of women from gaining senior positions in industries where the worker is likely to be exposed to hazardous chemicals, is increasing in industrialized countries. However, both male and female gametes are vulnerable to teratogenic effects in the workplace. Rather than excluding potentially talented workers and risking future litigation by fathers of occupationally damaged offspring, would it not be better to make the workplace safer and protect the health of all employees?

THE BIOCHEMISTRY AND PHYSIOLOGY OF DRUG ABUSE

Drugs are and always have been available in society, and the three most widely used, nicotine, ethanol and caffeine, are so common that they are not always considered to be drugs.

NICOTINE

Cigarette smoking (implicated in the development of lung cancer, heart disease, emphysema and related diseases) is the chief preventable cause of death in western societies yet, despite the known health hazards, more than 30% of adults smoke. In western societies smoking has steadily declined since 1965 but only due to a reduction in cigarette consumption by adult male smokers. Cigarette smoking among young women aged between 18 and 24 years, however, has steadily increased, with apprcximately 40% smoking. Approximately 27% of women between the ages of 25 and 34 smoke, and smoking among girls younger than 16 years is also rapidly increasing. Cigarette smoke contains from 2000–4000 different compounds, but the single component responsible for the popularity of smoking (and its addictive properties) is nicotine. Other toxins found in high concentration in cigarette smoke are carbon monoxide and hydrogen cyanide.

Nicotine is a natural alkaloid found in the leaves of the tobacco plant *Nicotiana tabacum*. In its pure form nicotine is colourless, odourless, highly water and lipid soluble and readily crosses cell membranes. Nicotine, from smoke in the lungs, rapidly enters the circulatory system and is preferentially taken up by the brain and adrenal glands. Subsequently the drug is distributed throughout the rest of the body and circulating levels decrease; it may accumulate in tissues such as body fat, where it can remain for extended periods. Nicotine is a CNS and adrenal gland stimulant affecting autonomic ganglion cell function and adrenaline/cortisol release. The major systemic responses to ganglion cell stimulation and catecholamine release are elevation of heart rate, blood pressure and oxygen consumption and vasoconstriction, the last including blood flow to the uterus and testes. Nicotine is catabolized in the liver to its major metabolites cotinine and nicotine-*N*-oxide, which are then excreted (together with a small percentage of unmetabolized nicotine) in the urine.

Maternal smoking and adverse effects on the fetus

Maternal tobacco use has been strongly implicated in many adverse outcomes of reproductive function including low birthweight, spontaneous abortion, early fetal death, increased risk of malformations and long-term defects of growth and behavioural development. Despite these serious and well established risks to the offspring and the wide public dissemination of warnings that smoking is harmful to the fetus, cigarette smoking during pregnancy is still prevalent. A 1983 survey in the USA revealed that 20–25% of the women who smoked before pregnancy continued to do so. Nicotine is freely transferred across the placenta and concentrates in the fetus, amniotic fluid, umbilical cord blood and term placenta, all of which retain higher nicotine levels compared to those in maternal blood. In the fetus itself the disposition of nicotine is not homogeneous as the brain, adrenal glands, heart, kidneys and spleen preferentially accumulate the drug.

Maternal metabolism and elimination of nicotine is fairly rapid since the average half-life ($t_{1/2}$) is between 30 and 140 minutes (cotinine has a longer $t_{1/2}$, ranging between 6 and 30 hours). In the fetus the $t_{1/2}$ is prolonged because the fetal liver, lacking a full complement of adult enzymes, is relatively incapable of metabolizing nicotine. Elimination of waste products including drugs such as nicotine takes place via the amniotic fluid and placental circulation; the waste re-enters the maternal circulation and is excreted by the mother. The fetus' inability to effectively metabolize nicotine (and other drugs) accounts for its longer drug exposure time and higher drug levels compared to that in maternal serum.

Intrauterine growth retardation (IUGR) and prematurity

Birthweight reduction occurring in infants born to mothers who smoke compared with babies of non-smokers is well established. The low birthweight is not usually attributable to a shortened gestation but to IUGR, and the exent of IUGR is directly proportional to the number of cigarettes smoked; such a retarded growth rate can persist after birth, sometimes for years. A correlation between maternal serum cotinine levels and reduced birthweight has been demonstrated and serum cotinine levels are conveniently used to assess and monitor the prevalence of cigarette smoking during pregnancy. In one study it was found that women with serum cotinine levels equal to or greater than 284 ng/ml had infants 441 g lighter, on average, than women with low serum cotinine levels of 24 ng/ml. Both light and heavy (that is, 10 cigarettes or more daily) women smokers have increased chances (53% and 130%, respectively) of giving birth to an infant weighing less than the average 2500 g.

The mechanisms by which cigarette smoking causes fetal growth retardation are not fully understood, but there are two factors of major importance in the etiology of IUGR. Firstly the uteroplacental circulation is impaired by the vasoconstricting effect of nicotine which results in decreased uterine blood flow, abnormal placental function, reduced oxygen tension, fetal hypoxia and reduced transfer of nutrients. Secondly, fetal hypoxia is exacerbated due to elevated carboxyhaemoglobin levels. Rapid transplacental transfer of carbon monoxide causes carboxyhaemoglobin levels to accumulate in the fetus and, as the $t_{1/2}$ of fetal carboxyhaemoglobin is three times longer than that of the maternal form, chronic fetal hypoxia is inevitable. In turn, impaired oxygen transport mechanisms impair fetal growth, development and well-being. In addition, the necessity for continual detoxification of inhaled smoking products increases maternal vitamin and nutrient requirements, further reducing their availability for fetal consumption. Nicotine also has direct effects on the fetal endocrine system. Increased catecholamine concentrations, resulting from the activation of the fetal adrenergic system, cause a cascade of secondary physiological changes.

Prematurity (duration of pregnancy less than 37 weeks) is also significantly more common in infants of smoking mothers compared to nonsmokers. As with IUGR, a dose–response effect is found between the incidence of prematurity and the number of cigarettes smoked daily (see section below).

Abortion and other fetal wastage

A significant reduction is found in the proportion of the live births among those who smoke five or more cigarettes daily, and the mortality rate progressively increases when pregnant women smoke more than ten cigarettes daily. All types of fetal wastage – abortion, stillbirth and neonatal death – increase with smoking. A significant proportion of fetal and neonatal deaths can be attributed to premature delivery as a result of a higher than normal incidence of abruptio placentae (premature placental separation) and placenta praevia (placenta located in the lower segment of the uterus). Both conditions are serious pregnancy complications which may result in maternal, as well as fetal, death; emergency care and access to a facility where Caesarean section can be performed is, therefore, needed. Premature and prolonged rupture of the fetal membranes is also common.

Histological investigations show that the placentae of smokers are commonly associated with abnormal insertions of the fetal membranes and umbilical cords. Ultrastructurally, the placental villi from heavy smokers are fewer in number and work inefficiently because the fetal capillaries have reduced luminal cavities, resulting from endothelial cell swelling and repeated vasoconstriction. A similar placental structure is found in women living under conditions of extreme hypoxia at high altitudes. As mentioned above, the anticipated outcome of cigarette smoking is catecholamine and cortisol release, uterine vasoconstriction, reduced placental perfusion and sustained high carboxyhaemoglobin levels in the fetal compartment. Doppler ultrasonography is able to graphically demonstrate that, approximately 20 minutes after maternal smoking, there is a significant increase in the resistance index of the umbilical and fetal anterior cerebral arteries which is seen to be accompanied by an increased fetal heart rate, all of which are responses to the increased placental resistance experienced as a result of vasoconstriction. Cyanide and other compounds from cigarette smoke additionally contribute to fetal toxicity *in utero*.

Congenital malformations

Several studies have reported a modest relationship between maternal smoking and an increased risk of birth defects. The specific abnormalities noted were cardio-vascular, urogenital, microcephalus, oral cleft defects (40–50% higher in one extensive study) and club foot. However, the association between smoking during pregnancy and the risk of congenital abnormalities in the offspring is difficult to assess accurately because of the presence of other risk factors and population heterogeneity. Further, smoking is also associated with an increased risk of spontaneous abortion so the magnitude of the relationship between maternal smoking and congenital malformations at birth is reduced. Additionally, malformations not apparent at birth were not included on the birth certificates from which the population-based data were collected.

Physical and mental development in later childhood

Increasing concern and the wish to prevent long-term neurological and intellectual handicaps has focussed attention on identification of biological and/or environmental factors which may adversely affect fetal differentiation. In general, the lowest risk of neonatal death and the greatest likelihood of optimal physical and intellectual development is in children weighing 3000 g or more at birth. Both animal experiments and human experience demonstrate a close correlation between the severity of intra-uterine hypoxia and the appearance of brain damage. Accidental exposure to high levels of carbon monoxide during pregnancy resulted in infants exhibiting severe mental retardation, motor disabilities, involuntary movements, hydrocephalus and a variety of other anomalies of development. Moderate exposure to carbon monoxide also has clinical implications. Heavy smokers, for example, have a carboxyhaemoglobin level of 10% or more. In a US survey of several thousand children followed from birth, physical and mental retardation at the ages of 7 and 11 years was attributable to maternal smoking; the retardation increased proportionately with the number of cigarettes smoked. After corrections for associated social and biological factors, 7-year-olds born to mothers who had smoked 10 or more cigarettes daily were, on average, 3 to 5 months retarded in reading, mathematical skills and general intellectual ability compared with the offspring of non-smokers.

Passive smoking

Smoking a cigarette produces two main kinds of smoke: mainstream smoke (that which the smoker draws into the lungs) and sidestream smoke which issues into the atmosphere when the cigarette is held smouldering. A cigarette produces about twice as much sidestream smoke as mainstream smoke and, since sidestream smoke is unfiltered, many of its constituents are present in higher concentrations than in mainstream smoke. For example, a cigarette generates up to 70 mg of carbon monoxide, of which some 50 mg is released in sidestream smoke. In smoky rooms such as pubs and conference rooms, the carbon monoxide level is usually about 10–15 p.p.m. but may be much higher, particularly if the air is recirculated without being adequately cleansed. In confined spaces such as in cars, up to 95 p.p.m. can be recorded in the vicinity of a person smoking a single cigarette. Protracted exposure to such elevated levels increases the concentration of carboxyhaemoglobin in the blood in those not smoking. Several studies have demonstrated that children whose parents smoke suffer more lower (cough, wheeze, bronchitis, pneumonia) and upper (colds, influenza, pharyngitis, tonsillitis) respiratory tract diseases than children whose parents do not smoke. Such diseases of passive smoking can be life-threatening because lower respiratory tract illness in the very young may later give rise to impaired lung function and chest disease in childhood and adulthood. The effect of passive smoking on the fetus is less well documented, although a Danish study reported that smoking by the father had a significant effect in reducing the birthweight. The effect of passive smoking was found to be greatest among populations of lower socio-economic status where paternal smoking was nearly as significant (66%) as maternal smoking in inducing IUGR.

Paternal smoking and adverse effects on the fetus

Chapter 5 briefly mentions that cytotoxic agents directly affect the germinal epithelium and sperm differentiation and that there is an association between cigarette smoking and testicular dysfunction. Morphological abnormalities are prevalent among individuals who smoke, and since smoke condensates are mutagenic these observations are not surprising and imply a logical mechanism by which male smokers may transmit terato-spermic effects to their offspring. In its 1986 report on environmental tobacco smoke, the US National Research Council reviewed several studies which showed that fathers who smoke produce smaller babies, with associated increased rates of perinatal mortality. One study showed that paternal smoking was more strongly associated with low birthweight than was maternal smoking, a finding consistent with the fact that men who smoke usually smoke much more than do women. These findings may also indicate a synergistic effect of paternal and passive maternal smoking. Another study revealed that, independently of parental age and social class, the frequency of severe malformations was doubled in children of non-smoking mothers when the children's father smoked more than 10 cigarettes daily. Childhood cancers are also significantly more common among children whose fathers smoked at the time of conception. Much of the damage to the testes caused by smoking is associated with oxidizing compounds (Chapter 8).

ETHANOL

The problem of alcohol abuse in present-day society has reached alarming proportions despite the increasing number of reports concerned with the medical, social and economical aspects of alcoholism. Inappropriately managed alcohol consumption places a heavy burden on society; immediately obvious are productivity losses, traffic deaths and morbidity, violent crime and alcohol-related illnesses. Less obvious are the reproductive burdens associated with alcohol consumption which may be of even greater magnitude because excessive drinking among men and women rarely renders them incapable of conceiving.

Metabolism of ethanol

Ethanol is a small, water- and lipid-soluble molecule which is directly absorbed from the stomach and small intestine. From the bloodstream it quickly diffuses into all tissues, readily crossing the blood–brain barrier and the placenta. In the fetus, the disposition of ethanol is rapid, with the liver, pancreas, kidney, lung, heart and brain reaching the highest levels. Most ethanol is oxidized by the liver's oxidative enzymes, in particular alcohol dehydrogenase which metabolizes 90–95% of the ethanol consumed to acetaldehyde. Acetaldehyde is then converted into acetyl coenzyme A which, in turn, is degraded into carbon dioxide and water. Other enzyme systems, such as the microsomal oxygenases, are also capable of metabolizing ethanol and a small amount is eliminated unchanged through the breath, sweat, urine and faeces. Alcohol acts as a CNS depressant and, by changing neurotransmitter release and neuronal function in the brain's most integrative areas (like the reticular activating system), causes central nervous dysfunction. Although chronic alcoholism produces permanent brain damage in adults, alcohol exerts its most profound CNS effects during the fetal period.

Because of low fetal hepatic alcohol dehydrogenase activity, it is the maternal elimination pattern of ethanol which determines the fetal exposure and is the rate-limiting step. Following maternal 'binge' type drinking, the amniotic fluid retains a reservoir of ethanol and acetaldehyde over relatively long periods. Since ethanol stimulates catecholamine release, the umbilical vessels respond by contracting and, in cases of chronic drinking, placenta villus deterioration follows repeated vasocontriction. Thus, ethanol-linked changes in placental physiology and structure adversely affect nutrient transport and gas exchange between the mother and her fetus. The deleterious effects of alcohol consumption on the fetus, in addition to specific CNS effects, are not unlike those described for nicotine: prenatal loss, fetal growth retardation and developmental defects. However, alcohol-related teratogenic effects are usually clustered and can be clinically recognized as the fetal alcohol syndrome.

Maternal ethanol consumption and the fetal alcohol syndrome

In 1968 a French pediatrician Lemoine and his colleagues first reported a distinct pattern of anomalies in babies born to families with a history of chronic alcoholism. The similarities included growth deficiency, microcephaly, a cluster of anomalous facial characteristics, cardiac defects, limb deformities, CNS dysfunctions resulting in hyperactivity, decreased attentiveness, delays in psychomotor and language development, poor visual memory and psychosocial maladjustment. The IQ of the children described was about 70; however, the severity of mental retardation and behavioural problems is related to the degree of alcohol exposure *in utero*. As similar clusters of malformations were subsequently observed and described, the full impact of ethanol on development became apparent and the category of anomalies was named the fetal alcohol syndrome or FAS. This is, of course, only recent history, as the detrimental consequences of alcohol consumption during pregnancy have been suspected since antiquity. Aristotle warned that women drunkards often gave birth to abnormal children; and the early Greeks, fearing the birth of a damaged child, prohibited drinking on the wedding night. The English gin epidemic lasting from 1720 to 1750 was another classic case which strengthened suspicions that alcohol could have deleterious effects on fetal development. The gin epidemic resulted from a large decrease in the price of gin associated with social helplessness, and alcohol abuse became a major health problem severely compromising the health and wellbeing of infants.

Of all the characteristics of FAS, mental retardation is the most damaging and consistent consequence, and alcohol is now the leading cause of mental retardation in the western world, followed by Down's syndrome and cerebral palsy. It has been estimated that the incidence, worldwide, of FAS is 1.9 cases per 1000 live births and, for the population most at risk, rises sharply to as high as 59 per 1000 live births. Apart from the personal tragedies, the costs of providing treatment to children with FAS (conservatively estimated for the USA at about US $322 million annually) is accelerating, particularly for those with IQ scores from 50 to 65. Despite the community effort many such children with special needs are still neglected because the requirements outstrip the available resources. These personal tragedies and the economic drain can, of course, be prevented by behavioural change.

Research on the possible teratogenic mechanisms involved in FAS has focussed on direct alcohol toxicity, acetaldehyde toxicity, fetal hypoxia, nutritional deficits and placental dysfunction. It is hypothesized that some of alcohol's effects may be due to the primary metabolite, acetaldehyde, which is a more potent teratogen than ethanol, but whether it crosses the placenta in sufficient concentration remains unclear. Since the placenta oxidizes ethanol to acetaldehyde, acetaldehyde can reach the human fetus by placental transfer and/or by production. Thus, by ethanol catabolism, the placenta may enhance the pathophysiology of ethanol-associated fetal injury. Other factors which frequently accompany alcoholism may also contribute to FAS. These include maternal smoking, caffeine and other drug consumption and paternal drug-consuming habits.

How much is too much?

Several problems are encountered when attempting to estimate the relationship between maternal alcohol consumption and the risk of FAS as there are critical periods during pregnancy when the fetus is particularly susceptible. Isolated facial anomalies are more likely when alcohol ingestion is confined to early pregnancy, whereas chronic alcohol consumption throughout pregnancy results in a wide variety of effects ranging from structural anomalies to growth retardation, neuroanatomical aberrations and compromised CNS function. Animal research suggests that consumption of large amounts of alcohol at one time (binge drinking) is more detrimental than one or two drinks daily, which is generally accepted as not being harmful. However, a recent study found associated musculoskeletal defects in babies in women who drank one or more alcoholic drinks per day, but not more than seven per week, during pregnancy. A number of other variables, including nutritional status, whether the woman was a heavy drinker preceding her pregnancy, multiple drug use, stressful life situations and genetic factors also interact with alcohol in the development of FAS. It is argued, therefore, that since no safe level of alcohol intake has been established, women should abstain completely from drinking during pregnancy and whilst attempting to conceive.

Postnatal neuroanatomical development may also be adversely affected by exposure to alcohol in breast milk. For instance, motor development, as measured by the Psychomotor Development Index was found to be significantly lower in 1-year-olds exposed to alcohol-contaminated breast milk.

Paternal ethanol consumption and adverse effects on the fetus

An effect of paternal drinking on the fetus has been suggested by epidemiological analyses. The relationship, as for nicotine, depends on alcohol-induced changes in sperm and relates to the mutagenic properties of the drug. Alcoholism has long been known to cause testicular atrophy, desquamation of the germinal epithelium, decreased sperm motility, oligo- and teratospermia, decreased testosterone production, loss of libido and impotency (brewer's droop!). Alcoholic males can influence reproductive outcome both behaviourally through decreased opportunities for fertilization and genetically through decreased quality of gametes.

CAFFEINE

Caffeine is the most widely consumed psychophysiologically active substance in the world. In the USA, for example, 92% of children aged 5 to 18 years were found to have consumed caffeine during a week in which they were asked to keep food records. Caffeine is similarly consumed in all other parts of the world, with its popularity reflected in trade statistics. For instance, it was estimated in 1984 that the industry employed more than 20 million people and was the most traded commodity after oil.

Caffeine is a trimethylxanthine alkaloid and is readily available in coffee (containing from 85 to 110 mg/cup), tea (about 50 mg/cup), cola beverages (30–45 mg/serving), cocoa (about 5 mg/cup), chocolate (25 mg/small bar), as well as preservatives, analgesics and other pharmaceutical preparations. The average daily caffeine intake of moderate to heavy consumers is about 463 mg/day or the equivalent of five or six cups of brewed coffee. In general, pregnant women reduce their caffeine intake; however, 90–95% still continue to consume caffeine throughout their pregnancies, with an estimated 18% consuming amounts equivalent to four or more cups of coffee daily.

Pharmacokinetics of caffeine

Pregnancy affects the pharmacokinetics of many drugs which need hepatic biotransformation prior to excretion due to competing demands on the metabolizing enzymes. For example, the mean caffeine plasma $t_{1/2}$ is prolonged in pregnant women and those using oral contraceptives, because the liver microsomal mixed-function oxidases metabolize a wide variety of xenobiotics and endogenous substrates including steroids. Normally the elimination of caffeine from the serum varies between 3 and 7 hours but the rate of elimination decreases progressively during the course of pregnancy, resulting in a two- to three-fold increase in half-life by the third trimester. Caffeine is first demethylated to the dimethylxanthine isomers theophylline, theobromine and paraxanthine, before being further processed to monomethylxanthines and monomethyluric acids. As described for nicotine and ethanol, caffeine and its metabolites are rapidly transferred across the placenta and, on account of slow fetal elimination and prolonged maternal $t_{1/2}$, caffeine accumulates in fetal tissues and amniotic fluid. There is the potential for high-level exposure since the human fetus and newborn infant are comparatively unable to metabolize caffeine. The mean plasma clearance time is about 100 hours in the neonate and progressively decreases only reaching the adult rate at 1 year of age. In view of this long caffeine $t_{1/2}$ in infants, continued exposure to caffeine through the breast milk may cause further adverse developmental effects.

The physiological effects of caffeine and its major metabolites

The major methylxanthines differ in their potency in the various physiological systems. Caffeine is a most powerful CNS stimulant, with theophylline (a major active component in tea) and theobromine (a major active component in cocoa beans) having less potent effects. Caffeine has a number of other general stress-like effects, such as catecholamine release, mobilization of free fatty acids, increased metabolic and heart rates and altered blood pressure. Structurally, methylxanthines are purine alkaloid analogues of the purines found in the nucleic acids RNA and DNA (Fig. 19.2). Because the methylxanthines are purine analogues, they interact chemically with DNA in different ways and may intercalate and/or disorganize its helical structure. In *in vitro* systems, the xanthine group of drugs causes chromosomal breaks and interferes with postreplication repair mechanisms. If these effects translate to *in vivo* systems (not conclusively demonstrated), they may result in the retention of faulty chromosomal rearrangements in sperm, maturing occytes and rapidly dividing embryos. A recent US report revealed that women who drank more than one cup of coffee daily were half as likely to conceive as those who drank less than one cup a day, and women who consumed more than

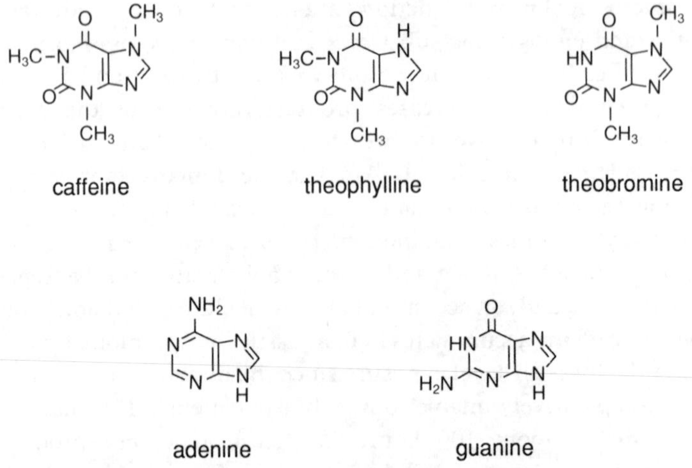

caffeine theophylline theobromine

adenine guanine

Fig. 19.2 The structures of the purine-derived methylated xanthines: caffeine (1,3,7-trimethylxanthine), theophylline (1,3-dimethylxanthine) and theobromine (3,7-dimethylxanthine) and the purine-derived nucleic acid bases adenine and guanine.

7000 mg caffeine monthly (about two and a half cups daily) are 4.7 times less likely to conceive than women who consumed less than five cups of coffee per month. Caffeine-induced subfecundity may well be the result of an increased incidence of unrecognized spontaneous losses of defective pregnancies, rather than a failure to conceive.

Effects on reproduction and prenatal toxicity

There is evidence that caffeine has a variety of teratogenic effects even at a moderate level of intake; for example, some studies report that women drinking more than three cups of coffee per day during the first trimester have an increased risk of miscarriage. Other studies have found that caffeine consumption prior to and during pregnancy is associated with prematurity, fetal IUGR, poorer neuromuscular development, poorer reflex functioning and, as a result of neonatal withdrawal symptoms, behavioural abnormalities. Since caffeine has not been studied as extensively as nicotine or ethanol, this drug's effects on human development are not fully known. A number of recent studies have, however, shown that as little as 150 mg caffeine per day during pregnancy can result in lowered infant birthweight. Infants exposed to high levels of caffeine *in utero* may suffer withdrawal symptoms (irritability, jitteriness and vomiting) after delivery. This withdrawal dysfunction may also be a potentiating factor in neonatal apnoea and sudden infant death syndrome (Chapter 11).

The association between maternal caffeine consumption and congenital abnormalities is equivocal because of the difficulty of dissociating methylxanthine consumption from other risk factors. However, caffeine is a strong potentiator of adverse reproductive effects when co-administered with other drugs. For example, caffeine exacerbates the effects of ethanol and a significant caffeine/cigarette interaction was found in an analysis of birth defects.

Paternal caffeine exposure and the fetus

There have been reports of a relationship between paternal caffeine exposure and adverse pregnancy outcome. This association is not surprising because the distribution of caffeine into semen rapidly equilibrates with that in the blood. A synergistic effect of caffeine consumption and cigarette smoking in relation to changes in semen quality has been reported. Chapter 5 discusses possible relationships such as drug-induced vasoconstriction and the etiology of testicular dysfunction (see especially Fig. 5.6, p. 91).

OVERVIEW

This chapter evaluated the, at least theoretically, preventable morphological, functional and behavioural effects of maternal and paternal drug abuse on the fetus, neonate and child. Because of the existence of many confounding factors influencing epidemiological assessments, mechanisms of teratogenesis, specificity, susceptibility and treatment insights have to come from the use of animal models. The section on maternal drug abuse, IUGR and sudden infant death syndrome in Chapter 11 is also relevant in the present context.

General references

Armstrong, B. G., McDonald, A. D. & Sloan, M. (1992). Cigarette, alcohol and coffee consumption and spontaneous abortion. *American Journal of Public Health*, **82**, 85–86.

Cohen, F. L. (1986). Paternal contributions to birth defects. *Nursing Clinics of North America*, **21**, 49–64.

Folb, P. I. & Dukes, M. N. G. (eds.) (1990). *Drug Safety and Pregnancy*. Elsevier, Amsterdam.

Gottesfeld, Z. & Abel, E. (1991). Maternal and paternal alcohol use and its effects on the immune system of the offspring. *Life Sciences*, **48**, 1–8.

James, J. E. (1991). *Caffeine and Health*. Academic Press, San Diego, CA.

Jauniaux, E. & Burton, G.J. (1992). The effect of smoking in pregnancy on early placental morphology. *Obstetrics & Gynecology*, **79**, 645–648.

Koren, G. (ed.) (1990). *Maternal-Fetal Toxicology: A Clinicians' Guide*. Marcel Dekker, New York.

Lemoine, P., Haronsseau, H. H., Borteryu, J. P. & Menuet, J. C. (1968). Les enfants de parents alcoholiques. Anomalies observés. A propos de 127 Cas. *Quest Medical*, **25**, 476–482.

Mendelson, J. M. & Mello, N. K. (eds.) (1992). *Medical Diagnosis and Treatment of Alcoholism*. McGraw-Hill, New York.

O'Rahilly, R. & Müller, F. (1992). *Human Embryology and Teratology*. Wiley-Liss, New York.

Paterson, D. C. (1977). Congenital deformities associated with Bendectin. *Canadian Medical Association Journal*, **116**, 1348.

Persaud, T. V. N. & Thomas, C. C. (1990). *Environmental Causes of Human Birth Defects*. Charles C. Thomas, Springfield, IL.

Schenker, S., Becker, H. C., Randall, C. L., Phillips, D. K., Baskin, C. S. & Henderson, G. I. (1990). Fetal alcohol syndrome: current status of pathogenesis. *Alcoholism: Clinical & Experimental Research*, **14**, 635–647.

Watson, R. R. (ed.) (1991). *Biochemistry and Physiology of Substance Abuse*. Vol. III. CRC Press, Boca Raton, FL.

Wigglesworth, J. S. & Singer, D. B. (eds.) (1991). *Textbook of Fetal Prenatal Pathology*. Blackwell Scientific, London.

Zhang, J., Savitz, D. A., Schwingl, P. J. & Cai, W. W. (1992). A case-control study of paternal smoking and birth defects. *International Journal of Epidemiology*, **21**, 273–278.

20 *Ethical aspects of human reproductive biology*

Rapidly evolving and complex new reproductive technologies have given hope to the infertile and engendered much controversy and social change. The treatment of human infertility was revolutionized in 1978 with the world's first IVF baby and many technical advances quickly followed. In the 1990s, these improved skills are contributing to human welfare, opening up prospects for the conservation of endangered species and upgrading animal production. Adaptations of the technologies central to controlling human fertility and infertility are also becoming central in the quest for increased food production and, together with advances in modern contraception, are a response to the worst legacies of increasing human population pressures. Modern reproductive technologies have, by increasing our autonomy from biological constraints, irreversibly altered long-held cultural concepts surrounding fertility, childbearing and child rearing. They have also raised many ethical issues.

The power of these newly developed reproductive technologies forces decisions in areas such as whether or not to prolong life and pain of a seriously ill infant, abort a supposedly defective fetus, use another person's gametes to create an embryo, or to use a surrogate to bear a child. We also often have to choose between the rights of the parents and those of the unborn child. The dilemmas created transcend the concerns of scientists, doctors, parents and other health carers and have become social issues with strong political and philosophical overtones. Decisions affecting the reproductive capacity of individuals, alternatives to neonatal intensive care and the consequences of prenatal testing and genetic counselling are made more difficult because they have to be made in the context of the same cost constraints that limit the application of all medical technologies.

SOCIAL IMPACT OF THE REPRODUCTIVE REVOLUTION

In spite of the benefits reproductive technology can offer, its application threatens traditional values and has, therefore, generated opposition from many quarters. The ethical debate surrounding surrogacy has become so political that governments have now to decide whether to intervene, regulate or prohibit it. Modern surrogacy challenges deeply ingrained assumptions and attitudes about parenthood which developed in our ancestors as part of the socialization process. Artificial reproduction procedures now make it possible to separate the various phases of the reproductive process. With gamete/embryo donation, the desires and needs of donor, recipient, wider kin and the general population have to be taken into consideration. In the past, genetic relatedness was of greater concern for men than for women but this may change now that modern technology can produce gestatory surrogates. The selective condemnation of surrogate technology may also reflect subconscious biological fears that if society officially condones women giving away their children, for whatever reason, traditionally held belief-systems about human roles in reproductive function may be undermined. The essential difference between the past clandestine use of surrogate mothers to overcome childlessness and modern surrogacy relates to powerful vested interests which are frequently executed by the use of legally binding contracts. An interesting example of selective discrimination was seen in Britain when the Warnock Committee's report was released. This committee approved of *in vitro* fertilization and artificial insemination as techniques for treating infertility (provided they are only used by 'stable' heterosexual couples) but disapproved of surrogacy, a sentiment mirrored in the official reports from many other countries. Feminist groups were late in entering the debate over reproductive technology, but this has now changed due to concerns about possible abuses, especially in the area of surrogacy. Opponents of the technologies argue that these represent examples of patriarchal exploitation which violate the rights of women; women who participate are considered as victims of male medical power. The surrogate mother is seen as a victim of commercial exploitation and surrogate motherhood has even been described as a new form of prostitution. However, to characterize women purely as victims denies them any degree of reproductive choice and fails to acknowledge that women themselves assist in the construction of the social reality within which reproductive technologies are utilized.

Anti-interventionist feminist concerns of the 1990s contrast with views held earlier this century when prominent feminists, such as Margaret Sanger in the USA and Marie Stopes in Britain, maintained that the development of birth technology was liberating for women because, by freeing them from compulsory motherhood, it provided choices of lifestyles. Sanger and Stopes successfully campaigned for legislative reform to permit the opening of medically supervised birth control clinics for the poor – a view in tune with earlier medical advances in contraception. This strategy also moved the sexual division of labour in the direction of equality. Feminism since the 1960s has moved away from the ties with biological thinking. Traditional motherly images have again become important as women claim a political identity as mothers and a wish to restore the sociobiological concept of woman-mother-nature. It is questionable whether feminist analysis can constructively develop further whilst it remains fragmented. Anti-interventionist policies may also inadvertently risk assisting extremist groups opposed to a large variety of reforms including female independence, abortion reform, use of contraceptives, divorce and continuing developments in reproductive technology. In the absence of financial independence, childbearing constitutes an investment activity for women. Dependent women, whose material well-being is determined either directly or indirectly by their men and children, have large families. Independent women, on the other hand, generally have smaller families. History has convincingly demonstrated that an advantageous shift in women's power relative to that of men, parents and children results in a fertility decrease. Ecological education and social change are of primary importance if population replacement or zero population growth is to be a viable option. Therefore, many more women in influencial and decision-making positions are required to ensure the success of a joint, and more just, future which includes the integration of the woman's unique talents and perspectives. The technological contributions in the areas of infertility research, contraceptive research, diagnosis of genetic aberrations and therapy have provided choices, accelerated social change and rewarded lower procreation through improved health. There is also the danger that restricted access to existing technologies may unwittingly assist our genes in their strategy of manipulating us to their advantage. Since genes have only one strategy, to replicate themselves, reducing flexibility and choices cannot be to our advantage in the struggle against over-population.

An exceptional effort is still needed to clarify the expansion of the concept of procreative rights. In dealing with reproductive choices, the emphasis so far has been, almost exclusively, on who may conceive, bear and rear a child (presently IVF-related technologies and adoption are restricted to couples; access to contraception and abortion is not general). Since any kind of combination of germ material is now possible and fetal transfer is a realistic alternative to abortion, the right of another woman to rear a child she biologically cannot conceive or bear also becomes open for discussion. Are all the applications of new technology to be available to those persons who desire and can afford them, and will those who do not participate be forced to bear the social consequences of any resulting mistakes? Commercialization of germplasm means the existence of a market place which can be shaped by consumer demand, state control or a combination of both. Its direction will depend on the power structure and whether individual states wish to take an even more active role in the reproduction of its population. In the final analysis, the major differences between countries will reflect political/social priorities rather than differences in technological sophistication.

Nevertheless, modern reproductive technology has developed too far to be politically suppressed, so potentially exploitative aspects need to be minimized by appropriate legal and institutional structures, especially in respect of our responsibilities to future generations. A basic difference between surrogacy and adoption which has not been adequately addressed is that surrogacy is aimed at the production of children for intending parents, whereas adoption is aimed at meeting the needs of existing children. Modern surrogacy has forced society to reconsider its definitions of motherhood, artificial insemination by donor has raised questions of fatherhood, and new aspects of teratology have raised questions of responsible parenthood. An increased awareness of environmental deterioration requires a moral responsibility for future generations of species, including human. This must become an integral aspect of any system of ethics. Simply to argue for the 'right to have children', unqualified by responsibility toward the unborn and the world they will exist in, reflects an attitude devoid of any moral consideration of the offspring's and the world's future.

HISTORY OF THE BIOETHICS MOVEMENT

Since World War II, phenomenal advances have occurred in all fields of human endeavour but probably most apparently in medical knowledge and scientific skills. The power of medicine to prevent nature from taking her course has been massively extended, and with that power has come a considerable extension of moral choice. For instance, once it became possible to keep patients alive by artificial respiration and to improve resuscitation techniques, doctors were faced with dilemmas caused by their new power. It was the civil rights movement, beginning with racial concerns but quickly spreading to demands for civil rights protection for other vulnerable groups, which insisted that individuals deserve respect as human beings. The emergence in the mid-1960s and the early 1970s of a number of technological advances in organ transplantation, long-term haemodialysis and ventilators sharpened our ethical dilemmas.

In 1967, Christiaan Barnard first transplanted the human heart. In 1968, the Harvard Medical School *ad hoc* committee published its report, delineating brain-death criteria so that patients being maintained on respirators could be pronounced dead and their organs used for transplantation. The Harvard report resulted in the passage of brain-death statutes in many states (such statutes were necessary because physicians risked being prosecuted for murder in organ donor situations).

The artificial kidney was the first life-prolonging, high-technology treatment to capture the public's attention. As a result, many articles appeared discussing how to choose patients for dialysis treatment and whether or not everyone who needed dialysis was morally entitled to it. Haemodialysis, more than organ transplants, increased public interest in the ethical dimension of health care problems because shortages of dialysis equipment made the ethical dilemma more obvious and poignant.

In 1976 the landmark decision made by the New Jersey Supreme Court in the matter of Karen Ann Quinlan, brain dead following a road accident, reintroduced public discussion of bioethics. The court ruled that if Quinlan's attending physician determined that there was no reasonable chance that she would ever return to a 'cognitive, sapient state', and if a hospital ethics committee agreed with that prognosis, then the life-support apparatus could be withdrawn at her guardian's or family's request. In 1983 the New York case of Baby Jane Doe brought the problem of treatment decisions for handicapped newborns before the public. In the USA, to protect anonymity, a boy or girl subject to court order is known as a John Doe or Jane Doe, respectively. This case involved parents who refused surgery, opting instead for conservative treatment including the use of antibiotics for their infant suffering with spina bifida and hydroencephaly.

Cases such as these demonstrate that both parents and doctors risk extensive legal involvement when treatment decisions, especially those withholding life-prolonging or aggressive procedures, are made. It was clear that those who held the moral responsibility for making these very difficult decisions needed as much help as they could get. In 1983 in the USA, the President's Commission for the Study of Ethical Problems in Medicine and Biomedical and Biobehavioral Research issued its report deciding to forego life-sustaining treatment in certain cases. The report included sections on making treatment decisions for incompetent patients and for seriously ill newborns. Other nations, including Australia, soon followed by producing and approving the release of discussion papers on the ethics of limiting life-sustaining treatment. Discussions were also extended to the problems presented by extremely low birthweight infants and the cost of intensive care for them, to infants with spina bifida and other birth defects, the sick elderly and the management of the terminally ill.

In the final analysis, individual rights, consumer and patient rights and bioethics movements all provided the impetus for the creation of a distinct transdisciplinary field of bioethics. Bioethics came into existence because there was a need to debate how research and health care decisions and regulations would be made, who should make them, and what the long-term implications would be. Ethics committees appeared to be a good approach because they were sufficiently similar to other ways of solving problems involving conflicting moral values and because they had the potential for providing public involvement. Since their introduction, medical/mental patients, and women and children in particular, have benefited from this new era of soul searching.

BIOETHICS COMMITTEES IN PRACTICE

Bioethics committees are now given institutional support and governmental encouragement, but each has to develop and refine its specific skills. Through their potential functions of education, policy recommendation and case reviews, hospital ethics committees can be an important step in raising everyone's consciousness about the values by which we live and die. Institutional Ethics Committees (IECs) should be composed of men and women reflecting different age groups and ideally include a combination of the following at some stage in their career.

(a) A layperson not associated with the institution
(b) An ethicist, philosopher and/or a minister of religion
(c) A lawyer
(d) A medical graduate with research experience if research decisions are to be made
(e) A nurse, social worker, physio/occupational therapist representing the patient care sector
(f) A gynaecologist/obstetrician/neonatologist or reproductive biologist if decisions involving reproductive technology are to be made
(g) Outsiders representing scientists, writers, professionals and others who can bring a fresh point of view.

The ethics committee model, however, is not a way of providing treatment decisons but rather a forum through which different values, perceptions and information about treatment decisions can be discussed, assessed and resolved by patients, their families and the health care team. Presently, a major function of ethics committees is in educating the hospital and the lay communities and arriving at appropriate policies and guidelines. These are then summarized and may become statements of policy which later still may be incorporated legally as part of a country's social policy.

Areas in which institutional ethics committees (IECs) operate

It is possible to strike a balance between the responsible and the irresponsible use of skills and technologies inherent in modern health care. Good IECs attempt to clarify medical and ethical issues and ensure that the patient's best interests have been properly considered. IECs, if properly constituted, can serve a useful purpose in guiding medical staff towards decisions that are sound in ethics as well as in medicine. A selection of the type of reproductive concern for which IEC opinion may be sought are as follows.

Gene transfer and fetal therapy A widely discussed issue is whether deliberate intervention in the human genome is acceptable under any circumstance. Generally, gene transfer to the human genome is considered acceptable, under current therapeutic standards, only if it provides benefit to the individual and the usual practices for innovative therapies are observed. Since somatic cell gene transfer has no role in producing a new generation, it is argued that it is not ethically different from other established forms of therapy. The use of chemo- or radiotherapy, for example, likewise alters the cellular genome of healthy as well as cancer cells. Reproductive policies are, however, the product of both scientific knowledge/technical capacity on the one hand, and social/cultural values of the other. Technical advances in germ line genetic modification raise the really difficult issue of eugenics – the controlled 'improvement' of human offspring – in unscrupulous hands. Ethical questions as to whether the techniques of assisted reproduction should be available to all to try to improve the health, the abilities or the physical traits of their offspring are increasing as the techniques of genetic engineering become more practicable. The demand for eugenics programmes is a reality. For example, parents have already sought to treat their children with growth hormone, now created in large quantities by genetic engineering, because they admire height and believe that being tall is advantageous in both economic and social terms.

Abortion and fetal right to life Fetal rights (unqualified by quality of life issues) have been among the most prominent and bitterly contentious of reproductive options. In the 1990s, life-sustaining technology to support artificial gestation is being used more frequently and with earlier fetuses which spend longer periods developing in the bodies of their dead mothers, or neomorphs. This procedure is entirely experimental since no-one knows what the long-term effects of artificial gestation will be on the offspring. In addition, there is the risk that the fetus may suffer intense deprivation or even torture *in utero* because a normal pregnancy is constantly stimulated by the movement of the mother, the myriad of sounds of her environment, her feelings and responses to those close to her and, above all, her communication with her growing child. Chapters 11 and 19 explore at-risk behaviours by parents which prevent the expression of the fetus' full genetic potential and legal perspectives arising from ethical considerations.

Embryo research The most important indication for embryo research from the clinical point of view is to improve the clinical results. Thus fundamental research and innovative medicine is not without moral restraint since those in the disciplines of science and medicine are themselves usually moral agents in the community – the first line in ethical guidance. In the event of moral breakdown, external controls, such as IECs and Courts of Law, are second lines of defence. Experimentation is necessary to provide the advantages of future human objectives; for example, as understanding of heredity has emerged through increasingly sophisticated scientific research, the experience accumulated has been beneficially integrated into various technological practices in both agriculture and medicine. Benefits from human pre-embryo and embryo research can be expected in the four major medical areas of infertility research, diagnosis of genetic aberrations and possible therapy, contraceptive research, and therapeutic use of embryonal tissue for transplantation in life-threatening medical conditions.

Conflict between parental and child's interests The paramount issue of what is in the 'best interest' of the child is sometimes lost in the public discussion of the 'absolute' right of the parents to determine these interests on their child's behalf. A good example was the case of Chad Green.[1] The

[1] Custody of a Minor, 379 N.E. 2nd 1053 (Mass. 1978), Custody of a Minor, 393 N.E. 2nd 836 (Mass. 1979).

parents refused to allow Chad Green, a leukaemic child, to be given cancer chemotherapy, bringing the right of the parents to choose medical treatment for ther child into direct conflict with the legal duty of the treating physician to provide accepted medical treatment. Western law stipulates that a child who is not receiving necessary medical care is neglected; to prove lack of parenting is, however, difficult. For the child to receive chemotherapy may depend on whether the treating physician can gain custody. Similarly, conflict between parental and fetal interests may arise which need external arbitration.

Assisted reproduction and informed consent To debate the morality of using medical knowledge to assist the infertile is very different from debating the morality of eugenics, although the two may intermingle. In North America, for instance, sterilizations have been forced on women whose fertility was regarded as a social danger by health officials. These operations were compulsorily carried out, early this century, on asylum inmates for the eugenic purpose of limiting reproduction of the 'unfit' and later, in the 1970s, secretly on poor women from ethnic minority groups seeking medical advice not related to sterilization. In the 1980s, Australian Aboriginal girls were, without their knowledge, injected with the contraceptive Depo-Provera for the sole purpose of restricting the black birth rate. Secret or coercive methods were similarly employed elsewhere, such as in India and China. As previously explained in this text, fertility rates fall when effective fertility control is socially grounded in rising living standards. Despite this documented understanding, priorities in the ethics of social inequality, child welfare and global population growth are still not adequately categorized.

Living wills and the right to die Living wills, also known as Advance Written Directions, are a means by which individuals may direct, in writing, that they do not wish to receive certain treatment if in the future they become terminally ill or severely impaired. The concept first received legal acceptance in the 1976 Californian Natural Death Act, which was passed as a consequence of a number of right-to-die cases, such as the Karen Ann Quinlan decision.

The issues raised above generally fall into two categories. The first deals with listing and defining the values that are challenged when human reproduction and quality of life become the target of intervention and conflict. The second seeks to provide a social and political process that can mediate among those involved. Reproductive technologies applied to humans have always been ethically more problematic than have other medical technologies, especially when the issues involve embryos. Conservatism, on the one hand, raises the question of how can science and technology continue to advance and successfully overcome deeply rooted, strongly defended values that originated in scientifically less knowledgeable times. Liberalism, on the other hand, does not guarantee that growing scientific understanding of human reproduction, with its attendant technologies, will deliver only benefits free from disadvantages and misapplication. Ideally, feared negative consequences should not cut off opportunities for responsibly gaining new fundamental knowledge; rather, general guidelines, expressive of the highest ethical standards, should be formulated to address control of unethical applications. Eventually a system of national and international institutionalized checks will evolve. The established public policy needs to be stable but also flexible enough to accommodate additional knowledge and provide continuing close scrutiny to maximize positive and minimize negative outcomes. In this respect, reproductive medicine remains close to the practice of general medical science within current ethical constraints.

ETHICAL PRINCIPLES IN HEALTH CARE

Ethical principles in health care stem from the assumption that there must be respect for persons, and there are three derived principles that ethics committee members must be concerned with in their work. These are autonomy, beneficence and/or nonmaleficence and justice.

The principle of autonomy has become extremely important in western society since World War II and is based on the principle that individuals should be permitted personal liberty to determine their own actions, choices and self-definition. We respect autonomy when we refrain from interfering with peoples' opportunities to control their own lives and when we accept the existence of individual values that do not need to be shared by others to be worthy of respect. In health care, patient autonomy is seen as the patient's right to receive accurate information about his or her health and treatment, including alternative procedures that might preserve life, prevent disease or relieve suffering, and to choose or refuse any or all of those treatments. In addition, patient autonomy is served by permitting patients to choose others, whether physician, family member, friend or attorney-in-fact to hold and receive information and to make decisions for them. With the doctrine of liberty replacing earlier authorities, there is greater scepticism about the omniscience of doctors and a stronger emphasis on patients' self reliance.

The principle of nonmaleficence (causing no harm) is not always easy to put into practice because in some situations doctors may be faced with a choice of harms, for example the use of pain-killing drugs has the known side effect of shortening life. Closely allied to this is the principle of beneficence (doing good). It is not enough for doctors not to do harm, they must increase the benefits to others which means acting in the patient's best interests. The problem is that the doctor decides what is in someone else's best interests, which brings the principle of beneficence into conflict with autonomy. According to some writers, the principle of beneficence is separate from the principle of nonmaleficence, that is, doing good is of equal importance to that of not causing harm. Others believe that beneficence is a fundamental principle that includes nonmaleficence as an aspect of the principle. A physician's obligation to preserve life may conflict with the obligation to do no harm or to do good. It may also conflict with the patient's liberty; for example, by refusing a request to end a suffering patients's life, the doctor is overriding the patient's right to choose death. Contradictory aspects may be ranked in importance so conflicts can then be resolved by that ranking. If one must choose between not causing harm or doing good, then one should choose not to cause harm. Similarly, preventing harm is preferable to removing harm.

Justice has many aspects, but in health care the aspect of concern is usually distributive justice, that is, how burdens and benefits should be distributed. The Hippocratic tradition stresses the importance of the doctor's duty towards the individual patient, but the community is composed of many patients – care for some may impinge upon the quality of care for others. Conflicting views about what is a fair distribution system and the most ethically appropriate way to spread limited resources throughout the community in itself poses problems. The idea that every individual has a right to health care often conflicts with the limited means society has to provide it. Therefore, fairness and justice are not interchangeable terms in all contexts, although they are often used that way. A choice might seem unfair to an individual but, in terms of broader social values, may be just. Affirmative action, for example, may seem just with respect to society even though it is unfair to some individuals. When there are insufficient resources, distributive justice is concerned with how the resources are to be allocated among those who are in need. The principle of distributive justice, under which those in greatest need have their needs met first, conflicts with the principle of autonomy, under which every individual is entitled to an equal share of the health care provided by the community.

In general, patient autonomy is considered to be the most important of the three principles in health care decision making at the patient care level. Justice or beneficence, however, may receive higher priority at governmental policy levels. The constitutional right of privacy and the informed consent doctrine are both based upon the principle of autonomy. Patient autonomy means that the preferences of competent patients are to be accepted as long as they understand the implications of their choices and the alternative choices that are open to them. This is not an absolute value, for there are times when respecting a patient's autonomy may cause great hardship to others, giving rise to questions of justice or beneficence. There have been instances, for example, in which the parents of young children have refused, for religious reasons, to consent to blood transfusions that they themselves needed. Courts have sometimes overruled patient preferences in such cases because harm would come to the minor children if their parent was allowed to die (nonmaleficence) or because the state's duty was to protect the minor children (beneficence). The state, as protector of the minor children, can make this decision. Similarly, as protector of all citizens, the state might choose to overrule patient preferences in such cases to protect society in general from having to take the responsibility for the care of these parents' children (justice).

The concept of the best interests of the child is now a widely recognised principle in areas such as adoption, custody and wherever the welfare of the child is at stake. It also logically follows that the best interests of children require a guaranteed future. The arrest of human population growth and preservation of as much as possible of the earth's remaining biodiversity and natural ecosystems are imperatives. It is already too late for us to leave restoration of the environment to natural forces, so all the available technologies should be employed in a responsible effort to establish global stability. If nothing is done we now know, as members of the most capable of species, that earth's continued existence will be free from humans. Our reproductive concerns have been centred largely at the level of the individual. We must hope that recent insights into individual rights will extend to other species and across generations so that the human species will remain an integral part of this biosphere.

General references

Bartels, D. M., Priester, R., Vawter, D. E. & Caplan, A. L. (eds.) (1990). *Beyond Baby M: Ethical Issues in New Reproductive Techniques*. Humana Press, New Jersey.

Baruch, E. H., D'Adamo, A. F. & Seager, J. (eds.) (1988). *Embryos, Ethics and Women's Rights – Exploring the New Reproductive Technologies*. Harrington Park Press, New York.

Blank, R. H. (1990). *Regulating Reproduction*. Columbia University Press, New York.

Bromham, D. R., Dalton, M. E. & Jackson, J. C. (eds.) (1990). *Philosophy and Ethics in Reproductive Medicine*. Proceedings of the First International Conference on Philosophical Ethics in Reproductive Medicine, Manchester University Press, Manchester, UK.

Cohen, J. & Hotz, L. R. (1991). Human embryo research: ethics and recent progress. *Current Opinion in Obstetrics & Gynecology*, **3**, 678–684.

Dans, P. E. (1992). Reproductive technology: drawing the line. *Obstetrics & Gynecology*, **79**, 191–195.

Filsinger, E. E. (ed.) (1988). *Biosocial Perspectives on the Family*. Sage, Newbury Park, CA.

Handwerker, W. P. (ed.) (1990). *Birth and Power – Social Change and the Politics of Reproduction*. Westview Press, Boulder, CO.

Merrill, D. C. & Weiner, C. (1992). Fetal medicine. *Current Opinion in Obstetrics & Gynecology*, **4**, 273–279.

Siedlecky, S. & Wyndham, D. (1990). *Populate and Perish: Australian Women's Fight for Birth Control*. Allen & Unwin, Sydney, Australia.

Singer, P., Kuhse, H., Buckle, S., Dawson, K. & Kasimba, P. (eds.) (1990). *Embryo Experimentation*. Cambridge University Press, Cambridge, UK.

Wajcman, J. (1991). *Feminism Confronts Technology*. Polity Press, Cambridge, UK.

Whiteford, L. M. & Poland, M. L. (eds.) (1989). *New Approaches to Human Reproduction – Social and Ethical Dimensions*. Westview Press, Boulder, CO.

Index